ND YAG LASER

Edited by **Dan C. Dumitras**

Nd YAG Laser
Edited by Dan C. Dumitras

Published by InTech
Janeza Trdine 9, 51000 Rijeka, Croatia

Publishing Process Manager Sasa Leporic

Technical Editor Teodora Smiljanic

Cover Designer InTech Design Team

Additional hard copies can be obtained from orders@intechopen.com

Nd YAG Laser, Edited by Dan C. Dumitras
p. cm.
ISBN 978-953-51-0105-5

For EU product safety concerns:
IN TECH d.o.o., Prolaz Marije Krucifikse Kozulić 3, 51000 Rijeka, Croatia,
info@intechopen.com or visit our website at intechopen.com.

Meet the editor

Professor Dr. Eng. Dan Constantin Dumitras graduated at the Faculty of Electronics, University Politehnica, Bucharest in 1970. He obtained his PhD at the Institute of Atomic Physics, Bucharest in 1978. Since 1970 he has been involved in research on laser physics and applications (frequency stabilization of lasers, photoacoustic spectroscopy, laser applications in medicine and biology, material processing and ultrashort pulse, high intensity lasers - extreme light) at the Department of Lasers, Institute of Atomic Physics and at the National Institute for Laser, Plasma and Radiation Physics, Bucharest. He works as a professor at Faculty of Physics, University of Bucharest and as a professor and a PhD supervisor at the Faculty of Applied Sciences, University Politehnica, Bucharest. He is the author and/or editor of 15 books and has published more than 120 papers in scientific journals. He held more than 200 presentations at the international conferences (30 invited lectures). Prof. Dumitras organized several international conferences: ROMOPTO 1997, ALT 2001 and ALT 2006, LEI 2009 and ICUIL 2012.

Contents

Preface

Discovered almost fifty years ago at Bell Labs (1964), the Nd:YAG laser has undergone an enormous evolution in the years, being now widely used in both basic research and technological applications. Nd:YAG laser is extremely versatile in its output characteristics, yielding emission from continuous wave to very large energy pulses or very high peak-power pulses. Also, it covers a large spectral range, namely near infrared, visible (second harmonic) and ultraviolet (third to fifth harmonic).

This book includes new results on current research and state of the art developments regarding Nd:YAG lasers and their applications. *Nd:YAG Laser* covers a wide range of topics, from new systems (diode pumping, short pulse generation) and components (a new semiorganic nonlinear crystal) to applications in material processing (coating, welding, polishing, drilling, processing of metallic thin films), medicine (treatment, drug administration) and other various fields (semiconductor nanotechnology, plasma spectroscopy, laser induced breakdown spectroscopy).

The editor would like to thank all the chapter authors for their effort in completion of this book.

Dan C. Dumitras
Senior Scientist, Department of Lasers,
National Institute for Laser, Plasma and Radiation Physics
Professor, Department of Physics, University of Bucharest and
Faculty of Applied Sciences, University "Politehnica" of Bucharest
Romania

1

Identification of Elements in Some Sudanese Gasoline Types Using Nd:YAG Laser Induced Breakdown Spectroscopy

Nafie A. Almuslet* and Ahmed Mohamed Salih
Institute of Laser, Sudan University of Science and Technology, Khartoum Republic of Sudan

1. Introduction

Laser-induced breakdown spectroscopy (LIBS) employs a low-energy pulsed laser (typically tens to hundreds of mJ per pulse) focused on sample to generate plasma that vaporizes a small amount of the sample. A portion of the plasma emission light is collected and a spectrometer disperses the light, emitted by excited atomic and ionic species in the plasma, a detector records the emission signals, and electronics take over to digitize and display the spectra [1].

LIBS is a type of atomic emission spectroscopy (AES). It has become a powerful analysis technique for both laboratory and field use. The main purpose of LIBS, like AES, is to determine the elemental composition of a sample (solid, liquid, or gas). The analysis can range from a simple identification of the atomic constituents of the sample to a more detailed determination of relative concentrations or absolute masses [2].

Examination of the emitted light provides the analysis because each element has a unique emission spectrum useful to "fingerprint" the species. Extensive compilations of emission lines exist [3 - 5]. The position of the emission lines identifies the elements and, when properly calibrated, the intensity of the lines permits quantification.

Q-switched lasers had the capability of producing high focused power densities from a single pulse of short duration sufficient to initiate breakdown and to produce analytically useful laser plasma.

Typically, however, the signals from many laser plasmas are added or averaged to increase accuracy and precision and to average out non-uniformities in sample composition. Depending on the application, time-resolution of the plasma may improve the signal-to-noise ratio or discriminate against interference from continuum, line or molecular band spectra [6].

Today LIBS is used to analyze gases, liquids, particles entrained in gases or liquids, and particles or coatings on solids. Liquids can be analyzed by forming the laser plasma on the liquid surface or on drops of the liquid [7, 8]. If the liquid is transparent at the laser wavelength, plasma can be formed in the bulk liquid below the surface [9].

* Corresponding Author

LIBS has the following advantages compared with some non-AES - based methods of elemental analysis [10]:

- Provide qualitative and quantitative information about the sample composition.
- Simultaneous multi-element detection capability.
- Simplicity.
- Provide real-time analysis.
- No need for sample preparation.
- Allows in situ analysis requiring only optical access to the sample.
- Adaptable to a variety of different measurement scenarios.

This work aimed to identify the elements in three different types of Sudanese gasoline. The bulk of a typical gasoline consists of hydrocarbons with between 4 and 12 carbon atoms per molecule (commonly referred to as C_4-C_{12}). Some of gasoline types contain small amounts of other elements, including sulfur, nitrogen, oxygen, and some trace metals. The identification of such elements is important in determination of gasoline quality due to their effects on combustion process.

The identification here is based on the use of laser induced breakdown spectroscopy (LIBS) technique, in which high peak power Q-switched Nd-YAG laser focused on the sample, to produce plasma emission of discrete lines.

These lines are the fingerprints of the atoms and ions constitute the sample where the line intensity is proportional to element amount in the sample. By recording these emission lines one can get qualitative information and identify the elements in each sample.

2. Materials and methods

A schematic diagram of the experimental setup used in this work is shown in figure (1).

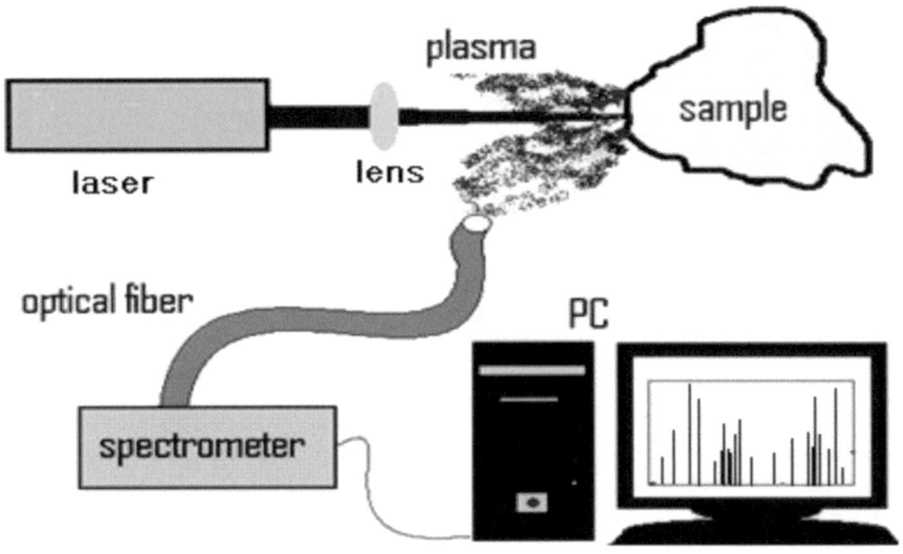

Fig. 1. Schematic diagram of the setup

The experimental setup consists of:-

i. Q-switched Nd: YAG laser supplied from XSD Hua Zhong Precision Instrument Factory, model LRH786T with 10 ns pulse duration and 0.5 Hz repetition rate. The pulse energy of this laser is adjustable starting from 12.57 up to 23.8 mJ.
ii. Spectrometer model USB 4000-UV/VIS, supplied from ocean optics - USA, attached with optical fiber and interfaced to computer with Windows operating system. The modular USB4000 is responsive from (170 -900) nm. The detector is Toshiba TCD1304AP linear CCD array detector, 3648 pixels, with a spectral response in the range (150 -950) nm.
iii. The software "SpectraSuite", supplied from Ocean Optics, was used in this work to control the Ocean Optics USB spectrometer.
iv. Samples: Three different types of Sudanese gasoline, collected from three different refineries, were investigated here. These are petroleum-derived liquid mixtures primarily used as a fuel in internal combustion engines. It consists mostly of aliphatic hydrocarbons obtained by the fractional distillation of petroleum, enhanced with iso-octane or the aromatic hydrocarbons toluene and benzene to increase its octane rating. Small quantities of various additives are common, for purposes such as tuning engine performance or reducing harmful exhaust emissions.

The experimental procedure was done as follows:

- Laser energy was adjusted to obtain sufficient peak power needed to form plasma.
- A laser pulse was fired on the sample cell without the sample and the plasma emission spectrum was recorded and saved as background.
- After that the plasma emission spectrum of every sample was recorded.
- The plasma spectrum was processed by subtracting the background.
- By referring to the atomic spectra database, the atoms and ions in the samples were identified.

3. Results and discussion

Figures (2), (3) and (4) show the emission spectra, in the range from 179 nm to 859 nm, of the three gasoline samples (1, 2 and 3), respectively. The laser energy per pulse was 23.4 mJ and the pulse duration was 10 ns that lead to peak power equals 2.34 $X10^6$ W. The power density was $1.32*10^8$ W/cm^2.

Atomic spectra database was used for the spectral analysis of the tested samples and the results are listed in table (1).

The three spectra of the samples show the essential atoms of the hydrocarbons that form the gasoline (C and H) with different amounts for each sample. There were also common relatively small amounts of other elements, including (Ar, Co, Fe, N, Ni, O, W, Si and Sc), found in all the three samples with different amounts. The Argon, Cobalt, Iron and Nickel were very low in sample 1 while larger amounts of them were found in samples 2 and 3. Silicon and Scandium were found nearly equal in all the three samples. The presence of all mentioned elements is due to the origin of the crude oil. The Oxygen atoms were found with relatively smaller amount in sample 1 compared with sample 2 and 3, it seemed that it was added in the refinery to reduce carbon monoxide and unburned fuel in the exhaust gas thus reducing smoke.

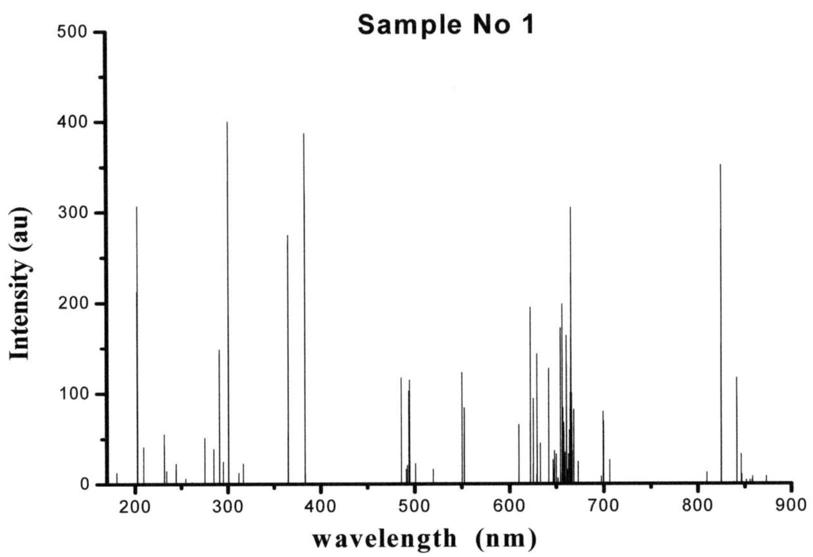

Fig. 2. LIBS emission spectrum of sample No 1

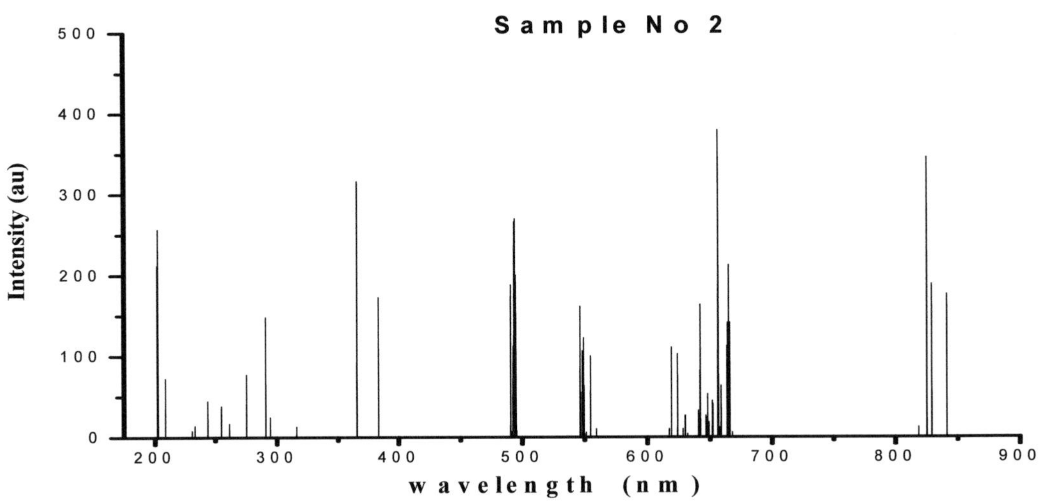

Fig. 3. LIBS emission spectrum of sample No 2

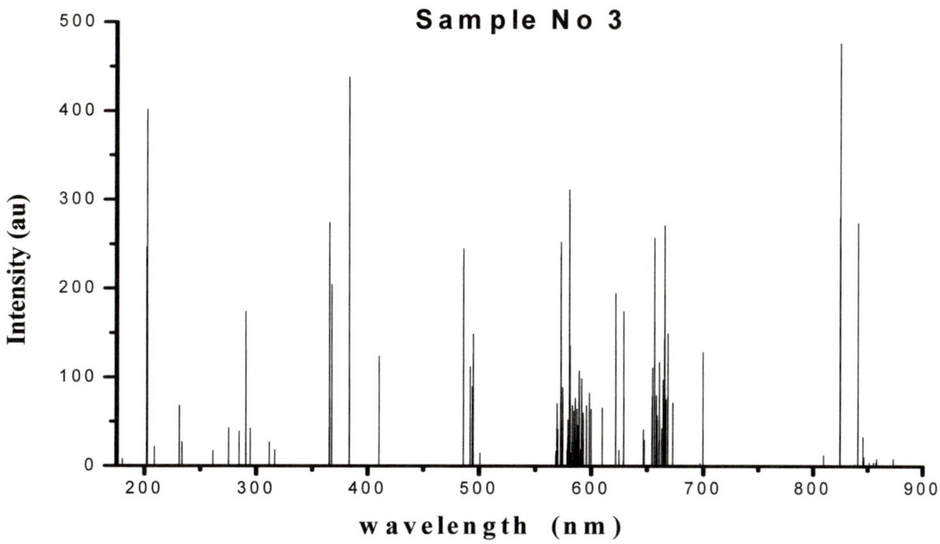

Fig. 4. LIBS emission spectrum of sample No 3

The appearance of Tungsten in the spectra was due to tiny fraction of the background lamp intensity that was not stored due to the change in its intensity.

The ratio of hydrogen to carbon determines the octane rating which is the measure of resistance of petroleum to engine knocking that damage it quite quickly. The octane rating decreases with carbon chain length and increases with carbon chain branching [11]. High rating means more hydrogen atoms with the same carbon atoms. That depicts that sample 1 has lower ratio (2.76) and lower rating, while samples 2 and 3 have higher ratios, 3.76 and 2.84, respectively.

Line Wavelength λ (nm)	Sample no. (1)		Sample no. (2)		Sample no. (3)	
	Element	Intensity (a.u.)	Element	Intensity (a.u)	Element	Intensity (a.u)
180.29	Rb II	12.29			Rb II	7.9
202.03	C II	212	C II	212	C II	247
202.64	O I	307.1	O I	256.9	O I	402.04
208.95	C III	40.96	C III	72.4	C III	24.5
231.43	O III	55	O III	8.19	O III	68.01
233.74	Na II	18.37	Na II	13.34	Na II	27.34
244.04	Fe II	22.53	Fe II	44.7		
254.94	O III	6.14	O III	38.4		
261.43	C III	2.05	C III	17	C III	42.76
275.18	C III	51.2	C III	77.24	C III	1721
290.88	C II	148.4	C II	148.4	C II	174.09
294.8	C III	24.58			C III	42.31
311.61	Hg II	12			Hg II	27.04
316.37	Na II	22.53	Na II	13.2	Na II	18
365.66	H I	274.5	H I	316.2	H I	387.1

Line Wavelength λ (nm)	Sample no. (1)		Sample no. (2)		Sample no. (3)	
	Element	Intensity (a.u.)	Element	Intensity (a.u)	Element	Intensity (a.u)
367.62					H I	204.1
383.54	H I	387.1	H I	173	H I	387.1
410.17					H I	123.84
486.13	H I	117			H I	245.13
491.74			B III	7.37		
492.34	O II	20.48	O II	49.15	O II	112.06
492.73			Fe I	112.64		
492.93			Na VI	36.86		
493.33			C I	266.24		
493.53			N I	83.97		
493.92			Fe I	270.33		
494.12	O IV	102.4			O V	89.51
494.32			K II	11.44		
494.71	O I	2.05	O I	200.7	O I	36
494.91	O II	114.69	O II	116.73	O II	148.72
500.85	O III	22.53			O III	14.52
545.22			W I	204.8		
546.39			Mg II	161.79		
546.97			W I	55.3		
548.33			Co I	106.49		
549.11			Kr I	122.88		
549.50			Ni I	63.49		
551.25			Ti I	6.14		
554.56			Xe III	100.35		
559.03	Ca I	3.97	Ca I	10.24		
568.33					V I	8.86
568.52					Th I	17.06
569.49					Ni I	70.3
569.88					Hg II	15.01
570.07					Ar I	41.63
572.97					S II	10.91
574.32					Fe I	88.73
575.29					Kr II	13.92
578.38					O II	17.06
579.15					Fe I	51.87
579.73					S II	33.44
580.5					C I	311.26
581.46					Xe I	15.01
581.65					Ne I	19.1
582.81					Fe I	25.25

Line Wavelength λ (nm)	Sample no. (1)		Sample no. (2)		Sample no. (3)	
	Element	Intensity (a.u.)	Element	Intensity (a.u)	Element	Intensity (a.u)
583					Xe I	68.25
583.19					K I	6.82
583.58					W I	6.82
585.12					Fe I	25.25
586.27					W I	29.34
587.43					He I	19.13
587.81					Co I	45.73
588.96					C I	84.64
589.16					C II	107.17
590.31					O II	6.82
590.69					Fe I	19.1
591.27					S II	92.15
592.81					N II	47.77
595.49					Kr I	68.25
598.37					Fe I	82.59
598.94					W I	0.67
599.9					N I	64.16
609.83	C II	65.53			C II	65.53
624.29			V I	102.4		
625.01	Ne II	94.21				
625.05					Ne II	18.1
628.84	Sc I	10.24	Sc I	10.24	Sc I	5.84
629.22	C I	143.36			C I	174.06
629.98			Rb I	26.62		
630.36			Kr II	26.62		
632.06			La II	4.1		
632.82	Ne I	45.06				
640.56			N I	32.77		
641.13			Eu I	26.62		
641.88	S III	126.97	S III	81.92	S III	79.4
642.26			N I	163.8		
646.59	O V	26.62			O V	41
646.78			Ni II	26.62		
647.34			Sc II	12.29		
647.53	Na II	16.38			Na II	29.22
647.72			Fe I	24.58		
648.1	Ar I	36.86	Ar I	53.25		
649.22			O II	18.43		
649.79	Nb I	32.77				
651.67	Ne II	6.14				

Line Wavelength λ (nm)	Sample no. (1)		Sample no. (2)		Sample no. (3)	
	Element	Intensity (a.u.)	Element	Intensity (a.u)	Element	Intensity (a.u)
651.85			Si I	45.06		
652.42			Si III	40.96		
654.37	O V	172.03				
654.67	Li II	11.59			Li II	78.56
655.61					Al III	6.82
656.29	H I	198.65	H I	379.8	H I	257.2
657.8	C I	67.58			C I	79.18
658.76	C I	26.62			C I	49.02
658.98	Si I	34.82			Si I	57.13
659.17			C IV	63.49		
660.57	C I	163.84			C I	117
661.41	Fe III	15.55				
662.73	O II	32.77			O II	42.54
663.84	Co I	15.86			Co I	97.1
664.03	S II	59.39	S II	112.64	S II	97.06
664.40	O IV	100.35	O IV	141.31	O IV	67.41
664.97	N I	161.79			N I	143.16
665.25			O II	212.99		
665.46	C I	305.15			C I	271.08
665.53	Mg IV	86.01			Mg IV	32.61
666.66	O II	100.35			O II	74.85
666.69	N I	36.86				
668.22	O V	81.92			O V	149.22
668.33			Th I	6.14		
672.61	N I	24.58			N I	71.21
697.08	Kr II	8.19			Kr II	3.02
699.10	W I	79.87				
699.47	S I	69.63			S I	128.6
706.28	Ni I	26.62			Ni I	43.01
824.99	H I	351.7	H I	346	H I	476.34
829.23			H I	189.16		
838.37					Cl II	2.05
841.33	H I	116.8	H I	177.03	H I	274.29
845.92	Ne II	32.77			Ne II	32.77
846.61	Fe I	10.24				
851.57	S II	4.1				
855.50	Hg I	4.1				
857.54	C II	4.1				
858.22	Fe I	8.19				

Table 1. Analysis of emission spectra for the three gasoline samples

Also there were relatively small amounts of neutral atoms which were not common for the three samples including (Ca, Eu, He, Hg, K, Kr, Nb, Ne, S, Th, Ti, V, Xe). This depicts that the original crude oils were brought from various fields and different refinery techniques were used to get the three types of gasoline, it is also may be due to additives added in order to enhance the gasoline quality.

Beside neutral atoms, ions of different amounts and ionization stages were found in the three samples like: Al III, B III, C II, C III, Fe II, Fe III, Hg II, K II, Kr II, Li II, Mg II, Mg IV, N II, Na II, Ni II, O II, O III, O V, S II, S III and Si III. Some of these ions were produced via the ionization of neutral atoms by the laser power density itself.

4. Conclusions

From the experimental results obtained in this work one can conclude that

- LIBS technique showed that the essential atoms forming the samples (Hydrogen and Carbon) appeared with higher amounts compared with other elements (like Calcium, Helium, Mercury, Potassium, Krypton, Neon, Niobium, Scandium, Silicon, Thorium, Titanium, Vandite) that were exist with very lower amounts.
- LIBS has the ability to detect, sensitively, almost all the elements and ions in gasoline samples.
- LIBS is a very good diagnostic technique that can be used in investigation of elements in liquid samples and to get the octane rating, in liquid fuels, precisely.

5. References

[1] David A. Cremers and Leon J. Radziemski “Handbook of Laser-Induced Breakdown Spectroscopy" John Wiley & Sons Ltd, England (2006).
[2] E. S. Dayhoff and B. Kessler, "high speed sequence photography of a ruby laser" Applied Optics 1 (1962) 339.
[3] J. Reader and C. H. Corliss, "Wavelengths and Transition Probabilities for Atoms and Atomic Ions Part II. Transition Probabilities", NSRDS-NSB 68, Washington DC: US Government Printing Office (1980).
[4] A. R. Striganov and N. S. Sventitskii, "Tables of Spectral Lines of Neutral and Ionized Atoms", New York: IFI/Plenum (1968).
[5] R. Payling and P. Larkins, "Optical Emission Lines of the Elements" Chichester: John Wiley (2000).
[6] G. M. Weyl, "Laser-Induced Plasmas and Applications", New York: Marcel Dekker (1989).
[7] J. R. Wachter and D. A. Cremers, "determination of uranium in solution using laser induced breakdown spectroscopy" Appl. Spectrosc., 41 (1987), 1042–1048.
[8] H. A. Archontaki and S. R. Crouch, "evaluation of an isolated droplet sample introduction system for laser induced breakdown spectroscopy" Appl. Spectrosc., 42 (1988), 741–746.
[9] D. A. Cremers, L. J. Radziemski and T. R. Loree, "theortical calculation of population of excited level of emitting atomic species in water" Appl. Spectrosc., 38 (1984), 721–729.

[10] Cristina Lo´pez-Moreno, Santiago Palanco, J. Javier Laserna, "Quantitative analysis of samples at high temperature with remote laser-induced breakdown spectrometry using a room-temperature calibration plot", Spectrochimica Acta Part B 60 (2005) 1034 - 1039.

[11] Francesco Ferioli, and Steven G. Buckley, "Measurements of hydrocarbons using laser-induced breakdown spectroscopy", Combustion and Flame 144 (2006) 435–447.

2

Nd:YAG Laser (1064 nm) in Management of Pilonidal Sinus

Ezzat A. Badawy
[1]Alexandria University
[2]Americal Safat Medical Center
[1]Egypt
[2]Kuwait

1. Introduction

Pilonidal sinus (PNS) is chronic inflammatory process of the skin in the cleavage between the buttocks (natal cleft). It is caused by keratin plug and debris clinically observed as a pit, having penetrated the dermis (1). The term pilonidal comes from a combination of Latin words meaning hair (pilus) and nest (nidal). PNS was first described by Hodges in 1880 (2).

PNS occurs ten times more often in men than in women (3). This may be attributed to the presence of more hair in men than in women. PNS usually occurs after puberty (4). PNS is more likely to occur in obese people and in those with thick, stiff body hair. Although the upper natal cleft is the most common site (Figure 1), PNS may occur in other sites especially where there is irregularity of skin surface with pressure applied to that region (5). The same condition can be seen in the clefts between fingers in those persons dealing with hair such as barbers and hairdressers (6, 7). It has been reported in other areas such as the umbilicus (8-10), the interdigital spaces of the foot (11), the finger pulp (12), the clitoris (13, 14), penis (15-17), the intermammary (18), the periungual (19) and the perineal regions (20).

2. Clinical picture

2.1 Etiology

The origin of the problem is not well understood. In the past, PNS was thought to be a congenital condition. However, there is nearly a consensus that it is an acquired condition. PNS has high incidence in certain communities which can be explained by the different hair distribution and its growth pattern. The condition is less common in Asians and Africans than in Caucasians (21). Nevertheless, most of the investigators suggest the acquired theory of PNS. Certain factors may be in favor of the occurrence of PNS. These factors are hormonal, presence of hair, friction, sitting or driving for long periods and infection. The risk factors of PNS were found in one study to be sedentary occupation (44%), positive family history (38%), obesity (50%) and local irritation or trauma prior to onset of symptoms (34%) (4). Akinci et al have studied the incidence of certain etiological factors in soldiers with and without PNS. They found that family history; obesity; being the driver of a vehicle;

and having folliculitis or a furuncle can be associated with the presence of PNS (22). Harlak et al have suggested that hairy people who sit down for more than six hours a day and those who take a bath two or less times per week have an increased risk for PNS than those without these risk factors (23).

Hormonal factor was thought to be one of the factors that can play a role in the etiology of PNS. At puberty, androgens are secreted to enhance the development of the pilo-sebaceous glands, which coincides with the onset of PNS (24). The early presentation of PNS is the visible pit in the midline of the natal cleft. This is actually the microscopic appearance of enlarged hair follicles. The weight of the buttocks can cause stretching of the follicular openings (25). It is explained by relatively high force of buttocks' weight applied over very small area of pilo-sebaceous follicle (about 1 mm^2), over the sacro-coccygeal joint. All the movements and sitting for long time can amplify the force applied to this area. When the force applied reaches a critical level, this can cause rupture of the follicle base; the weakest part. The friction between buttocks is another factor added to this force which can help for sucking keratin and hair into the distended follicles.

It has been postulated for long time that hair follicles were the source of PNS (25). Recently, studies have found that the specimens of excised lesions had the pits penetrating into the dermis but not all of them arose in hair follicles (26). Nevertheless, the hair has an important role in the development of inflammation. Sorts of treatments directed against hair follicles gave very good therapeutic effects.

2.2 Patho-physiology

In 1992, Karydkis has explained the patho-physiology of the PNS as an acquired condition. He called it "The hair insertion process". It consists of three main factors; the loose hair, the presence of vacuum force and the vulnerability of the skin (27). The entering hair causes an inflammatory foreign body reaction which can lead to the formation of multiple micro-abscesses. Eventually, that migrates further into the subcutaneous tissue forming a sinus which might complicate by abscess formation or creating more sinus tracts that open laterally forming fistulous tracts. However, during the surgical operations, only 50-70% of PNS had actually been found to contain hair (1, 28).

2.3 Symptoms and signs

Symptoms of PNS vary from a small asymptomatic dimple to a huge painful abscess with oozing pus (29). Early lesion appears as a pit or dimple which may be asymptomatic for the patient. However, PNS can cause discomfort and pain. Severe acute pain with or without a purulent discharge may be the presentation when an abscess is formed. Suppuration is very annoying to the patients causing staining of the underwear with unpleasant smell.

Chronic pain is the usual presentation of the chronic and complicated cases. Discharge, fistulae and disfiguring scarring can be seen in the neglected cases (4). On clinical examination, single or multiple sinuses can be seen. Discharge, debris and hair are seen arising from the sinus (Figure 1). Recurrent lesion is a common feature even after complete and wide surgical excision of PNS (Figure 2). Surgical treatment and recurrence lead to permanent scarring (Figure 3).

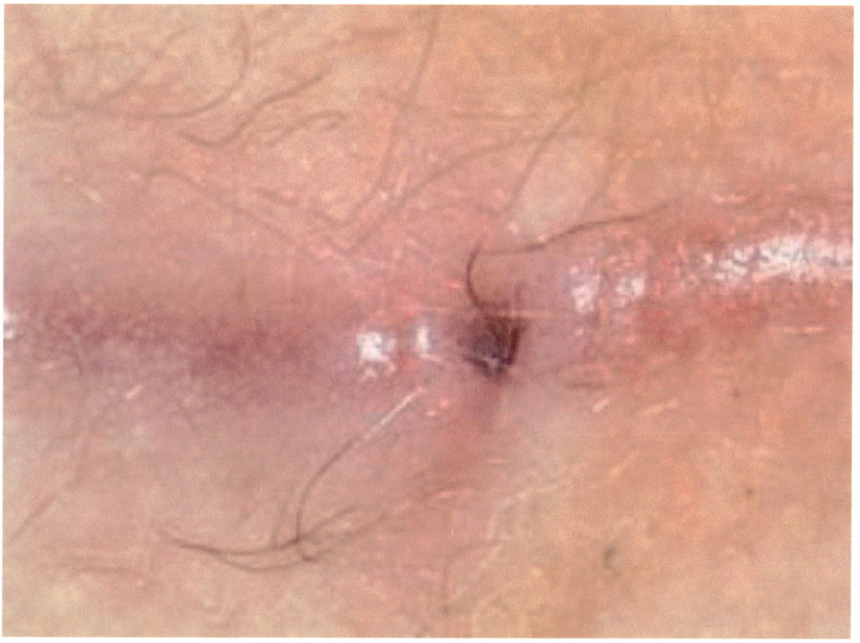

Fig. 1. PNS with hair protruding out of its opening. (Provided by Dr. Metwally Afifi)

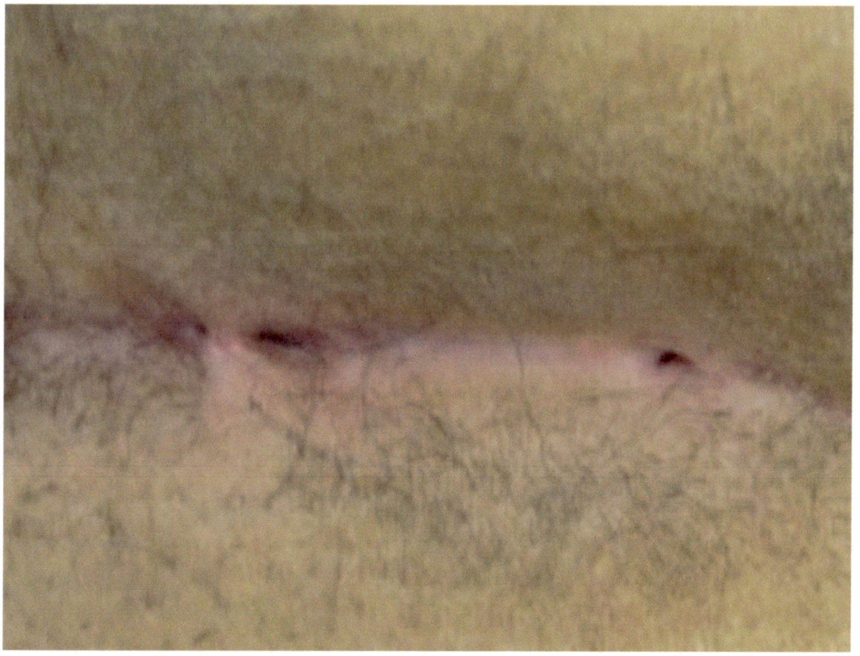

Fig. 2. Recurrent PNS. (Provided by Dr. Metwally Afifi)

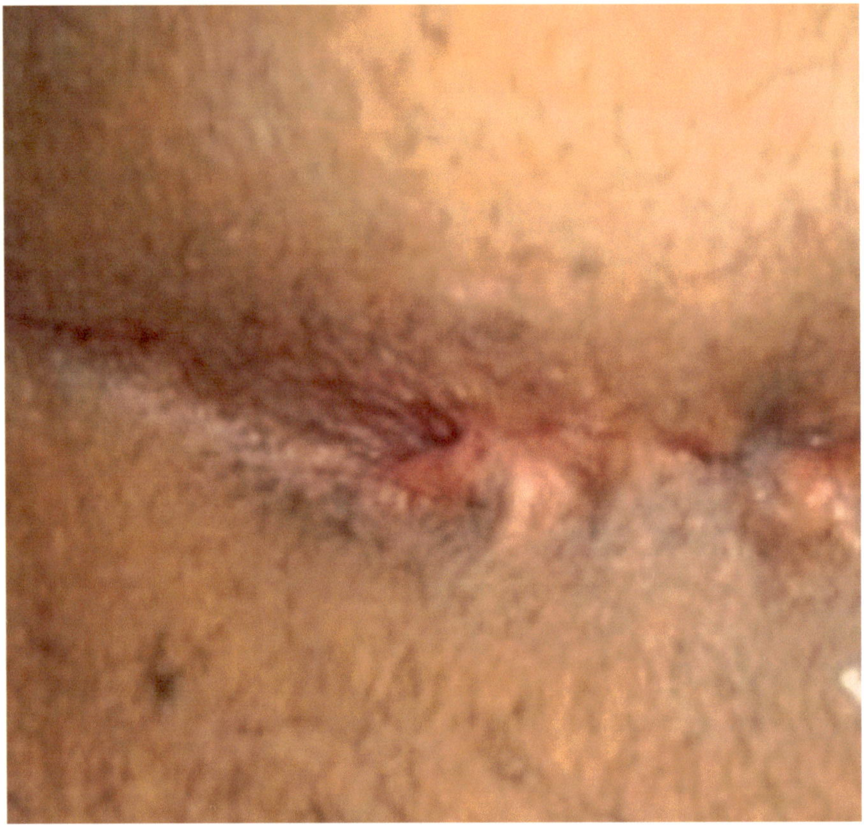

Fig. 3. Recurrent PNS with scarring and fistula. (Provided by Dr. Metwally Afifi)

2.4 PNS complications

Abscess formation is the most common complication of PNS. Recurrence rate of PNS (Figures 2 and 3) is very high whatever the sort of management (30). Chronicity of PNS can develop if the drainage at the abscess stage is not adequate. Development of fistulae is common due to repeated infection with suppuration with subsequent scarring (Figure 3). Systemic infection is a rare complication. Severe pyomyositis caused by methicillin-sensitive Staphylococcus aureus was reported secondary to an operated PNS in an apparently healthy young male patient (31). Moreover, septic arthritis was reported in a patient with rheumatoid arthritis after the resection of PNS (32). Although Actinomycosis associated with PNS of the penis is extremely rare, few cases have been reported (16, 17, 33). Neglected PNS can lead to unusual and life-threatening consequences. Necrotizing fasciitis and toxic shock syndrome have been developed in a previously healthy patient (34).

PNS rarely develops malignant changes (26, 35). If it happens, squamous cell carcinoma is the most common presentation (26, 36). Malignant changes usually occur in young men on very long standing antecedent recalcitrant PNS. Malignant changes of a chronic wound have worse prognosis than skin malignancy arising de novo (37). Although malignant change of PNS is rare, it carries bad prognosis.

2.5 Pathology

The pathological findings occurred in PNS were thought to be mainly due to the hair entry and its subsequent changes. The enlargement of follicles precedes the presence of hair inside the developed PNS (25). Being a foreign body inside PNS, hair is the main cause of the inflammatory reaction with subsequent edema, distension and obstruction of the pilo-sebaceous follicle. Infection can cause suppuration leading to abscess formation. If the patient is not probably treated, a chronic abscess will develop. Both aerobe and anaerobe can be detected in PNS (38, 39). The tissue can form a track to drain the abscess leading to sinus. Recurrence is a main feature of this condition. PNS rarely develops malignant changes (26, 35).

The histologic examination of PNS revealed the presence of chronic inflammation surrounding a sinus in the sub-cutaneous tissue. Hair in sinuses was found in three quarters of the specimens examined. Examination showed that hair entered via one of the created sinus openings. Pits were found to be indentations of the skin containing keratin plugs and debris. They may be isolated or connected with hair follicles (1).

2.6 Laboratory studies

The diagnosis of PNS is easy and depends mainly on clinical inspection and localization of the lesion. No specific laboratory tests are needed to confirm the diagnosis. Swab from oozing pus in acute abscess may be needed to do bacterial culture and antibiotic sensitivity. Excisional biopsy and histo-pathologic examination can be carried out for neglected cases and when malignancy is suspected.

3. Surgical management of PNS

Patients with mild form of PNS with no marked symptoms need no surgical intervention at this stage (26) but proper hygiene and hair removal are needed to prevent inflammation and complications. There are many types of surgical interventions to treat PNS (40-47). Nevertheless, all forms of surgical management have a significant failure rate, with a high incidence of recurrence (30). No single surgical technique can be relied upon to prevent recurrence (5). Different modalities of surgery have been used, but they often result in recurrence, and additional surgery is needed (30, 40). Conservative methods have been applied to treat PNS. The most commonly used conservative treatment is injecting phenol as a sclerosing agent (48 - 50).

The choice of the surgical intervention depends on the severity of signs and symptoms, being primary or recurrent and presence or absence of complications. If the patient presents with an abscess, primary incision and drainage of purulent discharge should be performed to relief symptoms. Complete drainage of the discharge with curettage to remove debris and infected tissue is important to improve the rate of healing (43). The application of platelet-derived growth factors to the post-operative PNS wound enhanced the healing process (51). Systemic antibiotics have been found to improve the healing rate after surgical excision (52). Otherwise, surgical excision of PNS may be followed by either primary closure or closure by secondary intention. Surgical unroofing of primary PNS gave a shorter time to heal and carried a lower complication rate compared to wide local excision (53). Recently, it is advocated to do surgical excision with primary closure (54).

Whatever the sort of management and surgical procedure in treating PNS, the recurrence rate is high (41-43). Reconstructive surgical procedures after excision of the PNS can be performed in cases of failure of the classic excision and for complex or complicated PNS (55-59).

4. ND:YAG laser in management of PNS

4.1 Role of Laser hair removal (LHR)

Management of PNS should be directed to treat the underlying etiological factors. Cubukcu et al had found that obese patients with high body mass index had a higher risk of recurrence of PNS after surgical intervention (60). Nevertheless, Cubukcu et al in another study have found that obesity alone is not an important factor in the etiology of PNS (61). Moreover, the presence of the hair seems to be the most important triggering factor in the pathogenesis and recurrence of PNS. By different ways of approach, many studies have confirmed the role of hair in the evolution and inflammation of PNS. Karydakis has concluded that hair insertion is the cause of PNS, and it prevents spontaneous recovery and delays healing of any wound in the depth of the natal cleft. He stated that hair is the cause of recurrence (27). Moreover, removal of hair can prevent further inflammation, progression and recurrence of PNS (62 - 64). As a conclusion of the extensive research carried out on the different types of PNS, management objectives should be directed not only at eradicating the obvious lesion present but also to prevent recurrence through managing the etiological factors. The re-accumulation or re-growth of hair may be the most important factors in the development of inflamed PNS (5). Odili and Gault have concluded that Laser hair removal (LHR) in the natal cleft is by no means a cure for PNS. Removal of hair by laser represents an alternative and effective method of hair removal. They have studied the effect of laser on the healing process of PNS. LHR was found to allow the PNS to heal rapidly. Odili and Gault have stated that LHR should thus be considered as an aid to an effective healing tool of the problem (65). As a matter of fact, removal of the hair in the natal cleft can prevent inflammation and recurrence of PNS. Nevertheless, regular razor epilation of hair increased the recurrence rate of PNS (66). Razor hair epilation can insert hair into skin and thus, it can aid in PNS recurrence and inflammation. Because LHR was found to be an effective adjunct in preventing the recurrence of PNS, many investigators suggested that LHR should be offered routinely to all patients (63, 64, 67 – 69). In a case report, a patient with recurrent PNS was satisfactorily treated by LHR using a ruby laser. The follow up period was 6 months (70). It is well documented now that eliminating the hair of the affected area could be a new approach to the treatment of patients with PNS. Six young men having PNS in a study done by Landa et al have also been treated by LHR. None of those patients experienced recurrence of PNS. Landa et al have concluded that LHR of the natal cleft should be considered a first choice treatment for recurrent PNS. Preventive LHR of the natal cleft in patients with recurrent folliculitis could avoid future surgery (71). Benedetto and Lewis have treated two patients with recalcitrant PNS by LHR using diode laser (72).

4.2 Advantages of Nd:YAG LHR

The destruction of hair follicles by laser is very efficient and popular method worldwide. Different types of laser and non-laser machines are used remove hair permanently (73), namely; normal mode ruby laser (694 nm), normal mode alexandrite laser (755 nm), diode lasers (800, 810 nm), Nd:YAG laser (1,064 nm), and intense pulsed light (IPL) sources (590-

1,200 nm). Moreover, Nd:YAG laser has peculiar advantages over the other devices. Nd:YAG laser was superior in hair reduction than Diode laser and it provided higher patient satisfaction (74). The wave-length of Nd:YAG-laser is 1064 nm. Consequently, the penetration of the laser beam through the skin is deeper than the other known lasers targeting the melanin and thus, it can treat the hair follicles which are present in the deeper part of skin. Because of its deeper penetration, it can reach also the base of PNS.

As general potential side effects, all laser types can potentially cause scarring, hypo, or hyperpigmentation especially in dark skin types. The risk of skin types V and VI to get either hypo- or hyper-pigmentation following laser treatment is very high. Nevertheless, this risk is very low with Nd:YAG laser. Nd:YAG laser can be used to treat all skin types with less risk of discoloration. Even skin type VI can be treated safely with this type of laser. Moreover, Nd:YAG laser has much less complications compared to other types of laser. Studies done by Goh (75) and Lanigan (76) have shown that Nd:YAG laser has been used successfully to treat abnormal hair in patients with skin types V and VI. In both studies, no risk of serious adverse side effects has been noted with the Nd:YAG laser (75, 76). However, pain is the most prominent side effect of Nd:YAG laser (62, 75, 76). Good cooling during treatment and proper choice of laser parameters can reduce the perception of pain.

4.3 Experience of Nd:YAG laser in PNS

We have carried out a study on twenty five patients to evaluate the effect of Nd:YAG LHR on the recurrence rate of PNS following surgical treatment. Fifteen of the patients have undergone LHR using Nd:YAG laser while ten patients did not have LHR during the study and follow-up period. All the twenty five patients underwent surgical excision to remove PNS prior to LHR sessions. None of the patients who performed LHR by Nd:YAG laser had shown recurrence of PNS during the study and follow-up period. Recurrence of PNS was seen in seven of those who had no LHR (62). Pain was the most prominent side effect seen in our study as it was a complaint of six patients out of fifteen. Nevertheless, the pain did not affect the patients passively. Two thirds of the patients were either satisfied or highly satisfied. All patients appreciated the improvement of hair density to different degrees. The degree of improvement was correlated with the number of LHR sessions. Avoidance of further surgical intervention was achieved by the use of Nd:YAG laser.

4.4 Protocol of Nd:YAG LHR in treatment of PNS

Generally, it is preferable to do primary closure of the excised PNS wound to assure rapid healing and to minimize the post-operative pain. LHR can be started after complete wound healing. Before doing LHR treatment, the area to be treated should be shaved and cleaned with local non-irritant antiseptic cleanser. Usually, no topical anesthetic cream should be applied before Nd:YAG laser treatment. However, sensitive patient can use topical EMLA cream one hour prior to treatment. Hair removal by Nd:YAG (1064-nm wavelength) laser is carried out in its normal mode. The usual spot size of the laser beam used in hair removal is ranged between 10 to 15 mm. The spot size more than 15 mm can be more painful. The range of effective laser fluence depends on the used laser spot size. Fluence of 35 to 50 J/cm2 can be used with 10 mm spot size laser beam. Less fluence can be used for larger spot size. In case of dark skin types, it is advised to use longer laser pulse width than that used for lighter skin. Skin type VI needs pulse width of 35 - 40 msec. or more. The laser treatment should be accompanied with a cooling method to alleviate the potential pain of laser. The cooling air is widely used and it is effective as well. It should be used during the whole laser treatment. The laser treatment is carried out on the area of natal cleft, extending at least 5 cm laterally in both sides, up to 5 cm above the PNS scar and down to the anal orifice. Larger area can be treated especially in hairy persons.

Mild to moderate erythema is seen immediately after treatment and it lasts few minutes up to few hours. Skin of the treated area becomes warm. Cooling of the skin is achieved by application of an ice-pack on the treated area immediately after treatment. Soothing gel and local antibiotic cream can be applied to reduce any unpleasant sensation and to avoid secondary infection. Non-infective folliculitis may be seen later on. If it happens, it lasts two to three days. Patients are examined clinically before each treatment in order to assess the degree of hair reduction, to exclude recurrence of PNS, to exclude any sign of infection, and to record any side effect. Treatments are usually performed at 6- to 8-week intervals. The intervals between sessions are becoming prolonged with advancement of treatment. That is because the hairs take more time to re-grow. Hair density in the treated area is becoming less with treatment and hair caliber is getting finer. Generally, the patients are requested to do session of LHR once they noticed regrowth of hairs. Fine hairs usually need more treatment sessions than coarse dark hairs.

4.5 Future treatment of PNS by Nd:YAG laser

Unpublished research has been carried out to evaluate the effect of Nd:YAG laser in treating the PNS itself without performing any surgical interventions. The underlying concept was to destroy the hairs inside the PNS itself in order to evaluate the subsequent effect on the inflamed PNS lesion. The preliminary results were very promising. The use of Nd:YAG laser can accelerate the healing of inflamed PNS. It helped in achieving marked improvement of mild to moderate lesions without carrying out surgical treatment. One of our treated patients had long history of unhealed post-operative wound which lasted few months. The unhealed wound showed rapid healing after Nd:YAG laser treatment. Other investigators have used Nd:YAG laser in the same way to treat PNS. Lindholt et al have succeeded in treating a case of PNS by Nd:YAG laser (77). Bashir and Kurwa have used the laser to destroy the hair within an active PNS in a 30-year-old white male patient, in order to remove the source of the foreign body reaction. They succeeded in treating the PNS which was finally healed (78).

Studies have shown that Nd:YAG laser is a good tool in different therapeutic aspects. It is very effective in hair removal, treating vascular lesions, for skin rejuvenation and as surgical ablation device. No risk of serious adverse side effects has been noted with the Nd:YAG laser (75, 76). Palesty et al have used it in pilonidal cystectomy. Patient postoperative satisfaction after laser excision was greater when compared with those who had traditional surgical excisions. Postoperative pain was less (79). Excision of PNS by laser seems to be promising therapeutic modality in the management of this condition.

5. Conclusion

Treatment of PNS is important, not only to alleviate pain and to reduce inflammation, but also to avoid long term complications. Malignant transformation is one of the important fatal complications (80) m60). Although the Malignant degeneration of PNS is a rare complication, it is associated with a high recurrence rate and poor prognosis (81) m61). The management of both asymptomatic non-inflamed, recurrent and inflamed PNS can be directed towards removal of the hair in the area by Nd:YAG laser. It can be done before the surgical intervention and post-operative to prevent recurrence. PNS itself can be removed by Nd:YAG laser and thus, patients can avoid risk of surgical complications. Nd:YAG laser may be a hope for this benign yet troublesome condition.

6. Acknowledgement

We kindly thank Dr. Metwally Afifi, consultant of general surgery, Al-Safat American Medical Center (Kuwait) for providing us with PNS photos.

7. References

[1] Sondenaa K, Pollard ML. Histology of chronic pilonidal sinus. APMIS (Acta Pathologica, Microbiologica, et Immunologica Scandinavica) 1995; 103(4): 267-72
[2] Hodges RM. Pilonidal sinus. Boston Med Surg J 1880; 103: 485-586.
[3] Dwight RW, Maloy JK. Pilonidal sinus: experience with 449 cases. N Engl J Med 1953; 249: 926-30
[4] Sondenaa K, Andersen E, Nesvik I, Soreide JA. Patient characteristics and symptoms in chronic pilonidal sinus disease. Int J Colorectal Dis 1995; 10(1): 39-42.
[5] Stephens FO, Stephens RB. Pilonidal sinus: management objectives. Aust N Z J Surg. 1995; 65(8): 558-60.
[6] Vergles D, Cupurdija K, Lemac D, Legac A, Kopljar M. Interdigital pilonidal sinus in a female hairdresser. ANZ J Surg. 2010; 80(11): 856.
[7] Aydin HU, Mengi AS. Recurrent interdigital pilonidal sinus treated with dorsal metacarpal artery perforator flap. J Plast Reconstr Aesthet Surg. 2010; 63(12): e832-4.
[8] Eby CS, Jetton RL. Umbilical pilonidal sinus. Arch Dermatol. 1972; 106(6): 893.
[9] Abdelnour A, Aftimos G, Elmasri H. Conservative surgical treatment of 27 cases of umbilical pilonidal sinus. J Med Liban. 1994;42(3):123-5.
[10] Abdulwahab BA, Harste K. Umbilical pilonidal sinus. Ugeskr Laeger. 2010; 172(41): 2848-9.
[11] O'Neill AC, Purcell EM, Regan PJ. Interdigital pilonidal sinus of the foot. Foot (Edinb). 2009; 19(4): 227-8.
[12] Grant I, Mahaffey PJ. Pilonidal sinus of the finger pulp. J Hand Surg Br. 2001; 26(5): 490-1.
[13] Palmer E. Pilonidal cyst of the clitoris. Am J Surg. 1957; 93(1): 133-6.
[14] Baker T, Barclay D, Ballard C. Pilonidal cyst involving the clitoris: a case report. J Low Genit Tract Dis. 2008; 12(2): 127-9.
[15] Rao AR, Sharma M, Thyveetil M, Karim OM . Penis: an unusual site for pilonidal sinus. Int Urol Nephrol. 2006;38(3-4):607-8.
[16] Chikkamuniyappa S, Scott RS, Furman J. Pilonidal sinus of the glans penis associated with actinomyces case reports and review of literature. ScientificWorldJournal. 2004; 4: 908-12
[17] Val-Bernal JF, Azcarretazábal T, Garijo MF. Pilonidal sinus of the penis. A report of two cases, one of them associated with actinomycosis. J Cutan Pathol. 1999; 26(3): 155-8.
[18] Sunkara A, Wagh D, Harode S. Intermammary pilonidal sinus. Int J Trichology. 2010; 2(2): 116-8

[19] Ngan PG, Varey AH, Mahajan AL. Periungual pilonidal sinus. J Hand Surg Eur 2011; 36(2): 155-7
[20] Testini M, Miniello S, Di Venere B, Lissidini G, Esposito E. Perineal pilonidal sinus. Case report. Ann Ital Chir. 2002; 73(3): 339-41
[21] Berry DP. Pilonidal sinus disease. J Wound Care 1992; 1(3): 29-32.
[22] Akinci OF, Bozer M, Uzunköy A, Düzgün SA, Coşkun A. Incidence and aetiological factors in pilonidal sinus among Turkish soldiers. Eur J Surg. 1999; 165(4): 339-42.
[23] Harlak A, Mentes O, Kilic S, Coskun K, Duman K, Yilmaz F. Sacrococcygeal pilonidal disease: analysis of previously proposed risk factors. Clinics (Sao Paulo). 2010; 65(2): 125-31.
[24] Price ML, Griffiths WAD. Normal body hair: a review. Clin Exp Dermatol 1985; 10: 87-97
[25] Bascom JU. Pilonidal disease: correcting over treatment and under treatment. Contemporary Surg 1981; 18: 13-28
[26] Lineaweaver WC, Brunson MB, Smith JF, Franzini DA, Rumley TO. Squamous carcinoma arising in a pilonidal sinus. J Surg Oncol 1984; 27(4): 39-42.
[27] Karydkis G E. Easy and successful treatment of PNS after explanation of its causative process. Aust N Z J Surg 1992; 62 (5): 385-9
[28] Bascom J. Pilonidal disease: origin from follicles of hairs and results of follicle removal as treatment. Surgery. 1980; 87(5): 567-72
[29] Solla JA, Rothenberger DA. Chronic pilonidal disease. An assessment of 150 cases. Dis Col Rec 1990; 33(9): 758-61.
[30] Bascom J, Bascom B. Failed pilonidal surgery: new paradigm and new operation leading to cures. Arch Surg 2002; 137: 1146–50.
[31] Lorenz U, Abele-Horn M, Bussen D, Thiede A. Severe pyomyositis caused by Panton-Valentine leucocidin-positive methicillin-sensitive Staphylococcus aureus complicating a pilonidal cyst. Langenbecks Arch Surg. 2007; 392(6):761-5
[32] Borer A, Weber G, Riesenberg K, Schlaeffer F, Horowitz J. Septic arthritis due to bacteroides fragilis after pilonidal sinus resection in a patient with rheumatoid arthritis. Clin Rheumatol. 1997; 16(6): 632-4
[33] Saharay M, Farooqui A, Chappell M. Actinomycosis associated with pilonidal sinus of the penis. Br J Urol. 1996; 78(3): 464-5
[34] Velitchkov N, Djedjev M, Kirov G, Losanoff J, Kjossev K, Losanoff H.Toxic shock syndrome and necrotizing fasciitis complicating neglected sacrococcygeal pilonidal sinus disease: report of a case. Dis Colon Rectum. 1997; 40(11): 1386-90.
[35] Tirone A, Gaggelli I, Francioli N, Venezia D, Vuolo G. Malignant degeneration of chronic pilonidal cyst. Case report. Ann Ital Chir. 2009; 80(5): 407-9.
[36] Bark T, Wilking N. Squamous-cell carcinoma in a pilonidal sinus. Case report. Acta Chir Scand. 1986; 152: 703-4.
[37] Trent JT, Kirsner RS. Wounds and malignancy. Adv Skin Wound Care 2003; 16(1): 31-34
[38] Miocinović M, Horzić M, Bunoza D. The prevalence of anaerobic infection in pilonidal sinus of the sacrococcygeal region and its effect on the complications. Acta Med Croatica. 2001;55(2):87-90.
[39] Søndenaa K, Nesvik I, Andersen E, Natås O, Søreide JA. Bacteriology and complications of chronic pilonidal sinus treated with excision and primary suture. Int J Colorectal Dis. 1995;10(3):161-6.
[40] Iesalnieks I, Furst A, Rentsch M, Jauch KW. Primary midline closure after excision of a pilonidal sinus is associated with a high recurrence rate. Chirurg 2003; 74: 461–8.

[41] Soll C, Hahnloser D, Dindo D, Clavien PA, Hetzer F. A novel approach for treatment of sacrococcygeal pilonidal sinus: less is more. Int J Colorectal Dis. 2008; 23(2): 177-80.

[42] Abdul-Ghani AK, Abdul-Ghani AN, Ingham Clark CL. Day-care surgery for pilonidal sinus. Ann R Coll Surg Engl. 2006; 88(7): 656-8.

[43] Berry DP. Pilonidal sinus disease. J Wound Care 1992; 1(3): 29-32.

[44] Al-Salamah SM, Hussain MI, Mirza SM. Excision with or without primary closure for pilonidal sinus disease. J Pak Med Assoc. 2007; 57(8): 388-91.

[45] Tejirian T, Lee JJ, Abbas MA. Is wide local excision for pilonidal disease still justified? Am Surg. 2007; 73(10): 1075-8.

[46] LORD PH, MILLAR DM. PILONIDAL SINUS: A SIMPLE TREATMENT. Br J Surg. 1965; 52: 298-300.

[47] Millar DM, Lord PH. The treatment of acute postanal pilonidal abscess. Br J Surg. 1967; 54(7): 598-9.

[48] Maurice B A, Greenwood R K. Conservative treatment of pilonidal sinus. Br J Surg. 1964;51:510–512.

[49] Sakçak I, Avşar FM, Coşgun E. Comparison of the application of low concentration and 80% phenol solution in pilonidal sinus disease. JRSM Short Rep. 2010; 1(1): 5.

[50] Kayaalp C, Olmez A, Aydin C, Piskin T, Kahraman L. Investigation of a one-time phenol application for pilonidal disease. Med Princ Pract. 2010; 19(3): 212-5.

[51] Spyridakis M, Christodoulidis G, Chatzitheofilou C, Symeonidis D, Tepetes K. The role of the platelet-rich plasma in accelerating the wound-healing process and recovery in patients being operated for pilonidal sinus disease: preliminary results. World J Surg. 2009; 33(8): 1764-9

[52] Bunke HJ, Schultheis A, Meyer G, Düsel W. Surgical revision of the pilonidal sinus with single shot antibiosis. Chirurg. 1995; 66(3): 220-3

[53] Lee SL, Tejirian T, Abbas MA. Current management of adolescent pilonidal disease. J Pediatr Surg. 2008; 43(6): 1124-7.

[54] Toccaceli S, Persico Stella L, Diana M, Dandolo R, Negro P. Treatment of pilonidal sinus with primary closure. A twenty-year experience. Chir Ital. 2008; 60(3): 433-8.

[55] Anderson JH, Yip CO, Nagabhushan JS, Connelly SJ. Day-case karydakis flap for pilonidal sinus. Dis Colon Rectum. 2008; 51(1): 134-8.

[56] Mentes O, Bagci M, Bilgin T, Ozgul O, Ozdemir M. Limberg flap procedure for pilonidal sinus disease: results of 353 patients. Langenbecks Arch Surg. 2008 ; 393(2): 185-9.

[57] Bessa SS. Results of the lateral advancing flap operation (modified Karydakis procedure) for the management of pilonidal sinus disease. Dis Colon Rectum. 2007 ; 50(11): 1935-40.

[58] Milito G, Gargiani M, Gallinela MM, Crocoli A, Spyrou M, Farinon AM. Modified Limberg's transposition flap for pilonidal sinus. Long term follow up of 216 cases. Ann Ital Chir. 2007; 78(3): 227-31.

[59] Mahdy T. Surgical treatment of the pilonidal disease: primary closure or flap reconstruction after excision. Dis Colon Rectum. 2008 Dec;51(12):1816-22.

[60] Cubukcu A, Gonullu NN, Paksoy M, Alponat A, Kuru M, Ozbay O. The role of obesity on the recurrence of pilonidal sinus disease in patients, who were treated by excision and Limberg flap transposition. Int J Colorectal Dis. 2000; 15(3): 173-5.

[61] Cubukcu A, Carkman S, Gonullu NN, Alponat A, Kayabasi B, Eyuboglu E. Lack of evidence that obesity is a cause of pilonidal sinus disease. Eur J Surg. 2001; 167(4): 297-8.

[62] Badawy EA, Kanawati MN. Effect of hair removal by Nd:YAG laser on the recurrence of pilonidal sinus. J Eur Acad Dermatol Venereol. 2009; 23(8): 883-6.

[63] Conroy FJ, Kandamany N, Mahaffey PJ. Laser depilation and hygiene: preventing recurrent pilonidal sinus disease. J Plast Reconstr Aesthet Surg. 2008; 61(9): 1069-72.

[64] Sadick NS, Yee-Levin J. Laser and light treatments for pilonidal cysts. Cutis. 2006; 78(2): 125-8.

[65] Odili J, Gault D. Laser depilation of the natal cleft—an aid to healing the pilonidal sinus. Ann R Coll Surg Engl 2002; 84: 29–32.

[66] Petersen S, Wietelmann K, Evers T, Hüser N, Matevossian E, Doll D. Long-term effects of postoperative razor epilation in pilonidal sinus disease. Dis Colon Rectum. 2009; 52(1): 131-4

[67] Schulze SM, Patel N, Hertzog D, Fares LG 2nd. Treatment of pilonidal disease with laser epilation. Am Surg. 2006 ; 72 (6): 534-7.

[68] Lukish JR, Kindelan T, Marmon LM, Pennington M, Norwood C. Laser epilation is a safe and effective therapy for teenagers with pilonidal disease. J Pediatr Surg. 2009; 44(1): 282-5.

[69] Oram Y, Kahraman F, Karincaoğlu Y, Koyuncu E. Evaluation of 60 patients with pilonidal sinus treated with laser epilation after surgery. Dermatol Surg. 2010; 36(1): 88-91.

[70] Lavelle M, Jafri Z, Town G. Recurrent pilonidal sinus treated with epilation using a ruby laser. J Cosmet Laser Ther 2002; 4: 45–7.

[71] Landa N, Aller O, Landa-Gundin N, Torrontegui J, Azpiazu JL. Successful treatment of recurrent pilonidal sinus with laser epilation. Dermatol Surg. 2005; 31(6): 726-8.

[72] Benedetto AV, Lewis AT. Pilonidal sinus disease treated by depilation using an 800 nm diode laser and review of the literature. Dermatol Surg. 2005; 31(5): 587-91.

[73] Haedersdal M, Haak CS. Hair removal. Curr Probl Dermatol. 2011;42:111-21.

[74] Wanitphakdeedecha R, Thanomkitti K, Sethabutra P, Eimpunth S, Manuskiatti W. A split axilla comparison study of axillary hair removal with low fluence high repetition rate 810 nm diode laser vs. high fluence low repetition rate 1064 nm Nd:YAG laser. J Eur Acad Dermatol Venereol. 2011 Sep 19. doi: 10.1111/j.1468-3083.2011.04231.

[75] Goh CL. Comparative study on a single treatment response to long pulse Nd:YAG lasers and intense pulse light therapy for hair removal on skin type IV to VI--is longer wavelengths lasers preferred over shorter wavelengths lights for assisted hair removal. J Dermatolog Treat. 2003; 14(4): 243-7.

[76] Lanigan SW. Incidence of side effects after laser hair removal. J Am Acad Dermatol. 2003; 49(5): 882-6.

[77] Lindholt CS, Lindholt JS, Lindholt J. Treatment of pilonidal cyst with Nd-YAG laser. Ugeskr Laeger. 2008; 170(26-32): 2321-2.

[78] Bashir S, Kurwa H. Successful resolution of a chronic pilonidal sinus with laser epilation. JAAD 2009; 60 (3 suppl 1): AB195.

[79] Palesty JA, Zahir KS, Dudrick SJ, Ferri S, Tripodi G. Nd:YAG laser surgery for the excision of pilonidal cysts: a comparison with traditional techniques. Lasers Surg Med. 2000;26(4):380-5.

[80] Mentes O, Akbulut M, Bagci M. Verrucous carcinoma (Buschke-Lowenstein) arising in a sacrococcygeal pilonidal sinus tract: report of a case. Langenbecks Arch Surg. 2008; 393(1): 111-4.

[81] de Bree E, Zoetmulder FA, Christodoulakis M, Aleman BM, Tsiftsis DD. Treatment of malignancy arising in pilonidal disease. Ann Surg Oncol. 2001; 8(1): 60-4.

3

Processing of Metallic Thin Films Using Nd:YAG Laser Pulses

Santiago Camacho-López et al.*
Departamento de Óptica, Centro de Investigación Científica y de Educación Superior de Ensenada, Carretera Ensenada- Tijuana, Zona Playitas, Ensenada, Baja California México

1. Introduction

Nd:YAG lasers are possibly the more widely used lasers either for basic research or for industrial and technological applications (Dubey, A. K. & Yadava, V. 2008). These lasers are also excellent pump sources for laser development, for instance Ti:sapphire ultrashort pulse lasers are based on CW Nd:YAG pumping. In particular, Nd-YAG lasers have been applied to study laser-induced oxidation in metals as titanium and chromium; semiconductors as silicon (Aygun, G. et al., 2006). (Perez del Pino, A. et al., 2004) demonstrated that the rutile phase of TiO_2 is obtained by laser oxidation in air of titanium films. Nd:YAG laser pulses have been used to laser-induce a phase transformation from W_3O thin films to WO_3 (Evans R., et al., 2007); laser ablation for micromachining of bulk metals as copper, bronze and aluminum has also been done using Nd:YAG nanosecond pulses (Maisterrena-Epstein R., et al., 2007); laser-induced oxidation and novel LIPSS formation in titanium thin films deposited on silicon substrates was demonstrated by using a single laser beam from a frequency doubled Nd:YAG nanosecond pulsed laser (Camacho-Lopez S., et al., 2008). Some works about pulsed laser oxidation have been reported (Dong, Q. et al., 2002). (Pereira, A. et al. 2004) have investigated the laser treatment in steel irradiating at various wavelengths by using different laser sources. In Table 1, we cited some works on the oxidation induced by pulsed laser irradiation in various metals. Recently, we have published results on fs-laser

---

*Marco A. Camacho-López[2], Oscar Olea Mejía[3], Rodger Evans[1], Gabriel Castillo Vega[1], Miguel A. Camacho-López[4], Manuel Herrera Zaldivar[5], Alejandro Esparza García[6] and José G. Bañuelos Muñetón[6]
[1]*Departamento de Óptica, Centro de Investigación Científica y de Educación Superior de Ensenada, Carretera Ensenada- Tijuana, Zona Playitas, Ensenada, Baja California, México*
[2]*Facultad de Química, Universidad Autónoma del Estado de México, Tollocan s/n, esq. Paseo Colón, Toluca, Estado de México, México*
[3]*Centro Conjunto de Investigación en Química SustenTable UAEM-UNAM (CCIQS), Facultad de Química, Universidad Autónoma del Estado de México, de la carretera Toluca-Atlacomulco, San Cayetano, México*
[4]*Facultad de Medicina, Universidad Autónoma del Estado de México, Paseo Tollocan s/n, esq. Jesús Carranza, Toluca, Estado de México, México*
[5]*Centro de Nanociencias y Nanotecnología, Universidad Nacional Autónoma de México, Carretera Tijuana-Ensenada, Ensenada, Baja California, México*
[6]*Centro de Ciencias Aplicadas y Desarrollo Tecnológico, UNAM, Apdo. Postal 70-186, México, DF, México*

oxidation in molybdenum thin films (Cano-Lara, M. et al., 2011). (Herman et al., 2006) have studied ablation of molybdenum thin films with short and ultrashort laser pulses. This work is the first study with Nd:YAG to investigate pulsed laser oxidation in molybdenum thin films.

Material	Research group Starting material Laser line		
Titanium	(Pérez del Pino et al. 2002) Titanium targets λ=1064 nm	(Lavisse, L. et al. 2002) Titanium targets λ=1064 nm	(Camacho-López, S. et al 2008) Titanium thin films λ=532 nm
Chromium	(Dong, Q. et al. 2002) Chromium films λ=1064 nm		
Steel	(Pereira, A. et al. 2004) Steel target 1064 nm 532 nm		
Molybdenum	Our work Molybdenum Thin Films λ=532 nm		

Table 1. Some works on Nd-YAG pulsed laser oxidation.

Some advantages of laser-induced metallic oxides are:

1. Oxidation can be performed in air, so that a controlled atmosphere is not necessary.
2. The time that it takes to achieve a given stoichiometry and crystalline phase is very rapid as compared to conventional thermal treatment.
3. A high spatial resolution patterning of metallic oxides is only possible by this method. Sizes are determined by the optical diffraction limit, therefore single "pixels" made of metallic oxide in the order of the laser light wavelength are possible.
4. In some cases, the laser-induced oxidation process is accompanied by laser-induced periodic surface structures (LIPSS) formation.

Titanium dioxide is an important material due to its wide range of applications. Titanium dioxide is a biocompatible material; in its thin film form, TiO_2 has applications as an antireflective coating, or anticorrosive coating. Additionally, titanium dioxide has applications as gas sensor material, in photocatalysis, among others (Linsebigler A. M. et al., 1995). It is well known that TiO_2 exist as a crystalline material in three phases: anatase, rutile and brookite (Beattie, I. R., Gilson, T. R. 1969).

Molybdenum oxides are attractive materials due to its potential technological applications. MoO_3 possesses photo-, electro- and gasochromic properties (Livage, J. & Ganguli, D. 2001). For instance, this material can be used in gas sensors, catalysis, smart windows, lithium microbatteries (Dieterle, M. 2001). MoO_2 has potential applications as a cathode material in the area of microbatteries; field emission, and also catalysis (Jun, Z. et al. 2003; Wang F. & Lu B. 2009; Mikhailova, D. et al. 2011).

The micro-Raman technique is very useful to obtain information about the composition and structure of the material into the laser irradiated zone. Spatially studies can be carried out since the laser beam can be Focalized down on the sample to 2 μm diameter (Witke, K. et al. 1998). For instance, we have obtained spatially resolved information about what kind of material is formed when a molybdenum thin film was irradiated with fs-laser pulses (Cano-Lara, M. et al., 2011).

In this chapter we present and discuss a series of experimental results on short (nanoseconds) and ultrashort (picoseconds) pulsed Nd:YAG laser processing of metallic thin films. Our selection of materials consisted of molybdenum (Mo) and titanium (Ti) thin films deposited on glass substrates and silicon wafers by the DC-magnetron sputtering technique. We studied the ablation features on the selected materials; once the ablation threshold fluence was determined, we carried out laser processing experiments setting our delivered fluence to a value well below ablation threshold. Under such a scenario we studied the following phenomena: laser-induced periodic surface structures (LIPSS) formation, and laser-induced oxidation on the metallic films. Our results show that it is possible to laser-induce MoO_2 and TiO_2 inside the irradiated zone; we also found that for certain laser irradiation conditions it is possible to obtain LIPSS formation driven by the polarization of the recording beam. The characterization of the laser irradiated metallic thin films consisted mainly on Atomic Force Microscopy (AFM), Scanning Electron Microscopy (SEM), and micro-Raman Spectroscopy.

2. Materials and methods

2.1 Deposition of titanium and molybdenum thin films

Titanium and molybdenum thin films were deposited using disks of titanium and molybdenum (99.9%, Lesker), respectively. Ar gas was used to sputter the targets by means of the DC-magnetron sputtering technique. Glass slides and silicon wafers were used as substrates. In Table 2, one can see the deposition parameters used to obtain each material.

Deposition parameters	Titanium/silicon	Titanium/glass	Molybdenum/glass
Discharge Power	30 W	100 W	150 W
Pressure	$1x10^{-3}$ mBar	$1.4x10^{-3}$ mBar	$1.4x10^{-3}$ mBar
Target-substrate separation	7 cm	7 cm	7 cm
Deposition time	40 min	16 min	10 min
Substrate	Silicon wafer(100)	Glass slide	Glass slide
Substrate temperature	room	room	room

Table 2. Deposition parameters for titanium and molybdenum thin films.

2.2 Laser processing of the metal thin films

The experiments were performed using a typical laser processing set up (see Figure 1) which consist of a computer controlled x-y-z translation stage, where the sample is conveniently hold; the laser beam from either a nanosecond (ns) or a picosecond (ps), frequency doubled, pulsed Nd:YAG laser was used to irradiate the samples at normal incidence; the laser beam can be either focused or non-focused onto the sample. The laser irradiation was carried out at 10 Hz repetition rate, using a single beam, taking care of using a per pulse laser fluence below the ablation threshold for the selected metallic films.

The delivered fluence is controlled by means of an attenuator made of a half-wave plate and a polarizer. An extra half-wave plate, or a quarter-wave plate, is used to change from linear polarization to linear polarization with another orientation, or from linear to circular polarization. The as-deposited metallic thin films were exposed to a series of thousands of pulses.

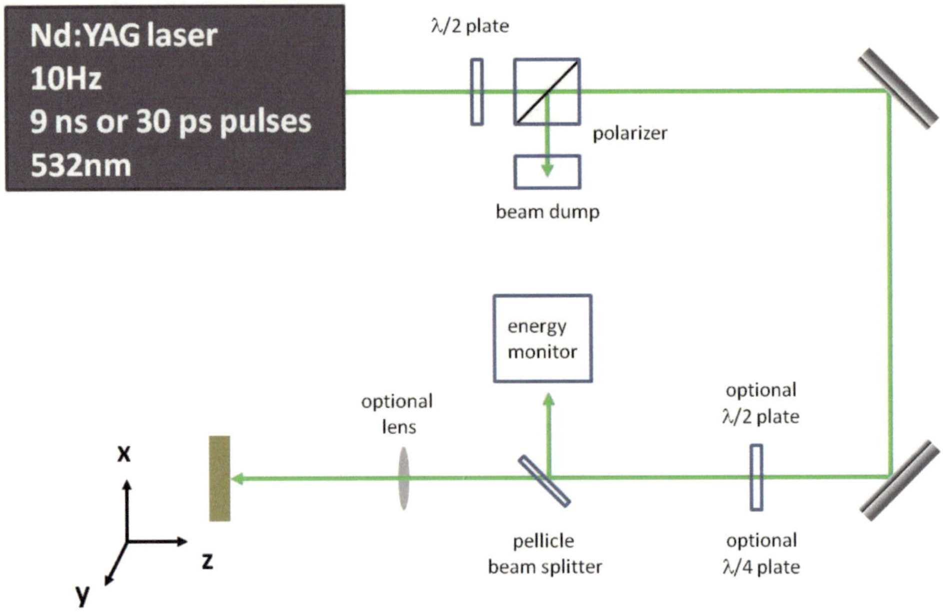

Fig. 1. LIPSS and laser-induced oxidation experimental set up.

Irradiation parameters	Titanium/silicon	Titanium/glass	Molybdenum/glass
Laser Fluence (per pulse)	0.24 J/cm2	0.08 J/cm2	0.08 J/cm2
Wavelength	532 nm	532 nm	532 nm
Repetition frequency	10 Hz	10 Hz	10 Hz
Pulse number	4000	4000 8000 12000	6000 12000 18000 10000 (0.16 J/cm2)
Atmosphere	air	air	air
Substrate temperature	room	room	room

Table 3. Irradiation parameters for the ns laser exposures

Irradiation parameters	Titanium/glass	Molybdenum/glass
Laser Fluence (per pulse)	0.24 J/cm2	0.24 J/cm2
Wavelength	532 nm	532 nm
Repetition frequency	10 Hz	10 Hz
Pulse number	2000, 4000, 6000 8000, 10000	2000, 4000, 6000 8000, 10000
Atmosphere	air	air
Substrate temperature	room	room

Table 4. Irradiation parameters for the ps laser exposures

The pair of frequency doubled Nd:YAG lasers: a Continuum, minilite II, 9 ns pulse duration; and a Ekspla, 30 ps pulse duration, were used to irradiate the metallic thin film samples in atmospheric air. The laser irradiation parameters we used in the experiments are presented in Tables 3 and 4.

2.3 Characterization of the as-deposited Titanium and Molybdenum thin films and the laser exposed sites

The as-deposited molybdenum and titanium thin films were characterized by X-Ray Diffraction (Bruker D8 Advance with Linxeye detector) with the Cu Kα radiation source (λ = 1.5406 Å) and a Scanning Electron Microscope (SEM). The SEM analysis was performed with a JEOL JSM-6510LV microscope in the high vacuum mode. The samples were characterized without any conductive coating with secondary electrons; the acceleration voltage was 20 kV. The AFM analysis was done using a Veeco CP-II in contact mode with a silicon nitride tip. The scanned size is 5x5 microns. The modified material which turns into metallic oxides during the irradiation process was characterized by microRaman spectroscopy. A micro-Raman system, LabRaman HR-800 of Jobin-Yvon-Horiba, was used to run and capture the micro-Raman spectra. The 632.8 nm line of a He-Ne laser was utilized to excite the material and the laser power at the sample was 5 mW. An Olympus BX-41 optical microscope was used to focus down the laser beam on the sample and to collect the scattered light. This was done using a 100X microscope objective lens. All the spectra are the result of 10 acquisitions of 60 s.

3. Experimental results

3.1 X-Ray Diffraction of the as deposited thin films

Figure 2 shows the XRD patterns for the sputtered deposited titanium and molybdenum thin films. The titanium XRD pattern (Figure 2a) contains peaks corresponding to the (002), (101), (102) and (103) reflection planes, that according to the literature correspond to the α-Ti phase. The diffraction peaks in the case of the molybdenum thin film (Figure 2b) correspond to the (110) and (220) reflection planes. This indicates that the molybdenum thin films grew preferentially acquiring the cubic phase.

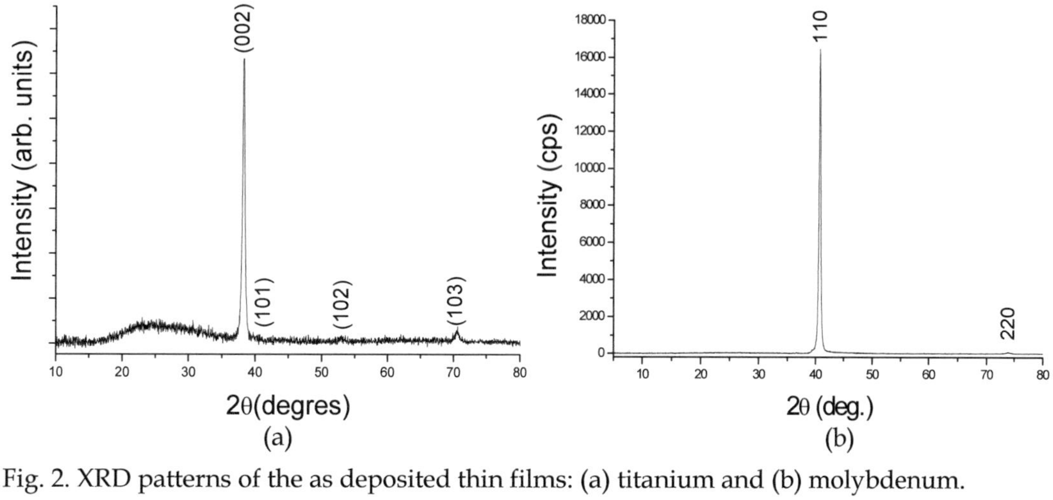

Fig. 2. XRD patterns of the as deposited thin films: (a) titanium and (b) molybdenum.

3.2 LIPSS formation characterized by AFM

3.2.1 Titanium / Silicon and ns pulses laser irradiation

A 320 nm thick Ti thin film deposited on silicon (100) was irradiated using a frequency doubled Nd:YAG laser, with pulses of 9 ns duration at 10 Hz repetition rate, and a per pulse laser fluence of 0.24 J/cm^2. The delivered fluence is well below the ablation threshold, which according to (Vorobyev, A. Y. & C. Guo, C. 2007) is in the order of 0.8 J/cm^2. Figure 3 clearly shows the formation of laser-induced periodic surface structures (LIPSS), such an effect takes place as a result of applying thousands of pulses. In this particular case, we are showing the formation of LIPSS for an exposure of 4000 pulses. Figure 3a, shows an AFM image from a non-irradiated zone in the as-deposited thin film, a homogeneously smooth surface can be identified, it is constituted by a compact layer of nanosized grains. Figures 3b-d show the results of the laser exposures on the Ti film; when either a linear or a circular polarized beam is used the film surface experiences significant changes. In Figures 3b-c, a linear polarized beam was used, notice that in those cases grating-like structures are formed, whose orientation follows the laser polarization direction (indicated with a blue arrow and a **E** which stands for the light electric field); we must observe that the grating-like structures are covered with quasi-periodically distributed craters. If we switch from a given laser linear polarization to the orthogonal linear polarization orientation the grating-like structure follows the polarization orientation. An interesting and expected result is shown in Figure 3d, where it is noticed that when circular polarization is used there is no LIPSS formation at all, however, a series of craters of a few hundred nanometers diameter are formed. Those craters are of the same nature than the ones formed on the grating-like structures under linear polarization exposures; as discussed in (S. Camacho-Lopez et al., 2008) those craters could be formed due to enhanced field effects, which would produce ablation at specific sites with sizes below the wavelength scale. Notice too that the initial nanosized grains that constitute the thin film seem to preserve well under circularly polarized light exposures.

In Figure 3e, we have a cross profile from one of the grating-like structures that result from the linear polarization exposures. It can be easily seen how the grooves periodicity is in the order of the laser wavelength (532nm); this fact is already well known in the LIPSS literature. However, we must note in this case that while in the great majority of the LIPSS reports in metals, the LIPSS formation comes from using laser fluences above the melting and even ablation thresholds (Sipe, J. E. et al., 1983; Young, J. F., et a., 1983); in our case the LIPSS features are not the result of melting or ablation, but the result of laser-induced oxidation of the Ti film. S. Camacho-Lopez et al., reported, back in 2008 for the very first time, LIPSS made of a metallic oxide. An interesting feature is that while in most of the reported work the LIPSS orientation is perpendicular to the laser polarization orientation, in the present case the LIPSS form consistently oriented parallel to the laser beam polarization.

On the optical side effects resulting of the LIPSS formation, we must mention that an angular selective reflectance was obtained when the processed sample is illuminated obliquely under white light and the sample is rotated around the normal to its surface (Camacho-Lopez S., et al., 2008). If the angle of incidence of the white light beam is varied a whole selection of colors is obtained by diffraction off the grating-like structures.

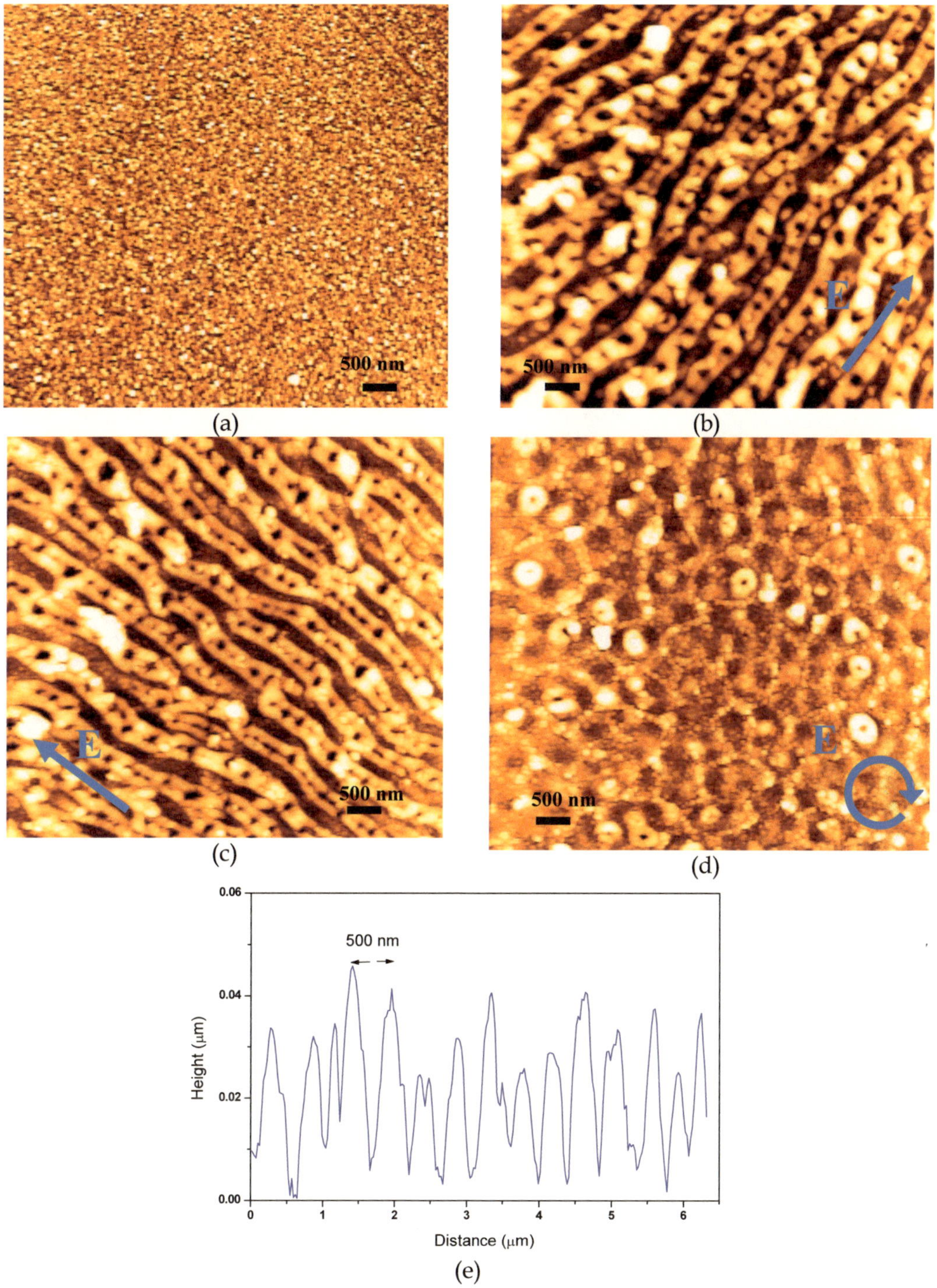

Fig. 3. AFM micrographs for a) an as deposited titanium film deposited in silicon and b-c) laser exposed spots to linear polarization and d) circular polarization; e) is a cross profile of the grating-like structures formed for linear polarization.

3.2.2 Molybdenum / glass and ps pulses laser irradiation

For the case of molybdenum, a thin film deposited on glass was laser irradiated with a frequency doubled Nd:YAG ps pulsed laser using the parameters already mentioned above (Table 4). LIPSS formation is easily obtained, as it shown in Figure 4, for a number of pulses as low as 2000. The LIPSS formation in this case is oriented perpendicular to the laser beam linear polarization (Figure 4a). Notice that craters as those showed in Figure 3 are not formed in this case. Another characteristic to notice here is the fact that the LIPSS periodicity is in the order of twice the laser wavelength (Figure 4b). It is important to point out that a sort of cone shaped structures can be identified to cover the main LIPSS formation; a 3-D AFM profile of such cone shaped structures is presented in Figure 4c, for the case of a spot that has been irradiated with 2000 ps laser pulses.

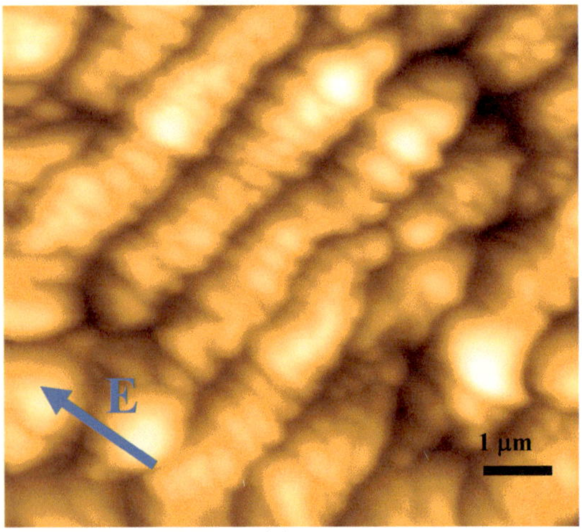

(a)

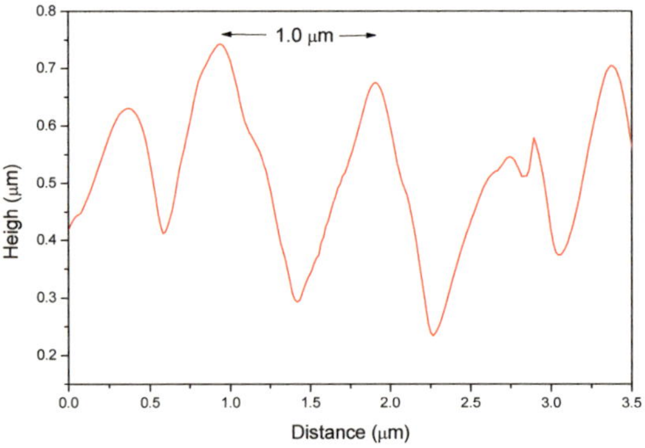

(b)

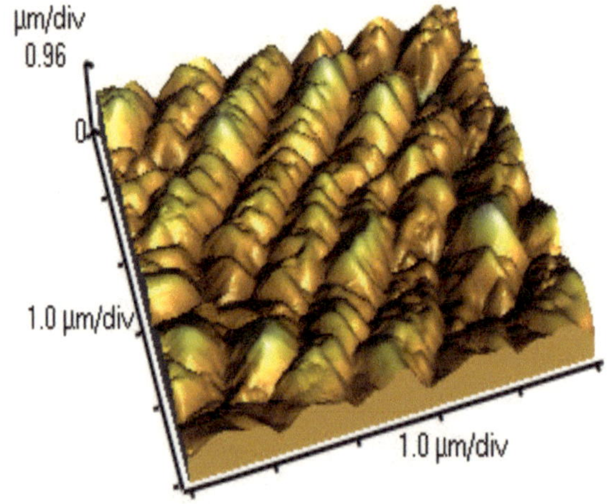

(c)

Fig. 4. AFM micrographs for a molybdenum thin film laser exposed to 2000 ps laser pulses. a) 2-D image of the LIPSS; b) cross profile that shows the LIPSS periodicity; c) 3-D micrograph of the cone shaped structures that cover the main LIPSS.

3.3 LIPSS formation characterized by SEM

3.3.1 Titanium / glass for ns and ps pulses laser irradiation

The as-deposited (on glass substrates) metallic thin films, for both selected metals (Ti and Mo), show smooth and homogeneous surfaces, this is seen for titanium in Figure 5a, and molybdenum in Figure 6a. In the specific case of the as-deposited titanium, the film shows very compact nanosized grain structure; while in the case of the as-deposited molybdenum, a nano-porous structure dominates the film surface texture.

The titanium thin film (500 nm thick) was irradiated using 9 ns laser pulses (see Table 3) and a per pulse fluence of 0.08 J/cm^2, at a repetition rate of 10 Hz; from Figure 5b we can observe that after 4000 pulses the surface of the metallic thin film, suffers almost no change in its morphology and texture. A few cracks developed but the nanosized grain layer still dominates the titanium film. We must mention though that no matter the lack of LIPSS formation, there is a definitely laser-induced oxidation effect on the Ti thin film, this is shown in the following section dedicated to micro-Raman characterization.

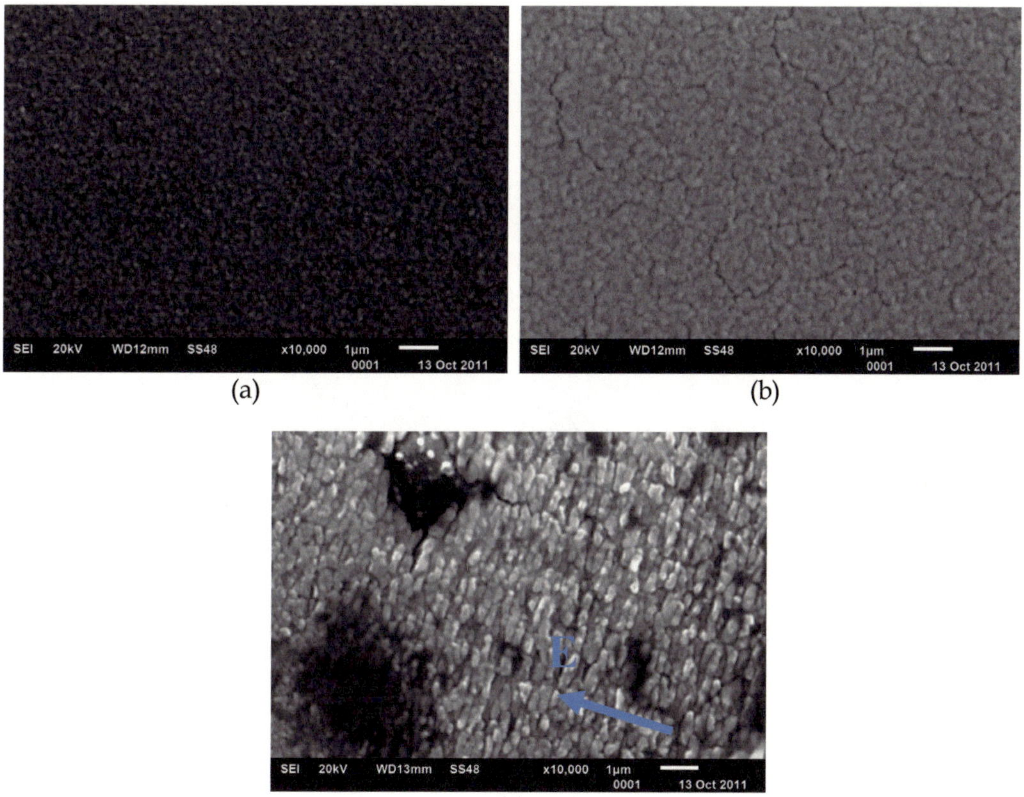

(a) (b)

(c)

Fig. 5. SEM titanium: a) as-deposited Ti thin film; laser irradiated with b) 4000 ns pulses, c) 2000 ps pulses.

When the titanium thin film (500 nm thick) was irradiated using a 30 ps pulse duration at 10 Hz repetition rate and a per pulse delivered fluence of 0.24 J/cm^2, and the number of pulses was set to 2000, the film is significantly modified. Figure 5c shows a typical SEM micrograph of the irradiated spot, where we can see LIPSS formation in a similar manner as we observed in the case of the titanium on silicon substrate irradiated with ns laser pulses. It must be noticed though that the LIPSS features in the ps laser irradiation case present some differences as compared to the ns laser irradiation results; we can see for instance that the oriented structures formed in a denser pattern, notice too that the craters formed in the case of ns laser pulses do not appear in the ps laser pulses case. A very relevant feature, however, is that the orientation of the formed LIPSS on the ps laser pulses case is orthogonal to the laser beam polarization. As expected, the grating-like structures periodicity is in the order of the laser wavelength, which is consistent to the known facts of the LIPSS formation phenomenon. The deep and dark areas in the micrograph most likely correspond to an inhomogeneous intensity distribution across the laser beam and therefore across the laser exposed area. Another feature we must notice is that the grating-like structures in this case are composed by rectangular platelets which actually flake off the sample.

3.3.2 Molybdenum / glass for ns and ps pulses laser irradiation

Figure 6 shows the as-deposited on glass substrate molybdenum thin film (Figure 6a); the ns pulses laser irradiated sample (Figure 6b) and the ps pulses laser irradiated sample (Figure 6c-d).

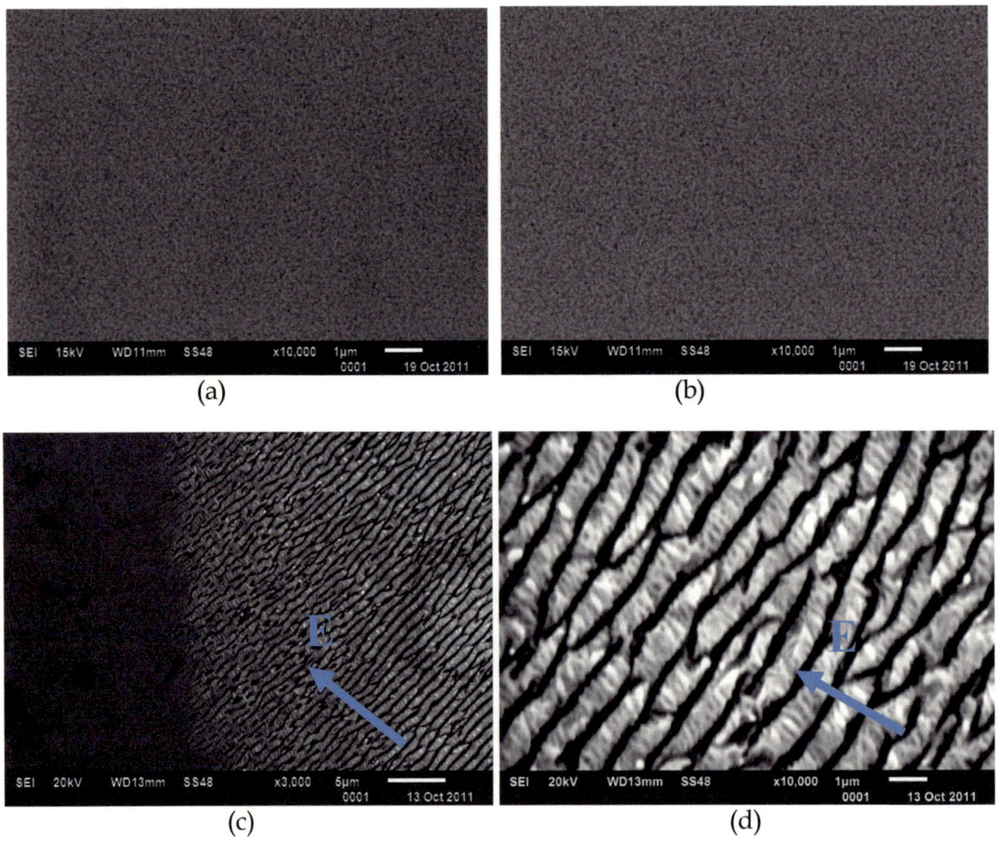

Fig. 6. SEM molybdenum: a) as-deposited thin film; laser irradiated with b) 18000 ns pulses, c) 2000 ps pulses, notice the sharp border between the laser affected and the non-affected zones, d) LIPSS zoom in.

As it occurred in the case of titanium, the ns pulses laser irradiation did not affected the molybdenum surface texture at all (Figure 6b) for up to 18000 pulses of a per pulse fluence of 0.08 J/cm^2, however, as it is shown in the following section dedicated to micro-Raman characterization, a laser-induced oxidation effect takes place as a result of the laser exposure.

For the ps pulses laser irradiation of the molybdenum thin film, as previously mentioned in the AFM section, the orientation of the formed LIPSS is perpendicular to the laser beam polarization. An interesting feature to be noted is a tendency to bifurcation across the grooves formed (Figure 6c-d). There are a few reports of LIPSS formed under femtosecond laser irradiation in metals, where the bifurcation effect is presented although it has not been explained yet. Notice too that craters as those showed in Figure 3 are not formed in this case; this is consistent with the results obtained on titanium deposited on glass when irradiated with the ps laser pulses.

3.4 Pulsed laser-induced oxidation characterized by micro-Raman spectroscopy

The micro-Raman spectra for the titanium thin film irradiated with ns is showed in Figure 7. Raman spectra are displayed between 200 to 800 cm^{-1}. The spectra 7(a-c) correspond to irradiated zones with 4000, 8000 and 12000 pulses at 0.08 J/cm^2.

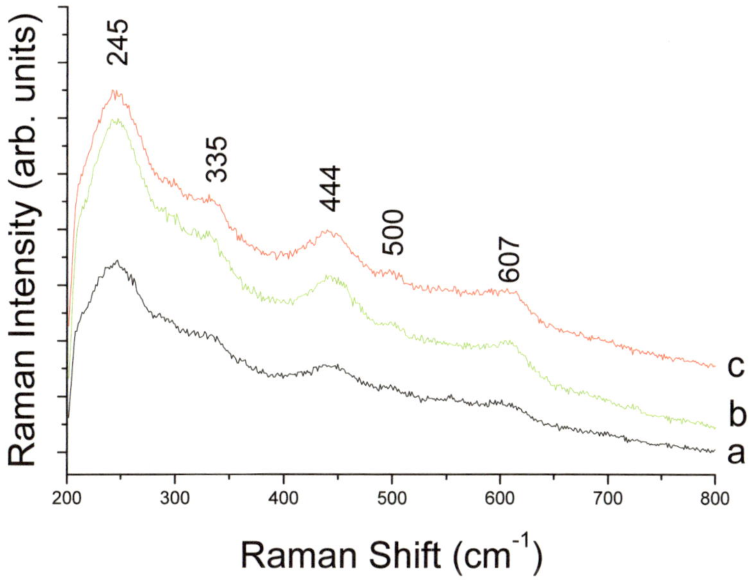

Fig. 7. micro-Raman spectra of a titanium film irradiated at : a) 4000, b) 8000 and c) 12000 ns pulses.

The micro-Raman spectra for the titanium thin film irradiated with ps are showed in figure 8. Raman spectra are displayed between 200 to 800 cm^{-1}. The spectra 8(a-d) correspond to irradiated zones with 2000, 4000, 6000 and 8000 pulses at 0.24 J/cm^2. It can be observed the presence of bands mainly at 442 and 610 cm^{-1}.

Figure 9 shows a set of micro-Raman spectra for the molybdenum thin film irradiated with ns laser pulses. The Raman spectra are displayed between 200 to 1000 cm^{-1}. For comparison purposes, the micro-Raman spectrum 9(a) of m-MoO_2 powder was included (Camacho-López M. A. et al.,). The spectra 9(b-d) correspond to irradiated zones with 6000, 12000 and 18000 pulses at 0.08 J/cm². The spectrum 9(e) corresponds to an irradiated zone with 10000 pulses at 0.16 J/cm². In all cases, micro-Raman spectra are constituted by peaks in the range 200-800 cm⁻¹.

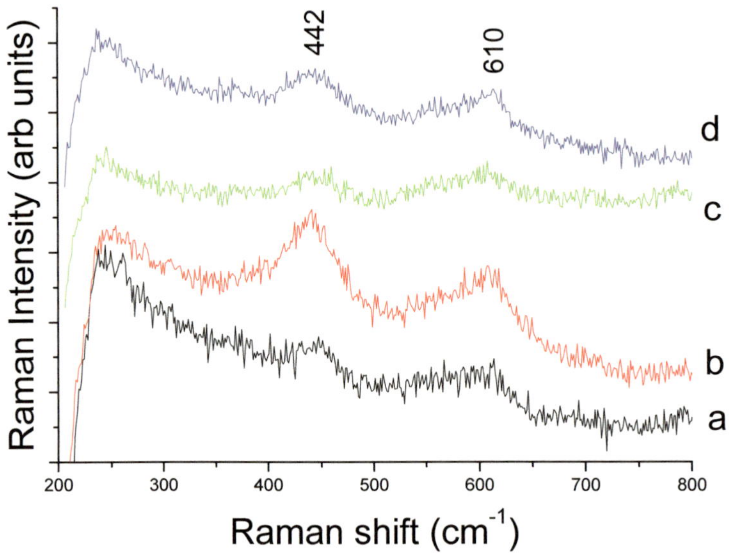

Fig. 8. Raman spectra of the titanium thin film irradiated with ps laser pulses: a) 2000, b) 4000, c) 6000, d) 8000 pulses

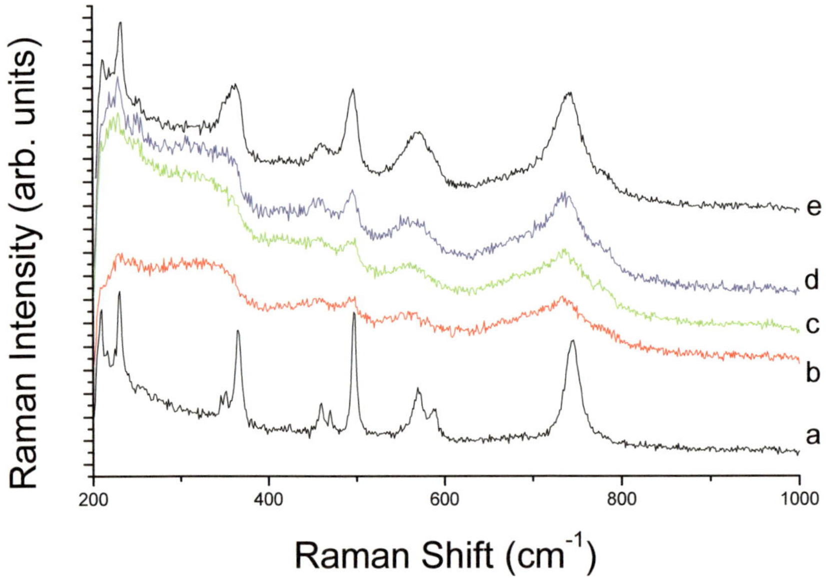

Fig. 9. Micro-Raman spectra of: a) m-MoO_2 powder; molybdenum thin film irradiated with ns pulses b) 6000, c) 12000 and d) 18000 pulses at 0.08 J/cm², e) 10000 pulses at 0.16 J/cm².

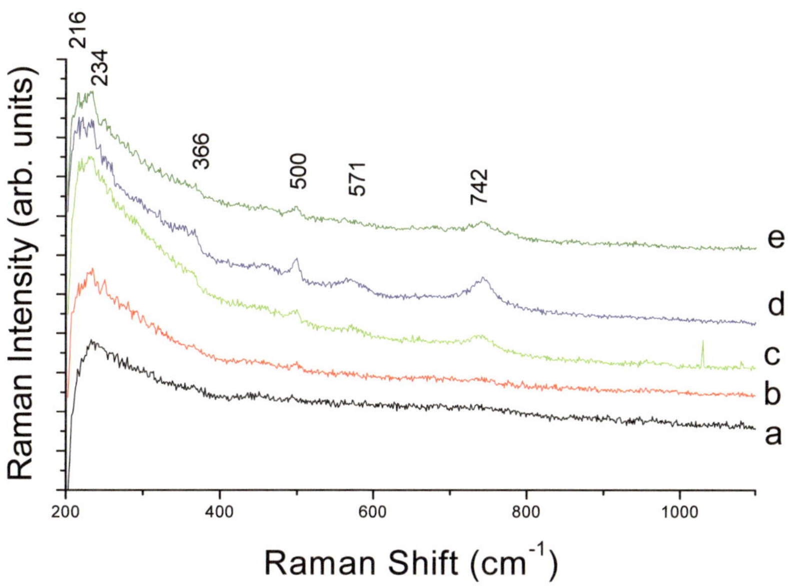

Fig. 10. Raman spectra of molybdenum irradiated with: a) 2000, b) 4000, c) 6000, d) 8000, e) 10000 ps laser pulses.

The micro-Raman spectra for the molybdenum thin film irradiated with ps are showed in Figure 10. As before, the micro-Raman spectra are displayed in the range 200 to 1000 cm^{-1}. Five zones of the thin film were irradiated with 2000, 4000, 6000, 8000 and 10000 ps laser pulses. Each spectrum in Figure 10 corresponds to each irradiated zone.

4. Discussion

4.1 LIPSS formation

The LIPSS phenomenon has been well studied in bulk metals for laser fluences above ablation threshold, but there are no reports of this effect on metallic thin films when irradiated at laser fluences well below ablation threshold. On top of that, there are no reports at all of such LIPSS formed during the laser-induced growth of metallic oxides. In the present work, we have shown different features that can be achieved on the LIPSS formation depending on pulse duration. We selected two metals as titanium an molybdenum to show some of the main characteristic transformations, that occur when those metals in their thin film form are laser irradiated with a frequency doubled Nd:YAG laser in two distinct pulse duration regimes. For the case of titanium and ns laser pulses, it was easy to obtain LIPSS formation for samples deposited on a Silicon substrate; on the contrary it was not possible to obtain LIPSS formation when the titanium is deposited on a glass substrate, although we did get laser-induced oxidation. We did not try molybdenum deposited on Silicon. As for molybdenum deposited on glass it was neither possible to form LIPSS under ns laser irradiation, but laser-induced oxidation is still possible. Remarkably, the situation changes significantly when the titanium and molybdenum thin films deposited on glass substrates are laser irradiated with a frequency doubled Nd:YAG ps laser. For both cases, we obtained LIPSS formation with very distinct features; while titanium forms compact and dense LIPSS with periodicity in the order of the wavelength, molybdenum forms LIPSS covered by cone shaped structures of a few hundred nanometers size. The periodicity of the LIPSS in the case of molybdenum is twice the laser wavelength, which is somehow unexpected.

4.2 Pulsed laser-induced oxidation

4.2.1 Titanium oxide induced by pulsed laser irradiation

From the literature it is well known that the Raman spectrum for the titanium rutile phase, in the range 200 to 800 cm^{-1}, is constituted by three bands located at 236, 444, 609 cm^{-1} (Porto et al., 1967; Beattie I. R., Gilson T. R. 1967; Escobar-Alarcón et al., 1999). There is good agreement between the Raman peak positions obtained for de irradiated zones with ns and ps pulses (Figures 7 and 8) and those reported in the literature for the rutile phase. This indicates that the irradiated material suffered an oxidation passing from Ti to TiO_2 in its rutile phase. This transformation was obtained by (Pérez del Pino et al., 2002) irradiating titanium targets by using the fundamental line (1064 nm) of a Nd:YAG laser. Additionally to the rutile phase, they found another phases like β-Ti_2O_3 and TiO into the irradiated zones. Analyzing our spectra for the ns pulses, we can observe two bands at 445 and 335 cm^{-1} that could be assigned to Ti_2O_3. Therefore, the Raman results indicate that a mixture of TiO_2 and Ti_2O_3 is present in the irradiated zones. It is worth noting that in the case of ps pulses, the Raman spectra do not present the bands corresponding to the Ti_2O_3 phase. In this case only the rutile phase of TiO_2 is obtained in the irradiated zones.

4.2.2 Molybdenum oxide induced by pulsed laser irradiation

From the literature, only a little amount of work on the Raman features for the m-MoO_2 can be found. For instance (R. Srivastava, R. & Chase L. L. 1972) reported the Raman spectrum for a single crystal of MoO_2. (Spevack, P. A.; McIntyre, N. S. 1992, 1993) published the Raman spectrum of MoO_2 for a powder and for thin films too. (M. Dieterle, 2001) reported some molybdenum oxide Raman spectra, in particular for MoO_2. (A. Blume 2004) has extensively studied the Raman spectra of a variety of MoO_x ($2 \leq x \leq 3$). Recently, (Camacho-López et al., 2011) studied the transformation of m-MoO_2 to MoO_3 by micro-Raman spectroscopy. The following Table presents the Raman frequencies obtained for the m-MoO_2 phase by several research groups. Analyzing the Raman frequencies from the literature (Table 5), some differences are observed. More detailed Raman studies on MoO_2 are necessary to determine the spectral changes related to substoichiometry.

Table 6 presents the Raman frequencies obtained for the molybdenum thin films irradiated with ns and ps pulses. Comparing the frequencies obtained in the spectra 9(b-d) and 10(c-e) with those reported in the literature for the m-MoO_2 phase, we can point out that the position of the majority of the Raman peaks (for instance spectrum 9e) have a reasonable agreement with the characteristic Raman spectrum for the m-MoO_2 phase. However, the peaks of the spectra 9(b-d) and 10(c-e) are wider than those for the spectrum 9(a). In particular, the FWHM for the peak (744 cm^{-1}) was indicated in Table 6. This result indicates that the crystallinity of the material in the irradiated zone is not optimized for a laser fluence of 0.08 J/cm^2. When the laser fluence is increased (0.16 J/cm^2), the peak position shifts towards higher frequencies and the FWHM decreases as it is observed in the spectrum 9(e). As a reference value, the FWHM of the peak centered at 744 cm^{-1}, for a crystalline MoO_2

powder (Table 6) and the 1D nanorods (Latha, K. et al. 2007) is 20 cm^{-1}. For a laser fluence of 0.08 J/cm^2 the FWHM is 150 cm^{-1}, while for 0.16 J/cm^2 the FWHM is 50 cm^{-1}. This indicates that crystallinity can be improved with the increase in the laser fluence. The shift in the position of the Raman peaks could be related to substoichiometric molybdenum oxide induced in the irradiated zone of the thin film. It must be noted that 6000 pulses are sufficient in the two cases (ns and ps) to induce the molybdenum oxide. Optically, the molybdenum thin film did not suffered any changes.

Single crystal MoO_2 (R. Srivastava et al. 1972)	m-MoO_2 (Spevack, P. A. &McIntyre, N. S. 1992)	Powder m-MoO_2 (Camacho-lópez, M. A. et al. 2011)	m-MoO_2 (Blume, A. 2004)	Commercial Powder MoO_2 (Latha, K. et al. 2007)
	203	203 208	 208	229
	228	229	232	226.3
	345	346 350	 353	346.2
	363	365	370	360.2
	461	459 469	448 463 473	458.2
505	495	496	501	494.1
	571	569	572	567.2
595	589	588	590	584.4
760	744	744	748	738.5

Table 5. Raman frequencies for m-MoO_2 reported in the literature.

m-MoO_2 (spectrum 9a)	Our work Nanosecond pulses Spectrum 9e	Our work Nanosecond pulses Spectra 9(b-d)	Our work Picosecond pulses Spectra 10(c- e)
203 208	 211	 218	 216
229	232 251	228 248	234
346 350	349		
365	362		366
459 469	461	457	
496	496	495	500
569	570	565	571
588			
744 (20 cm^{-1})	742 (50 cm^{-1}) 780	678 737 (150 cm^{-1}) 778	742

Table 6. Raman frequencies obtained in our work

5. Conclusions

We have demonstrated that it is possible to create laser-induced periodic surface structures (LIPSS) on metallic (titanium and molybdenum) thin films, by irradiating the thin film with either a nanosecond or a picosecond, frequency doubled, Nd:YAG pulsed laser. We found that in the nanosecond regime the LIPSS formation seems to be influenced by the substrate type (silicon or glass); although the delivered laser fluence could also be a factor. A very interesting fact is that in the case of the irradiation with nanosecond laser pulses on titanium /silicon, the LIPSS orientation forms parallel to the laser beam polarization; while in the case of the irradiation with picosecond laser pulses on titanium / glass, the LIPSS orientation switches to perpendicular to the laser beam polarization. The last also holds for the irradiation with picosecond laser pulses on molybdenum / glass. We also demonstrated that it is feasible to obtain TiO_2 in its rutile phase and MoO_2 in its monoclinic phase, by using low repetition rate Nd:YAG pulsed lasers in the short pulse regime (nanoseconds) and the ultrashort pulse regime (picoseconds). The laser-induced metallic oxides TiO_2 and MoO_2 synthesize in very specific crystalline phases which depend on the laser irradiation parameters.

6. Acknowledgements

The authors acknowledge partial support to this work from AFOSR-CONACyT Grant FA9550-10-1-0212. We are grateful to Dr. Ivan Garcia for his support on the characterization work.

7. References

Aygun, G.; Atanassova, E.; Kostov K.; Turan, R. (2006). *XPS study of pulsed Nd:YAG laser oxidized Si,* Journal of Non-Crystalline Solids, Vol. 352 (August 2006) pp. 3134–3139, ISSN 0022-3093.

Beattie, I. R., Gilson, T. R. (1969) *Oxide phonon spectra,* J. Chem. Soc. A: Inorg. Phys. Theor. (1969) pp. 2322-2327.

Blume, A. *Synthese und strukturelle Untersuchungen von Molybdän-, Vanadium- und Wolframoxiden als Referenzverbindungen für die heterogene Katalyse.* PhD Tesis, Technischen Universität Berlin (2004).

Camacho-López, S.; Evans, R.; Escobar-Alarcón, L.; Camacho-López, M. A.; Camacho-López, M. A. (2008). *Polarization-dependent single-beam laser-induced grating-like effects on titanium films.* Appl. Surf. Sci., Vol. 255 (December 2008), pp. 3028, ISSN 0169-4332.

Camacho-López, M. A.; Escobar-Alarcón, L.; Picquart, M.; Arroyo, R.; Córdoba, G.; Haro-Poniatowski, E. (2011). *Micro-Raman study of the m-MoO_2 to a-MoO_3transformation induced by cw-laser irradiation. Opt.* Mater., Vol. 33 (January 2011) 480, ISSN 0925-3467.

Cano-Lara, M.; Camacho-López, S.; Esparza-García, A.; Camacho-López, M. A. (2011). *Laser-induced molybdenum oxide formation by low energy (nJ)–high repetition rate (MHz) femtosecond pulses.* Optical Materials, Vol. 33 (September 2011) pp. 1648–1653, ISSN 0925-3467.

Dieterle, M. *In situ resonance Raman studies of molybdenum oxide based selective oxidation catalysts*. PhD Tesis, Technische Universität Berlin (2001).

Dong, Q.; Hu, J.; Lian, J.; Guo, Z.; Chen, J.; Chen B. (2003). *Oxidation behavior of Cr films by Nd:YAG pulsed laser*. Scripta Materialia, Vol. 48 (May 2003) pp. 1373–1377, ISSN 1359-6462.

Dubey, A. K.; Yadava, V. (2008) *Experimental study of Nd:YAG laser beam machining – An overview. Journal of materials processing technology*. Vol. 1 9 5 (January 2 0 0 8), 15–26. ISSN 0924-0136.

Escobar-Alarcón, L.; Haro-Poniatowski, E.; Camacho-López, M. A.; Fernández-Guasti, M.; Jímenez-Jarquín, J.; Sánchez-Pineda, A. (1999) *Structural characterization of TiO_2 thin films obtained by pulsed laser deposition,* Applied Surface Science Vol. 137 (January 1999) pp. 38–44. ISSN 0169-4332.

Evans, R.; Camacho-Lopez, S.; Camacho-Lopez, M. A.; Sanchez-Perez, C.; Esparza-Garcia, A. (2007). *Pulsed laser-induced effects in the material properties of tugnsten thin films.* J. of Phys. CS., Vol. 59, pp. 436-439, ISSN 1742-6596.

Hermann, J.; Benfarah, M.; Coustillier, G.; Bruneau, S.; Axente, E.; Guillemoles, J. F.; Sentis, M. Alloncle, P. ; Itina, T. (2006). *Selective ablation of thin films with short and ultrashort laser pulses*. Applied Surface Science. Vol. 252, (April 2006), pp. 4814-4818. ISSN 0169-4332.

Jun, Z.; Ning-Sheng, X.; Shao-Zhi D., Jun, C.; Jun-Cong S.; Zhong-Ling W. (2003). *Large-Area Nanowire Arrays of Molybdenum and Molybdenum Oxides: Synthesis and field emission properties*. Adv. Mater. Vol. 15 (November 2003), pp. 1835-1840.

Latha, K.; Yuan-Ron M.; Chai-Chang T.; Yi-Way L.; Sheng Yun W.; Kai-wen C.; Yung L. (2007) *X-ray difracction and Raman scattering studies on large-area array and nanobranched structure of 1D MoO_2 nanorods*, Nanotechnology Vol. 18 (March 2007) pp. 1-6. ISSN 0957-4484.

Lavisse, L.; Grevey, D.; Langlade, C.; Vannes, B. (2002). *The early stage of the laser-induced oxidation of titanium substrates*. Applied Surface Science, Vol. 186 (January 2002), pp. 150-155, ISSN 0169-4332.

Linsebigler A. M., Lu G. , Yates J. T., Jr.(1995). *Photocatalysis on TiO_2 Surfaces: Principles, Mechanisms, and Selected Results. Chem. Rev. Vol. 95 (July* 1995) pp. 735-758.

Livage, J., Ganguli, D. (2001). *Sol gel electrochromic coatings and devices: A review*. Solar Energy Materials & Solar Cells Vol. 68 (June 2001) pp. 365-381 ISSN 0927-0248.

Maisterrena-Epstein, R.; Camacho-Lopez, S.; Escobar-Alarcon, L,; Camacho-Lopez, M. A. (2007). *Nanosecond laser ablation of bulk Al, Bronze, Cu: ablation rate saturation and laser-induced oxidation*. Superficies y Vacio, Vol. 20 (September 2007), pp. 1-5, ISSN 1665-3521.

Mikhailova, D.; Bramnik, N. N.; Bramnik, K. G.; Reichel, P.; Oswald, S.; Senyshyn A.; D. M. Trots.; H. Ehrenberg. (2011). Layered $LixMoO_2$ Phases with Different Composition for Electrochemical Application: Structural Considerations, Chem. Mater. (July 2011), Vol. 23, pp. 3429–3441.

Pereira, A.; Delaporte, P.; Sentis, M.; Cros, A.; Marine W.; Basillais, A.; Thomann, A. L.; Leborgne, C.; Semmar, N. Andreazza P.; Sauvage T. (2004). *Laser treatment of a steel surface in ambient air*. Thin Solid Films Vols 453-454, (April 2004), pp. 16-21. ISSN 0040-6090.

Pérez del Pino, A.; Serra, P.; Morenza, J. L. (2002). *Oxidation of titanium through Nd:YAG laser irradiation*. Applied Surface Science, Vol. 197-198 (September 2002), pp. 887-890, ISSN 0169-4332.

Porto, S. P. S.; Fleury, P. A.; Damen, T.C. (1967) *Raman Spectra of TiO_2, MgF_2, ZnF_2, FeF_2, and MnF_2*, Phys. Rev. Vol. 154 (February 1967) pp. 522-526.

Sipe, J. E.; Young, J. F.; Preston, J. S.; Van Driel H. M. (1983). *Laser-induced periodic surface structure. I. Theory*. Phys. Rev. B Vol. 27 (January 1983), pp. 1141-1154. ISSN 1098-0121.

Spevack, P.A.; McIntyre, N. S. (1992). *Thermal reduction of molybdenum trioxide*. J. Phys. Chem., Vol. 96 (October 1992) pp. 9029–9035. ISSN 0022-3654.

Spevack, P. A.; McIntyre, N. S. (1993). *A Raman and XPS investigation of supported molybdenum oxide thin films. 1. Calcination and reduction studies. J. Phys. Chem.*, (October 1993), *97* (42), pp 11020–11030. ISSN: 0022-3654.

Srivastava R.; Chase L. L. (1972) *Raman spectra of CrO_2 and MoO_2 single Crystals*, Solid State Communications Vol. 11 (July 1972) pp. 349-353. ISSN 0038-1098.

Vorobyev, A. Y.; C. Guo, C. (2007). *Residual thermal effects in laser ablation of metals* J. Phys. CS Vol. 59 (2007) 418-423.

Young, J. F.; Preston, J. S.; van Driel, H. M.; Sipe, J. E. (1983). *Laser-induced periodic surface structure. II. Experiments on Ge, Si, Al, and brass*. Phys. Rev. B 27 (January 1983), pp. 1155-1172. ISSN 1098-0121.

Wang F. ; Lu B. (2009). *Well-alignedMoO$_2$ nanowiresarrays: Synthesis and field emission properties*. Physica B Vol. 404 (December 2009) 1901–1904. ISSN 0921-4526.

Witke, K.; Klaffke, D.; Skopp, A.; Schreckenbach, J. P. (1998). *Laser-induced transformation as a tool for structural characterization of materials by Raman spectroscopy*. J. Raman Spectrosc., Vol 29 (May 1998) pp. 411-415. ISSN 0169-4332.

4

Pulsed Laser Welding

Hana Chmelíčková and Hana Šebestová
*Institute of Physics of the Academy of Sciences of the Czech Republic,
Joint Laboratory of Optics of Palacký University and Institute of
Physics of the Academy of Sciences of the Czech Republic
Czech Republic*

1. Introduction

Metal joining by means of components heating to the melting temperatures was known thousands years ago in old age Greece. Heat sources have been developed from forging furnace to modern methods of plasma arc welding, electric resistance welding, oxy-fuel welding or laser welding. Laser, as a source of intensive light beam, starts to be implemented into industrial welding systems due to its advantages in comparison with classic methods, for example narrow heat affected zone, deep penetration, flexibility and many others. Besides the welding of compatible metals it is also possible to weld plastics.

The principle of "light" welding is the same for all suitable laser wavelengths. The absorption of laser radiation in thin work piece surface layer leads to the temperature rise to the melting or vaporization point. Due to the conduction of generated heat to the surrounding material volume sufficient weld pool is melted. However heat conduction also causes essential energy losses.

Laser welding important processing parameters are laser beam properties (power, beam quality and diameter, wavelength, focusing lens length), weld conditions (focus position towards the material surface, relative motion of the work piece towards the laser spot, weld type, processing gas) and physical properties of welded material and work piece dimensions. Typical power densities applied in laser welding lie between 10^5 - 10^7 W.cm^{-2}.

Lasers with active media formed by CO_2, semiconductors, Yb:YAG, Nd:YAG crystals or Ytterbium doped fibres can be used in industry for various components welding. High power gaseous continual CO_2 laser with wavelength 10 600 nm and wall-plug efficiency about 10 % has excellent beam quality, near Gaussian mode and high depth of focus, thus it is suitable for deep penetration welding. Far infrared radiation of this laser type cannot be transferred by means of optical fibre. Therefore typical CO_2 laser welding system is equipped with fixed processing head and work piece positioning mechanism. A mixture of helium and nitrogen is recommended as a processing gas to suppress plasma shielding effect.

Solid state pulsed Nd:YAG laser's wavelength 1 064 nm is suitable for fibre guiding from resonator to the processing head, fixed on a robot arm. This allows welding few meters far from the laser source which is especially suitable for large or in shape complicated components. Laser beam has a multimode profile with quality 20 – 30 mm.mrad, wall – plug efficiency of the flash-lamp pumped Nd:YAG lasers reaches only about 3%. In recent years new and much more efficient laser types were developed – diode lasers with beam quality up to 30 mm.mrad, diode pumped disc Yb:YAG lasers and Ytterbium doped fibre lasers with excellent beam quality, both in near infrared region. With more compact design, higher efficiency 30 % (fibre laser) to 50 % (diode laser) and lower running costs these systems are going to replace above mentioned Nd:YAG and CO_2 lasers (Němeček & Mužík, 2009).

Nevertheless many pulsed laser welders are already installed in small and middle enterprises and various materials are being processed. Processing parameters optimization is a goal of many research projects to improve productivity and decrease occurrence of defects that results in lower production costs.

2. Processing parameters optimization

To achieve fully penetrated high quality weld it is necessary to set optimal combination of processing parameters. They involved three different groups: laser beam properties (wavelength, power, diameter, divergence) material properties (density, thermal conductivity, specific heat, latent heat of melting and vaporization, thickness, joint configuration) and very important interaction parameters (welding speed, focusing element length, focus plane position towards the material surface, shielding gas direction and flow, absorptivity of material surface) (Duley, 1998). At room temperature almost all metals have absorptivity about 10 % - 20 %. It increases during material heating and leaps to 80 % - 90% when metal melting point is reached.

Heat Q necessary to melt material mass m is given by the well known equation

$$Q = m[c(T_m - T_0) + L_m] \tag{1}$$

where c is specific heat, T_m melting temperature, T_0 initial temperature and L_m latent heat of melting. In the case of continual laser welding this equation can be transformed to the following form

$$\frac{Q}{t} = \rho D h v[c(T_m - T_0) + L_m] \tag{2}$$

where Q/t represents power required for melting, ρ volume density, D spot diameter, h penetration depth and v welding speed. Then, penetration depth can be expressed as follows

$$h = \frac{\frac{Q}{t}}{\rho D v[c(T_m - T_0) + L_m]}. \tag{3}$$

Including surface absorption and heat conduction losses, penetration depth can be roughly estimated

$$h = K\frac{P}{Dv} \tag{4}$$

where P is laser power and K constant resulting from material physical characteristics including surface reflectivity and other energy losses. Thus, penetration depth is proportional to the applied laser power and inversely proportional to the spot diameter and welding speed.

P/v ratio defines the heat input to a unit length. Melted area cross section linearly rises with the heat input (Ghaini et al., 2007). In the case of continual laser welding processing parameters like laser power and welding speed are simply entered into the welding device control system pursuant to the material thickness and physical properties. Much more complicated setting of pulsed laser parameters is outlined in paragraph 2.1.

Surface power density or laser beam intensity is defined as a portion of laser power and laser spot area on material surface. Three welding modes are used in praxis, heat conduction mode, penetration mode and deep penetration (keyhole) mode.

Heat conduction welding is characterized by power density in the interval 10^4 - 10^6 W.cm^{-2} that causes only surface melting up to 1 mm. The weld is wide and shallow with aspects ratio about 2:1. Only laser beam without sharp intensity peak can be used, or the focal plane must be shifted some millimetres above the material surface. In the case of pulsed laser, pulse length 1 ms - 10 ms is used. When power density balances around the critical point 10^6 W.cm^{-2}, produced welds are deeper than in the case of the conduction welding. Aspect ratio is about 1:1 that indicates penetration mode welding (Lapšanská et al., 2010).

Keyhole welding mode starts when energy density exceeds 10^6 W.cm^{-2}. Laser beam is focused on the material surface and the fusion zone rapidly heats up to the boiling point. Melted material begins to vaporize at the centre of the weld spot and creates a blind hole (keyhole) in the centre of the weld line (Fig. 1). The pressure of hot metal vapour keeps the hole open during the welding. Presence of the keyhole allows the laser energy to reach deeper into the fusion zone and consequently to achieve deeper weld with lower aspect ratio (Kannatey-Asibu Jr., 2009). Keyhole mode welding is a typical application of high power continuous lasers or high energy pulsed lasers.

During the deep penetration laser welding plasma can be generated above the keyhole. Ionised metal vapour and shielding gas absorb laser light, change its direction and cause lower process efficiency. Inert gases such as argon, nitrogen, helium and their special mixtures are used for plasma reduction. On the other hand, thanks to the plasma plume presence, welding process can be controlled by means of plasma intensity measurement (Aalderink et al., 2005).

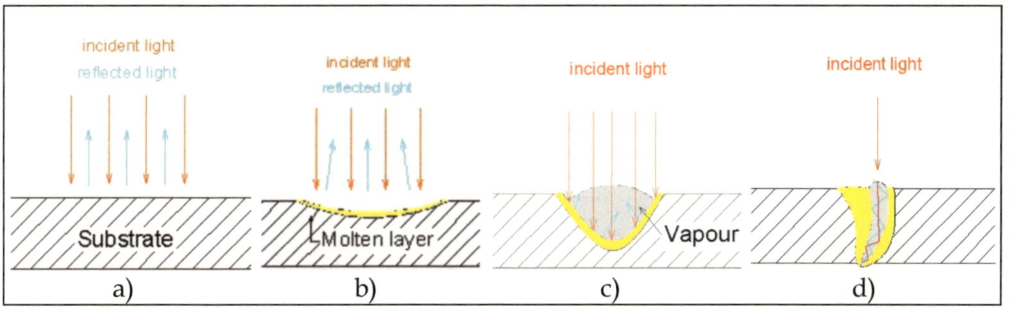

Fig. 1. Keyhole formation, a) surface irradiation, b) surface melting, c) vaporisation and cavity formation and d) light absorption inside the keyhole.

2.1 Pulsed laser welding parameters

In the case of pulsed laser, more parameters are involved. Three basic parameters that must be set at laser source control panel are frequency f (Hz), pulse length t (ms) and flash lamp charging voltage U (V). These parameters define actual pulse energy E (J). Peak power P_{peak} (kW) is defined as a portion of energy and pulse length

$$P_{peak} = \frac{E}{t}. \tag{5}$$

Peak power determines interaction intensity of laser beam with material for given spot size. According to the material thickness and welding mode peak power values 0.2 kW to 5 kW are recommended in the operation manuals of laser welders.

Laser average power P (W) is given as a product of actual energy and pulse frequency

$$P = Ef \tag{6}$$

and determines the welding speed.

There are two possible welding methods using pulsed laser. The first and simpler one is a spot welding which often replaces resistance welding nowadays. One or more pulses land material surface to reach required penetration depth in spot welding. No mutual motion between the processing head and material is applied. Spot welding is also often used for rough fastening of components to be subsequently seam welded using either pulsed or continual laser. This procedure reduces final distortions resulting from high thermal gradients corresponding to high value of applied power densities.

To achieve continuous tight welds using a pulsed laser, pulse overlap must be applied. This is realized using suitable combination of processing parameters. Pulse overlap PO is defined as follows

$$PO = 1 - \frac{v}{Df} \tag{7}$$

To achieve hermetic tight joints pulse overlap is recommended to be 80 %.

In comparison with conventional welding methods and continual laser welding, higher peak power densities in laser pulse mode causes higher heating and cooling rates which can result in weld defects and inhomogeneous microstructure. Many experimental works have been realised to optimise pulsed laser welding parameters for different kind of metals with goal to eliminate defects. For instance (Ghaini and al., 2006) studied overlap bead on plate welding of low carbon steel, (Tzeng, 1999) made successful welds without gas formed porosity in lap joints of zinc-coated steel. Another important parameter was introduced in these studies which is effective peak power density.

Effective peak power density $EPPD$ is defined as a product of peak power density PPD and pulse overlapping index F

$$EPPD = F \times PPD \tag{8}$$

where

$$F = \frac{1}{1 - PO}. \quad (9)$$

This parameter was introduced to better formulate the real power reaching material surface. When seem welding is required more pulses land material surface (Fig. 2) and their contribution adds. Thus the effective peak power density can be used to compare energy requirements of welds accomplished with different pulse overlap.

Fig. 2. Top view and longitudinal weld cross section.

2.2 Experimental work

Pulsed Nd:YAG laser system LASAG KLS 246 - 102 with maximal average power 150 W and beam parameter product 22 mm.mrad was used to carry out experiments focused on the study of the effect of processing parameters on weld dimensions and its surface character. Material to be welded was 0.6 mm thick stainless steel AISI 304.

So as the results were not affected by an accidental misalignment of components to be butt joined, by possible contamination or presence of surface defects of contact areas, or other unsuitable initial conditions influencing weld properties, it was decided not to join two sheets but to make a deep remelting of one sheet, which is, in fact, bead-on-plate welding. This strategy ensures that only the effect of energy changes will be studied. Laser welding is very demanding on pieces to be welded preparation, especially when narrow laser beam is used in near focal position. Therefore, highly precise prepared edges and minimal gap between the components to be joined are always supposed, which are conditions that need to be fulfilled in every precise butt joint laser welding application to prepare a high-quality weld. Then, the results of bead-on-plate experiments can be applied to the real sheet welding at conditions suitable for a high-quality weld joint preparation. This simplification can be used for the effect of processing parameter changes study in different laser applications.

Welding itself was realized 4 mm under the focal plane to ensure the sufficient beam diameter on the specimen which was 0.85 mm. Focussing lens with 100 mm focal length was used. Cleaned degreased weld pieces were clamped in a mounting jig. Pure argon gas at coaxial 8 l.min^{-1} flow rate was used to protect the weld pool against its oxidation. In each set of experiments, only one parameter was changed keeping all the other parameters constant to be able to identify the effect of the one examined parameter. Table 1 presents processing parameters of each set of experiments.

series	E (J)	P (W)	P_{peak} (kW)	t (ms)	v (mm.s^{-1})	f (Hz)
1	3.5	45.5	1.59 - 0.8	2.2 - 4.4	4.0	13.0
2	3.5 - 6.5	45.5 - 80.6	1.03 - 1.82	3.4	4.0	13.0
3	5.0	45.5	1.47	3.4	1.5 - 6.0	13.0
4	5.0	15.0 - 35.0	1.47	3.4	4.0	3.0 - 7.0
5	1.8 - 6.2	23.4 - 80.6	1.2	1.5 - 5.1	4.0	13.0

Table 1. An overview of processing parameters of realised experiments

2.2.1 Pulse length effect

The first set of experiments was focused on the effect of pulse length which was changed in the interval from 2.2 ms to 4.4 ms. To keep constant beam energy 3.5 J charging voltage had to be decreased when pulse length was increased (Fig. 3). Thus the average power remained constant and peak power decreased (Fig. 4).

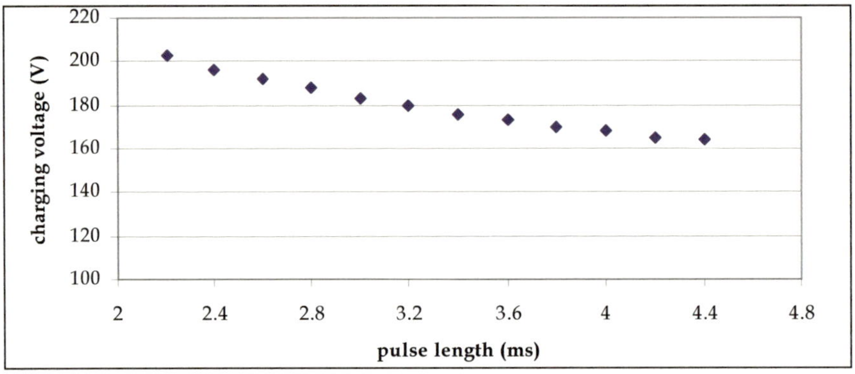

Fig. 3. Charging voltage vs. pulse length keeping constant pulse energy.

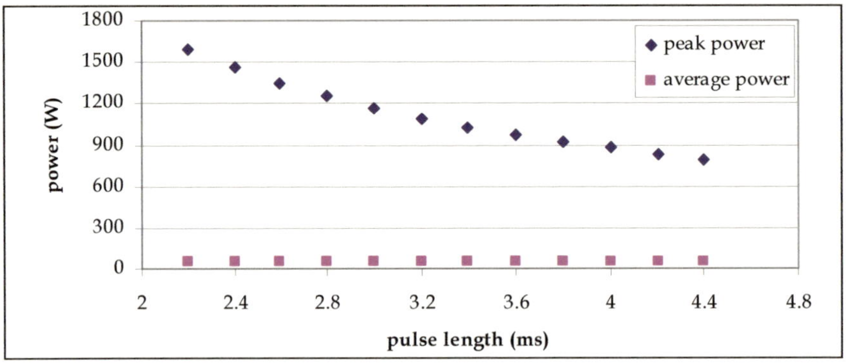

Fig. 4. Average and peak power vs. pulse length keeping constant pulse energy.

Pulse length increase led to the decrease of penetration depth (Fig. 5) that corresponds to the decrease of peak power which seems to be a critical parameter.

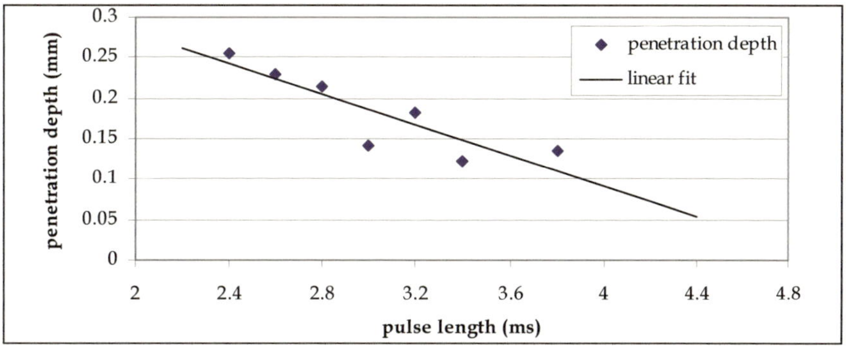

Fig. 5. Penetration depth vs. pulse length.

These results also showed that applied parameters were not sufficient for the full penetration of 0.6 mm metal sheet (Fig. 6). However the effect of pulse length is evident.

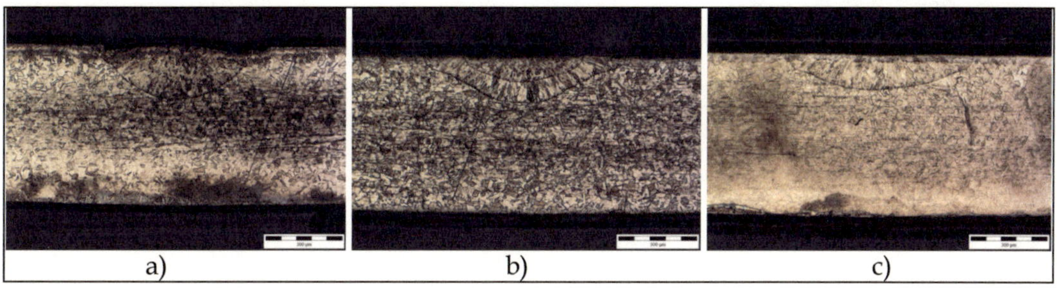

Fig. 6. Perpendicular cross sections of welds at a) 2.8 ms, b) 3.2 ms and c) 3.8 ms.

Laser scanning confocal microscope LEXT OLS 3100 was used to image and analyse laser weld surfaces. Spot diameter varied in the interval from 0.704 mm to 0.759 mm and its slightly decreasing tendency with increasing pulse length was detected (Fig. 7). Average value of spot diameter reached 0.73 mm. According to the 0.025 mm deviations measured within the each sample, no definite relationship between spot diameter and pulse length can be determined in the investigated region.

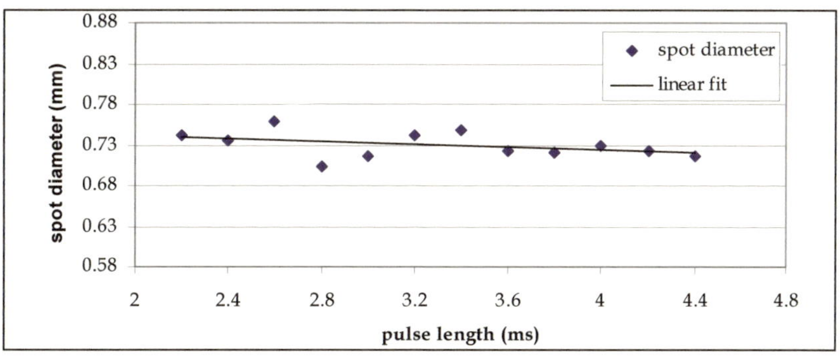

Fig. 7. Spot diameter vs. pulse length.

Fig. 8 presents an example of surface longitudinal central profile. In technical practise, strict claims on weld surface quality are often posed. Suitable combination of welding parameters can minimise post processing mechanical treatment. Fig. 9 presents surface images of selected samples. Pulse overlap 63 % remained constant during all the experiments of this set.

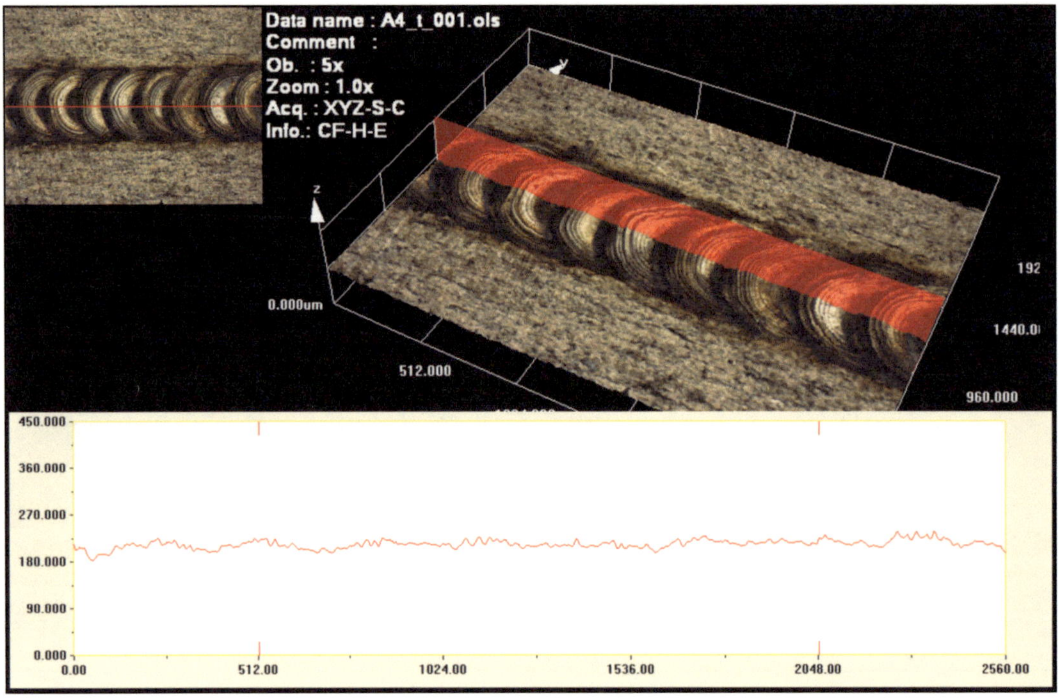

Fig. 8. 3D surface reconstruction and central profile of the weld realised at 3.8 ms.

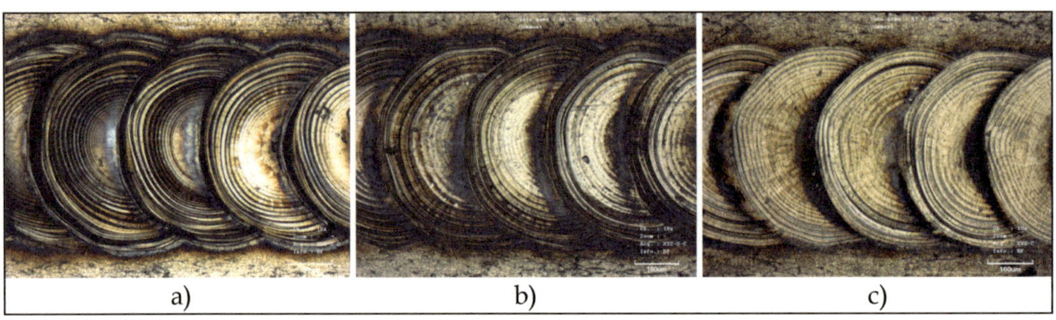

Fig. 9. Weld surface in case of pulse length a) 2.6 ms, b) 3.4 ms and c) 4.4 ms.

2.2.2 Pulse energy effect

The second series of experiments studied the effect of pulse energy which was set in the interval from 3.5 J to 6.5 J via charging flash lamp voltage changes (Fig. 10). Increasing energy naturally increases average as well as peak power (Fig. 11).

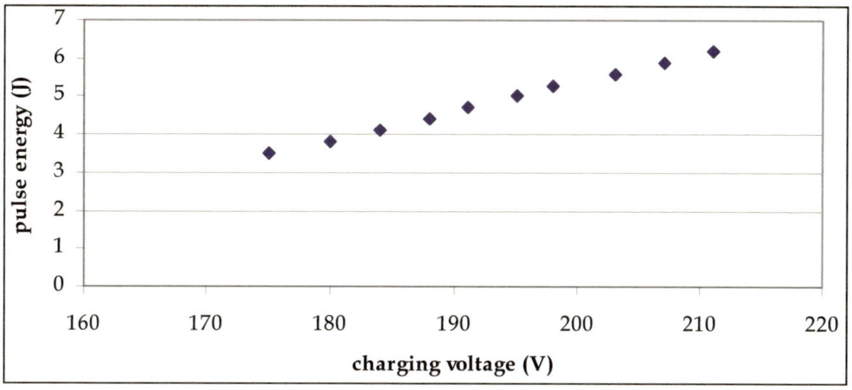

Fig. 10. Pulse energy vs. charging voltage.

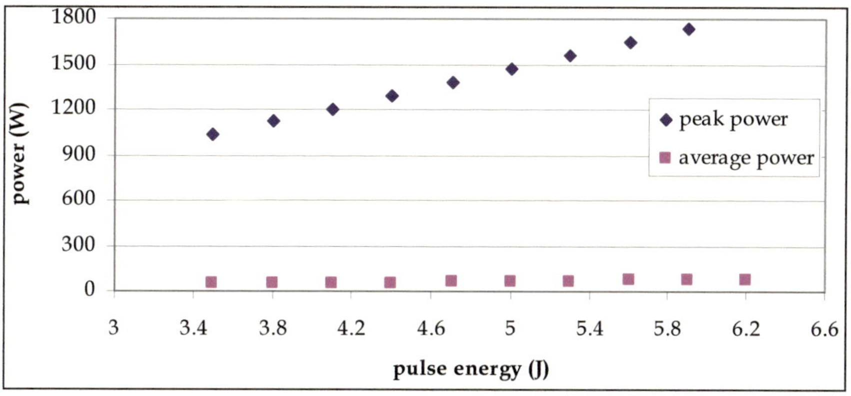

Fig. 11. Average and peak power vs. pulse energy.

Higher laser power leads to the deeper penetration which corresponds to the higher applied power resulting in higher heat input. Nevertheless penetration depth evolution is not linear (Fig. 12).

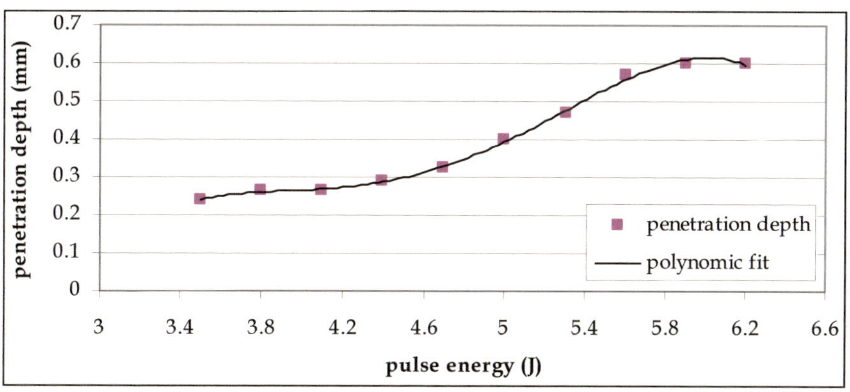

Fig. 12. Penetration depth vs. pulse energy.

Fig. 13 presents reached aspect ratios (penetration depth to weld width) as a function of effective peak power density. In this case pulse overlap 63 % was applied. Thus pulse overlapping index was 2.7. It means that power really touching a unit of material surface is 2.7 times higher in comparison with laser output power assuming no energy losses between laser output and material surface.

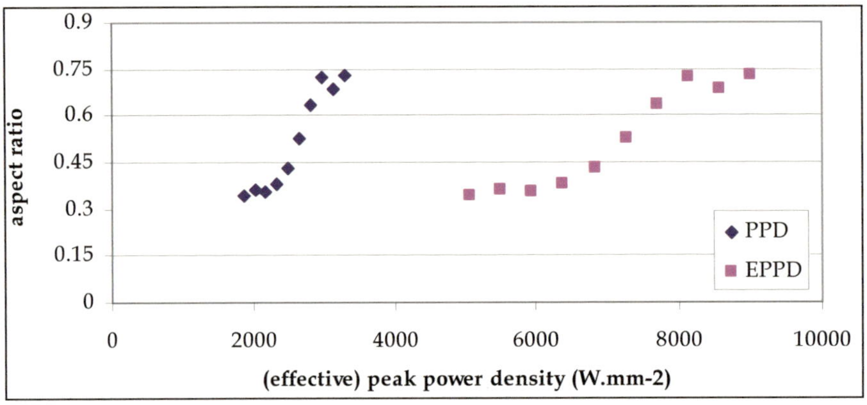

Fig. 13. Aspect ratio vs. effective peak power density.

Penetration depth slightly increases with pulse energy until about one half of the sheet thickness is penetrated. Then, when the formation of the keyhole starts (4.7 J), it increases steeply until the full penetration is reached at 5.9 J (1.7 kW). Keyhole formed at lower peak powers is not stable enough to establish true keyhole welding, and penetration welding mode can be observed (Fig. 14).

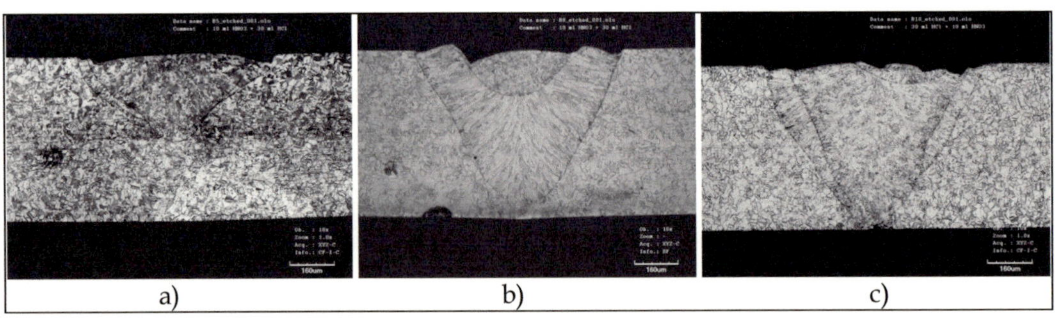

Fig. 14. Perpendicular cross sections of welds at a) 3.8 J, b) 4.7 J and c) 6.2 J.

Aspect ratio increases with increasing effective peak power density until the full penetration is reached at about 8.6 kW.mm^{-2}. At the moment of full penetration spot diameter is maximal, and aspect ratio decreases. Once the full penetration has been achieved, increasing effective peak power density does not lead to another weld width grow. Rather, conversely, the width decreases because the beam can escape from the melt pool because of the full penetration and because less energy is absorbed. That is why the aspect ratio increases again. These data corresponds to the weld cross-section fusion area measurement (Fig. 15).

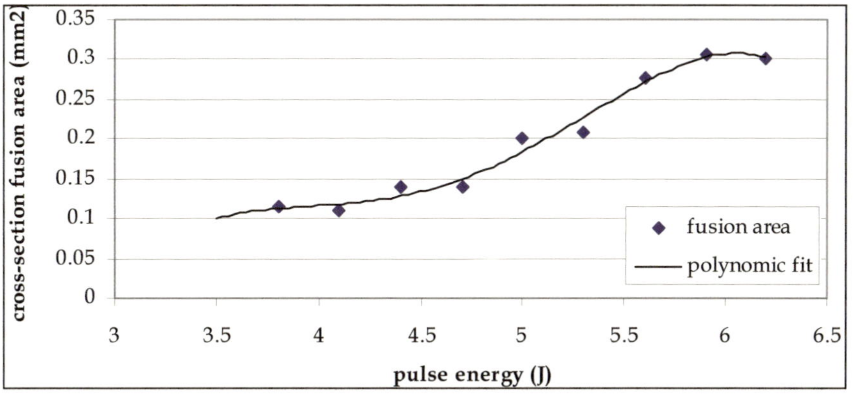

Fig. 15. Cross-section fusion area vs. pulse energy.

Fig. 16 presents the effect of beam energy on spot diameter. Beam diameter on the specimen was 0.84 mm for all carried out experiments. Spot diameter does not reach this value for almost all spots. This fact corresponds to the relative low energy application, non-uniform heat distribution, resulting from characteristic profile of beam intensity, and is supported by relatively low thermal conductivity of welded material.

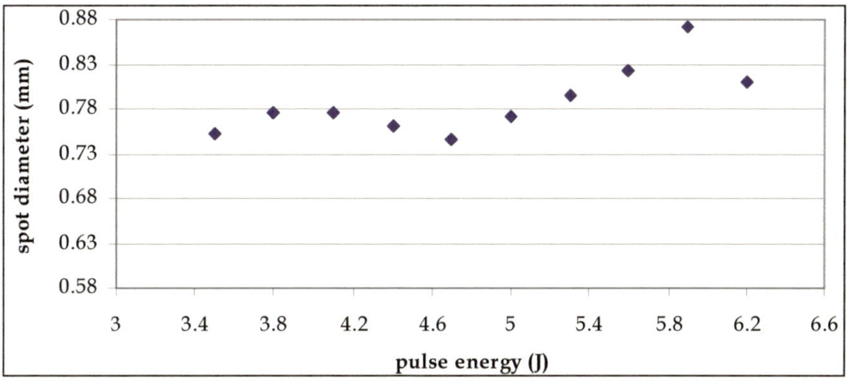

Fig. 16. Spot diameter vs. pulse energy.

Spot shape is also very demanding on pulse energy. The volume of melted material is higher at higher energies which leads to the higher deformation of the spot shape (Fig. 17).

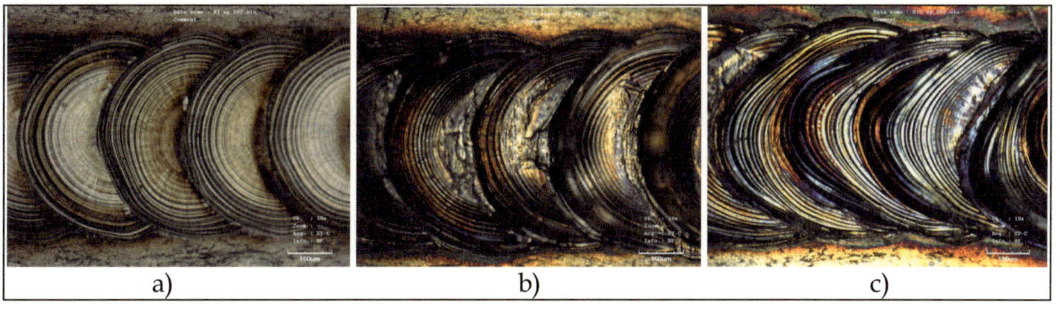

Fig. 17. Weld surface in case of pulse energy a) 3.5 J, b) 5 J and c) 6.2 J.

2.2.3 Welding speed effect

The third experiment was focused on the effect of welding speed. Welding speed was changed from 1.5 $mm.s^{-1}$ to 6 $mm.s^{-1}$.

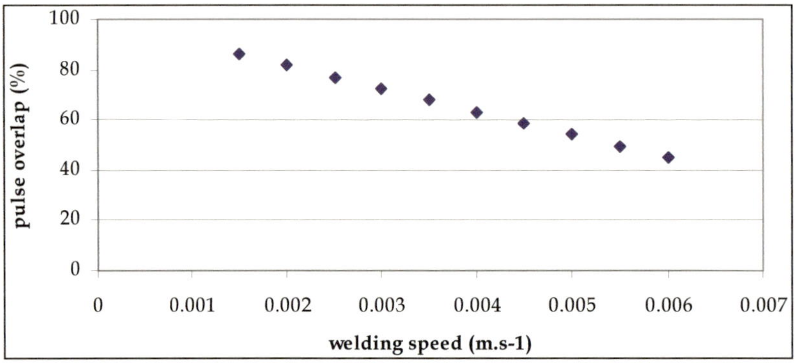

Fig. 18. Pulse overlap vs. welding speed.

Speed changes correspond to the pulse overlap change from 86 % to 40 % (Fig. 18, Fig. 19).

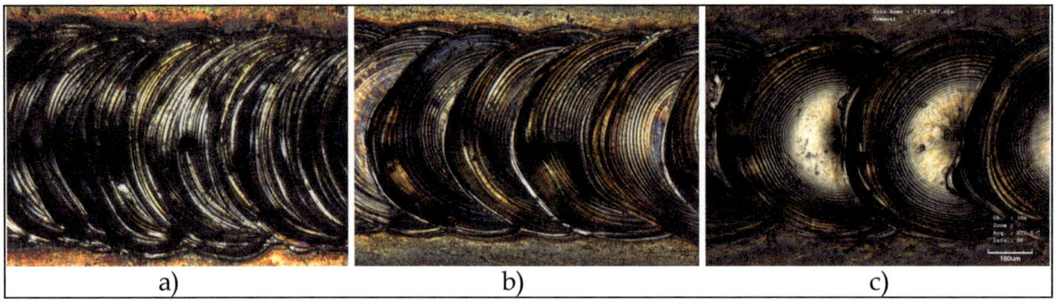

Fig. 19. Weld surface in case of welding speed a) 1.5 $mm.s^{-1}$, b) 3.5 $mm.s^{-1}$ and c) 6 $mm.s^{-1}$.

These experiments did not prove any significant change in penetration depth (Fig. 20). On the other hand, weld width decreased with increasing welding speed (Fig. 21). These give us positive information that slight required changes of pulse overlap via welding speed do not significantly influence penetration depth.

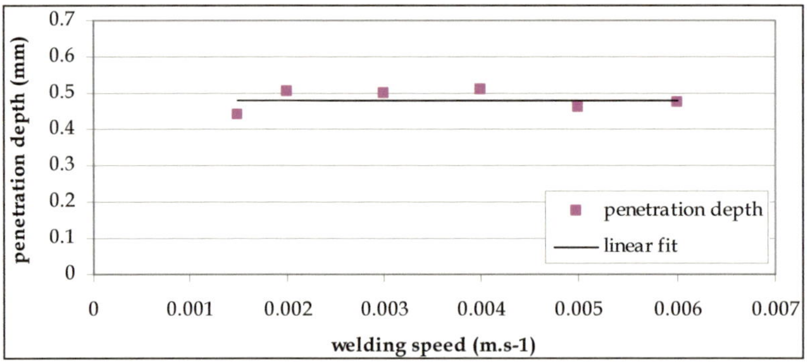

Fig. 20. Penetration depth vs. welding speed.

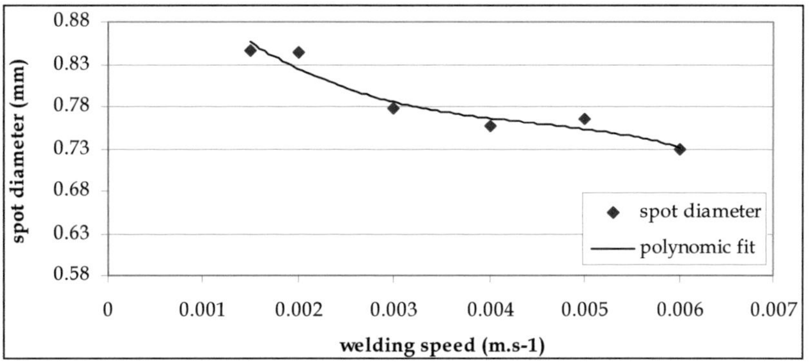

Fig. 21. Spot diameter vs. welding speed.

2.2.4 Pulse frequency effect

Another experiment concerned with pulse frequency effect. Pulse frequency was changed from 9 Hz to 17 Hz. In this case peak power remained constant, while the average power increased (Fig. 22).

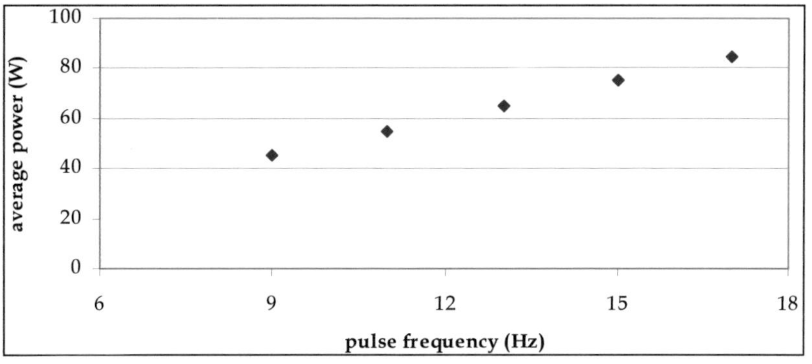

Fig. 22. Average power vs. pulse frequency.

Frequency increase naturally increases pulse overlap (Fig. 23, Fig. 24). Pulse overlap must be high enough so as also bottom side of the sheet is continuously penetrated (Fig. 25).

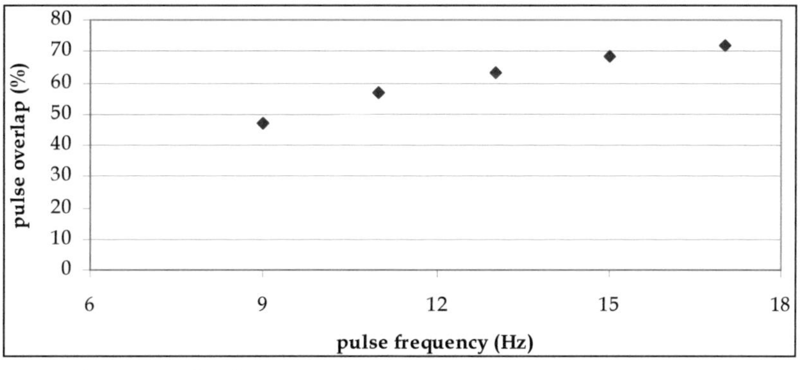

Fig. 23. Pulse overlap vs. pulse frequency.

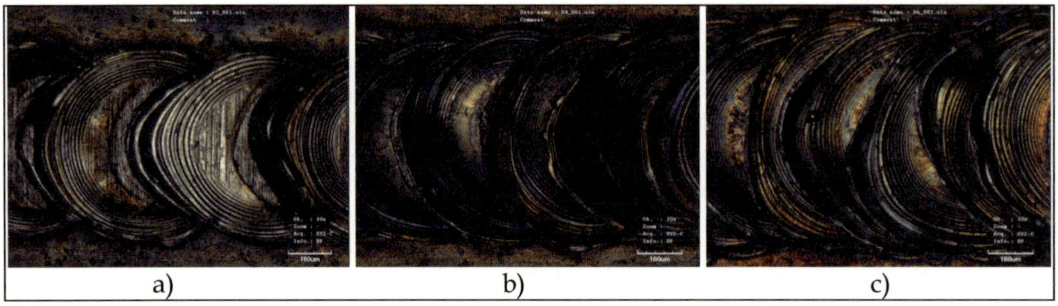

Fig. 24. Weld surface in case of frequency a) 9 Hz, b) 13 Hz and c) 17 Hz.

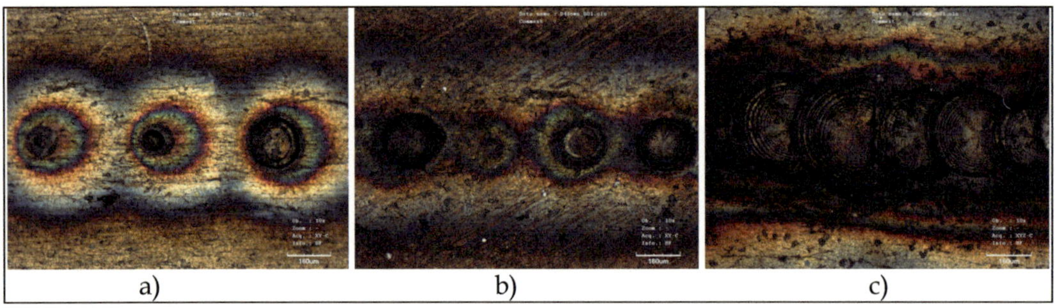

Fig. 25. Weld bottom side in case of frequency a) 9 Hz, b) 13 Hz and c) 17 Hz.

Power increase led to the increase of penetration depth (Fig. 26) as well as the weld width. Frequency 11 Hz was sufficient for the full penetration. Spot diameter slowly increases with increasing frequency (Fig. 27).

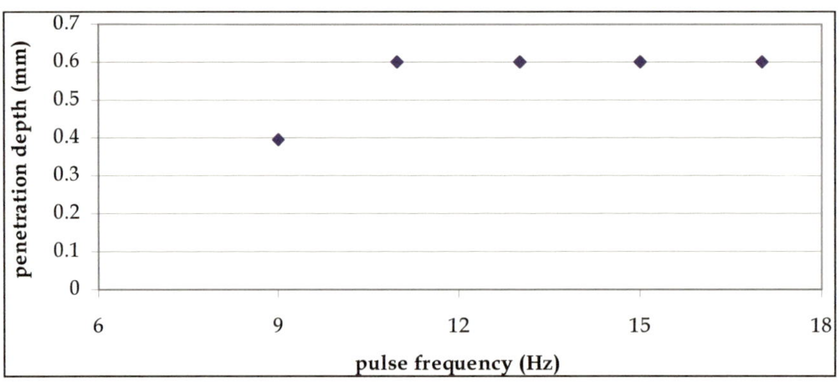

Fig. 26. Penetration depth vs. pulse frequency.

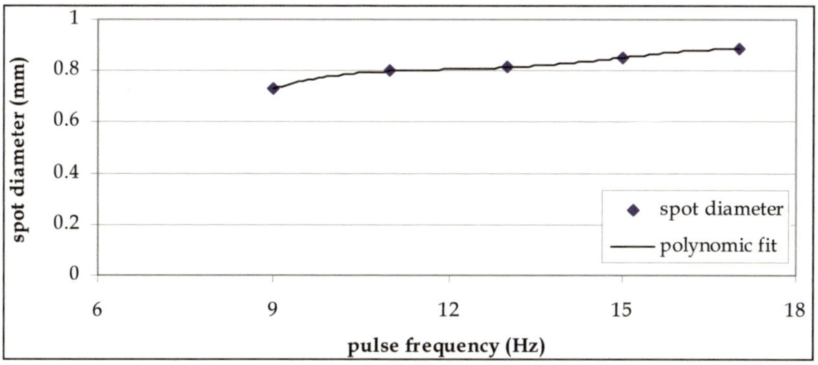

Fig. 27. Spot diameter vs. pulse frequency.

2.2.5 Combination of pulse energy and pulse length effect

In the last experiment beam energy and pulse length were changed simultaneously to keep constant peak power (Fig. 28). Although the peak power is still the same average power increases (Fig. 29).

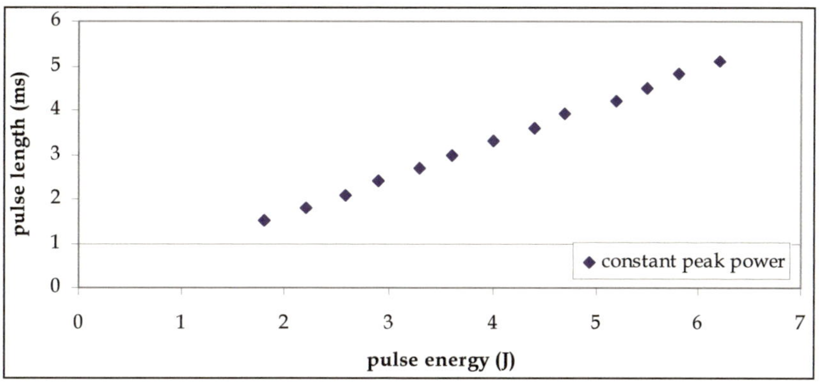

Fig. 28. Pulse length vs. pulse energy keeping constant peak power.

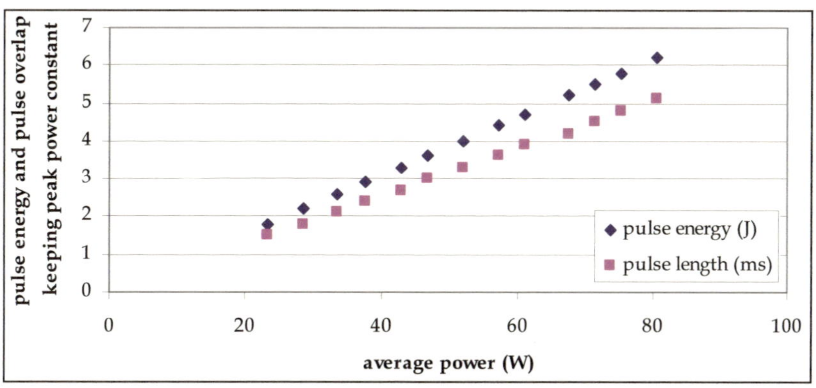

Fig. 29. Combination of pulse energy and pulse length (keeping constant peak power) vs. average power.

The higher energy and pulse length the deeper penetration depth (Fig. 30). This result points out the fact that even peak power can not be used as a definite indicator of penetration depth in pulsed laser welding.

Fig. 31 presents a comparison of these results with results of the effect of pulse energy (chapter 2.2.2). It is obvious that pulse energy is a critical parameter since it defines the volume of melted material.

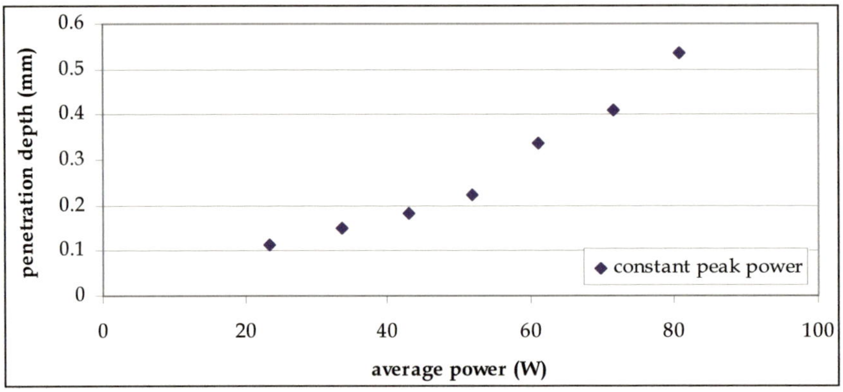

Fig. 30. Penetration depth vs. average power keeping constant peak power.

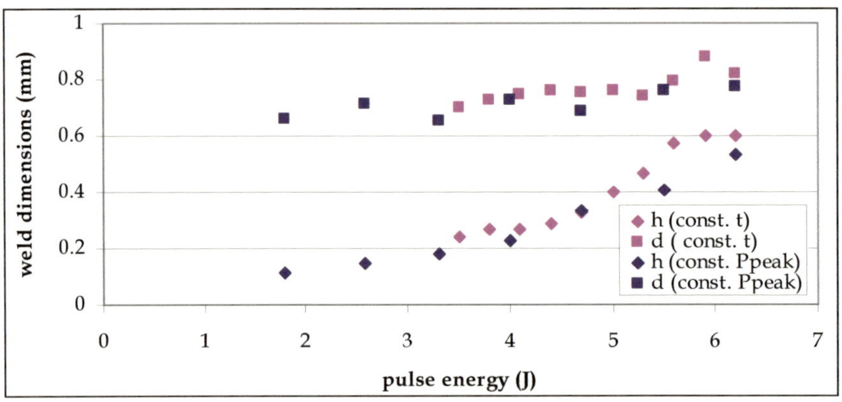

Fig. 31. Weld width *d* and penetration depth *h* vs. pulse energy for constant and variable pulse length.

3. Conclusion

High power Nd:YAG lasers with millisecond pulses are used in industry for spot or overlap welding. Many research papers reporting on pulsed laser welding parameters optimisation leading to the production of sufficiently deep welds without defects have been published. Usually more parameters were simultaneously changed in such optimisation processes. The aim of our research work was to identify the effect of each parameter separately.

Pulsed overlap bead-on-plate welding of 0.6 mm thick AISI 304 stainless steel was realised in our laboratory. Flash lamp pumped pulsed Nd:YAG laser KLS 246-102 with multimode beam profile was used for five series of welding experiments. The influence of different process parameters - pulse length, pulse energy, welding speed, pulse frequency and combination of pulse energy and pulse length on weld dimensions was examined in five separate experiments. Weld cross sections and laser spots overlap on samples surfaces were observed and measured by means of laser scanning confocal microscope LEXT OLS 3100. The effect of penetration depth and surface spot diameter on applied parameters was outlined.

Following from the above mentioned results, each processing parameter more or less influences weld characteristics. It is obvious that the knowledge of only one parameter, for example beam energy or average power, is not sufficient for the prediction of weld dimensions in pulsed laser seam welding. Peak power, pulse length and frequency in combination with processing speed are also very important. Suitable combination of processing parameters must be always found.

Experiments with flash lamp pumped solid state laser with very low efficiency will be followed by new studies on modern laser systems - diode and fibre lasers.

4. Acknowledgment

The Academy of Sciences of the Czech Republic supports this work under the project no. KAN301370701.

5. References

Aaldering, B. J.; Aarts, R. G. K. M.; Jonker, J. B. & Meijer, J. (2005). Weld Plume Emission During Nd:YAG Laser Welding, *Proceedings of the third International WLT - Conference on Lasers in Manufacturing*, Munich, Germany, June 13-16, 2005, Available from http://doc.utwente.nl/52666/2/weld_plume.pdf

Duley, W. W. (1998). *Laser welding*, A Wiley-Interscience publication, ISBN 978-0-471-24679-4, Hoboken, New Jersey, USA

Ghaini, F.M.; Hamedi, M. J.; Torkamany & Sabbaghzadeh, J. (December 2006). Weld metal microstructural characteristics in pulsed Nd:YAG welding, In: *Scipta Materiala 56 (2007)*, 955 – 958, Elsvier Ltd., Available from http://www.sciencedirect.com/science/article/pii/S1359646207001194

Kannatey-Asibu Jr., E. (2009). *Principles of Laser Material Processing*, A John Wiley & Sons, Inc., Publication, ISBN 978-0-470-17798-3, Hoboken, New Jersey, USA

Lapšanská, H.; Chmelíčková H. & Hrabovský, M. (2010). Effect of Beam Energy on Weld Geometric Characteristics in Nd:YAG Laser Overlapping Spot Welding of Thin AISI 304 Stainless Steel Sheets, *Metallurgical and Materials Transactions B*, Vol. 41, No. 5 (2010), pp. 1108-1115. ISSN 1073-5615

Němeček, S. & Mužík T. (2009). Laser Material Processing - Hardening and Welding, *Proceedings of METAL 2009 18th International Conference on Metallurgy and Materials*, Hradec nad Moravicí, Czech Republic, May 19-21, 2010, Available from http://www.nanocon.cz/data/metal2009/sbornik/List/Papers/014.pdf

Tzeng, Y. F.(1999). Pulsed Nd:YAG Laser Seam Welding of Zinc-Coated Steel *Welding research supplement,* Vol.6, No.4, (July 1999), pp. 238-s -211-s, ISSN 1729-8806

5

Laser-Assisted Cold Spray (LACS)

Dimitris K. Christoulis[1], Michel Jeandin[1],
Eric Irissou[2], Jean-Gabriel Legoux[2] and Wolfgang Knapp[3]
[1]*MINES-ParisTech, Centre des Matériaux/CNRS.-U.M.R. 7633,*
C2P-Competence Center for spray Processing
[2]*National Research Council Canada - Industrial Materials Institute, Montreal, QC*
[3]*CLFA-Fraunhofer-ILT, Paris*
[1,3]*France*
[2]*Canada*

1. Introduction

Thermal spray processes are used for the formation of metallic, ceramic and composite coatings. In thermal spray techniques, atomization jets or process gasses are used to melt or partially melt and to accelerate the feedstock powder or droplet material towards a prepared substrate (Davis, 2004). The feedstock itself may be in powder, rod or wire form. Upon impact, a bond forms with the surface, with subsequent particles causing thickness build-up and forming a lamellar structure. Thermal sprayed coatings are applied in various industrial sectors: automotive industry (Barbezat, 2005; Barbezat, 2006), marine industry (Wood et al., 2004), biomedical applications (Guipont et al. 2002, Liu et al., 2009) and aeronautical/aerospace applications (Dorfman et al., 2004; Evans et al., 2006) to name a few. Nevertheless, the deposition of coatings by thermal spraying technologies presents several disadvantages: high oxygen concentration, high porosity, non-uniform microstructure due to unmelted particles, quenching and thermal stresses, complex processes difficulties to retain the initial microstructure of the sprayed powder, low deposition efficiency. In order to minimize these defects and improve the properties of the coatings, techniques which control accurately the spray conditions (pressure and composition of the spraying chamber atmosphere, surrounding gas, temperature, power, etc.) have been developed (Sarafoglou et al., 2007).

However, these problems of conventional thermal sprayed coatings can be minimized or eliminated by employing Cold Gas Dynamic Spraying (Cold Spray), which is the newest process of the thermal spray techniques. Cold Spray is a solid state process where powder particles are accelerated by a low temperature supersonic gas stream and are plastically deformed upon impact on a substrate to form a coating. The formation of coatings by means of cold spray, compared with the conventional sprayed coatings has the major advantage of the absence of macroscopic melting of the sprayed particles. This leads to coatings which do not present quenching and thermal stresses. Also, cold-sprayed coatings retain the initial microstructure of the powder.

In thermal spray, coating adhesion strength is paramount since advanced coatings must remain bonded to the substrate under various and severe conditions. The same requirements of high coating adhesion hold true for cold spray coatings. However, the bonding mechanism of cold sprayed particles has not been completely clarified. Experimentally, it has been shown, that the particles can bond and form a coating only if they exhibit a material-dependant critical velocity (Schmidt et al., 2009). This critical velocity depends also on surface conditions (Ajdelsztajn et al., 2006; Golesich et al., 2008). The appropriate preparation of substrate surface prior to cold spraying contributes to the formation of high adhesive coatings. The adhesion of cold sprayed coatings can be increased by means of conventional methods such as grit-blasting and chemical solvents etching. However, grit-blasting may provoke contamination of the substrate by grit inclusions, which in turns can decrease the fatigue resistance while chemical solvents etching generates hazardous waste and is environmentally unfriendly (Garcia-Alonso et al., 2011).

Cold spray was coupled with a nano-pulsed Nd-YAG laser leading to the development of the Laser-Assisted Cold Spray (LACS) process which is a novel process, for the pre-treatment of the substrate and the coatings build-up. This leads to the production of advanced protective coatings of high adhesion and cohesion strength. At this pioneering stage, LACS was used for the formation of aluminum cold-sprayed coatings substrates onto aluminum alloy substrates (AISI-2107) and Ni-20Cr coatings onto Inconel718 substrates. Coatings adhesion strength was measured by using another Nd-YAG laser (i.e. LAser Shock Adhesion Testing, namely LASAT®). LASATesting® was shown to be a powerful tool for studying local adhesion of coatings and for determining the influence of metallurgical and morphological interface features on adhesion

2. Cold Spray technology

Cold Gas Dynamic Spraying (Cold Spray) is the newest process of thermal spray techniques. The cold spray process was initially developed in the mid-1980s at the Institute for Theoretical and Applied Mechanics of the Siberian Division of the Russian Academy of Science (Irissou et al., 2008a) but was introduced in laboratory level to North America and Europe only in the 2000s. Contrary to the conventional thermal spray technologies (plasma spraying, HVOF, arc-spraying etc), Cold Spray is characterized by a low spraying particle temperature and high spraying particle velocity (Figure 1). The main advantage of the Cold Spray technology, compared with the conventional thermal spraying techniques, is the absence of the macroscopic melting of the spraying particles. In Cold Spray technology, the sprayed particles are heated in temperature below their melting point and they are accelerated to supersonic velocities. Thus, the sprayed particles are plastically deformed upon impact on a substrate to form a coating. This leads to coatings where common problems of conventional thermal spray technologies, like high-temperature oxidation, evaporation, melting, crystallization, residual stresses, debonding, gas release, and other common problems for conventional thermal spray methods are minimized or eliminated (Papyrin et al., 2008).

In cold spray, fine powders in the 5 and 50 μm range are generally used. However, in some cases, coarser powders can be cold sprayed under appropriate conditions (Van Steenkiste et al., 1999; Van Steenkiste et al., 2002; Christoulis et al., 2011; Li et al., 2008)

In practice, the principle of the cold spray process can be described as follows. The powder is supplied to the nozzle by a high pressure carrier gas. The kinetic energy of the particles is increased by the propellant gas which can be heated between 200°C and 850°C. In the most of the cases, the feeding and the propellant gases are of the same nature. The gases which are used are nitrogen (N_2), Helium (He), compressed air, or mixtures of them (Christoulis et al., 2011; Wong et al., 2011; Gartner et al., 2006; Sundararajan et al., 2010; Chavan et al., 2011].

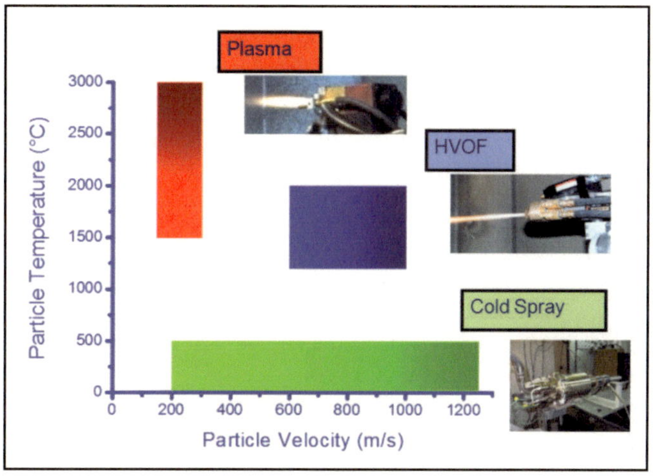

Fig. 1. Cold Spray technology and other conventional thermal spray techniques.

The geometry of the De Laval nozzle (Figure 2) results to the acceleration of the mixture propellant gas-powder at supersonic velocities at the exit of the nozzle. Although the gas temperature may seem high at first glance, the divergent of the nozzle causes expansion of gases and therefore a significant lowering of temperature. The powder particles, which also have a very limited residence time in the hot gas flow, in any case remain in a solid or slightly viscous (surface heating).

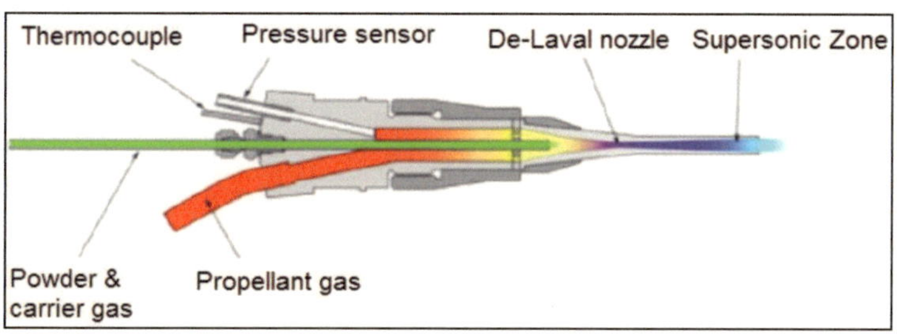

Fig. 2. Gold Spray Gun

3. Surface pre-treatment

The formation of advanced coatings is strongly depends on the proper substrate surface pre-treatment prior to the thermal spraying process. Preparation steps must be undertaken correctly in order the coatings to perform the design expectations, otherwise, a total failure could occur (Davis, 2004).

Coating adhesion is connected directly with the surface pre-treatment of the substrate. In thermal spray, coating adhesion strength is paramount since coatings can only fulfill its function if it remains bonded to the substrate under various and severe conditions. Many studies are devoted to the bonding mechanisms and to the optimization and improvement of coating bond strength in thermal spray (Greving et al., 1994; Brandt, 1995 ; Lu et al., 2002 ; Araujo et al., 2005 ; Day et al., 2005).

The same requirements of high coating adhesion hold true for cold spray coatings. However, the cold spray process differs from other thermal spray processes in its bonding mechanisms since there is no macroscopic melting of the feedstock powder. It is often concluded that the cold-sprayed particles adhere only on "nascent" surfaces (Ichikawa et al., 2008) produced by the impacts of high-velocity sprayed particles which provoke the fracture of the pre-existing oxide layers of the substrate. However, the "cleaning" of substrate by sprayed particles can be characterized as an accidental method which induces the waste of the costly powders.

The major surface preparation techniques which are currently carried out in thermal spraying technology are presented in the next two paragraphs. The first paragraph includes the main conventional techniques, while the second one presents the most innovate laser pre-treatment technologies.

3.1 Conventional techniques

In thermal spraying technology, coating adhesion can be increased by means of conventional methods (i.e grit-blasting, chemical etching, formation of bond-coatings). It should be mentioned that the decreasing of the substrates surface is important and it should be realized before grit-blasting. The decreasing of the substrate is performed to remove contaminants such as oil, grease, paint, rust, scale, and moisture. The decreasing of the surfaces is carried out either by using chemical solvents (aceton, MEEK etc) either by using hot vapor (Davis, 2004).

Grit-blasting is the most common technique which is used as a preparation method in conventional thermal spray processes (Bahbou et al., 2004; Peredes et al., 2006 ; Bobzin et al., 2010 ; Gonzalez-Hermosilla et al., 2010). In the case of cold spray process, it is often claimed that the substrates do not require any surface preparation; the substrates can be used in their as-received state after the necessary cleaning (Kroemmer et al., 2006; Hartmann, 2010). However, the grit-blasting of the substrates is accomplished in various application of cold spraying process in order to enhance the adhesion of the coatings (Richer et al., 2006; Koivuluoto et al., 2007; Bala et al., 2010 ; Koivuluoto et al., 2010). Experimental results of cold-sprayed copper onto copper substrates have shown that the coatings adhesion strength is significantly higher on grit-blasted substrates (36.0 MPa) compared to the coatings adhesion strength onto as-received substrates (5.4 MPa) (Makinen et al., 2007). Contrary, in the case of cold sprayed titanium, experimental results have shown that the bond strength of titanium cold sprayed coatings is higher onto polished substrates compared to grit-blasted substrates (Marrocco et al., 2006). Generally, in thermal spraying technology the increase of the substrate roughness via grit-blasting is a common procedure. Usually coating adhesion is correlated with the mean arithmetic roughness (R_a) of the substrates. However, it seems that (R_a) is not enough (Zecchino, 2003) for the characterization of a roughened surface and also it has still not been clarified which characteristics of the roughness influence the coatings' adhesion (Fukanuma et al., 2003). Furthermore, grit-blasting may provoke contamination of the substrate by grit inclusions which can be catastrophic for some applications. In automotive industry, which is a predominant sector for the application of thermal spray technology (Barbezat, 2005; Barbezat, 2007), for the production of engine blocks with thermally sprayed cylinder, other techniques should be used for the preparation of the substrate since the grit can remain within the many internal passages of the engine block. The grit can come loose during operation and later cause engine breakdowns (Schlaefer et al., 2008). Also, it is should be referred that grit-blasting is environmentally unfriendly and also it can be unhealthy for the operators of grit-blasting apparatus since the grits have been correlated with serious diseases such as silicosis, aluminosis, lung scarring, pneumoconiosis, or emphysema (Petavratzi et al., 2005).

Other techniques which are used for the pre-treatment of the substrate are the water-jet, the chemical etching (Pawlowski, 2008) and the macroroughening (Davis, 2004). Water-jet pretreatment has been used in the case of shrouded plasma sprayed MCrAlY coatings onto nickel superalloys (i.e Inconel 718, Rene 80 and Mar-M 509) (Pawlowski, 2008). The surface of the substrate is roughened by the water-jet, and it was used since the morphology of a water-jet treated surface is much finer than that of a sand-blasted one (Pawlowski, 2008).

Chemical etching with various acids (sulfuric, nitric, or hydrochloric acid) is another pre-treatment method of the metallic substrates, which was used to modify the roughness of the substrate and thus to increase the adhesion of the thermal sprayed coatings. However, it is not used so often any more due to environmental reasons.

Also, it should be mentioned that the application of a bond-coatings is also another pre-treatment method of the substrate. Of course the bond coats applied over grit-blasted surfaces (Davis, 2004). However bond coatings are applied in order to enhance coatings adhesion, in the cases that the coating cannot adhere directly to the substrate. In gas turbines technology, bond coatings (MCrAlY) are applied before the formation of the thermal barrier coatings (Padture et al., 2002; Saral et al., 2009; Zhang et al., 2009)

3.2 Laser techniques

Thermal spray and laser processing can be considered as half brothers since they show many common features due to the use of a (more or less) high-energy source for both (Jeandin et al., 2010). Their combination can therefore be very fruitful and prominent to achieve coatings, which results in their most recent and advanced applications (Jeandin et al., 2010).

Laser thermal spray hybrid processes have been developed as a result of the successful combination of laser and thermal spray technologies. Various types of laser have been combined with the guns of thermal spray process. Combining laser processing to thermal spray resulted in a major improvement for thermal spray in 3 sub-areas, i.e. that of pre-treatment, that of post-treatment and that of simulation of thermal and kinetic phenomena. In the next two paragraphs (§3.2.1 and §3.2.2), advanced laser pre-treatment methods are presented while section 4 is devoted to laser techniques which have been developed for the simulation of thermal and kinetic phenomena appeared in thermal spray technology. The post-treatment methods are not examined in the current chapter.

3.2.1 Laser heating and/or cleaning

The laser pretreatment of the substrates can provoke the ablation and/or the heating of the substrates. The interaction mechanisms which will appear between laser and matter (substrate) depend on the substrate material, the interaction time and the power density of the laser (Garcia-Alonso et al., 2011). The cleaning of the substrate form dust, oil, and contaminants is achieved by the ablation. Also, ablation can eliminate the superficial native oxide layers of the substrate which prevent the metallurgical bonding between the sprayed particles and the substrate allowing better particles/substrate contact which results in improved interfacial adhesion.

Laser pretreatment can provoke also the preheating of the substrate. Experiments have shown that the increased temperature of the substrate can lead to formation of coatings with improved properties (Gartner et al., 2006; Sundararajan et al., 2010). The preheating of the substrate can be achieved also by conventional methods (furnace or by passing the thermal spray gun over the substrate). However, the conventional methods are not fully controlled and also can provoke undesired phenomena (intensive oxidation or distortions of the material).

In early 2000's Zieris et al. developed the Laser Assisted Atmospheric Plasma Spraying (LAAPS) (Zieris et al., 2003). LAAPS has been developed by the combination of the gun of the Atmospheric Plasma Spraying with diode or Nd-YAG laser.

In LAAPS technology, three different arrangements between laser beam and plasma torch are possible (figure 3). Depending on the application the appropriate arrangement is selected. In the first case (figure 3a) the laser beam precedes the plasma torch and it acts as a pre-treatment method. The laser creates a molten pool crater at the substrate surface and thus the sprayed particles impinge on the molten substrate and form a coating that is metallurgically bonded to the substrate (Zieris et al., 2004). In the second arrangement (figure 3b), the laser beam is concentric with the plasma torch while in the third arrangement (figure 3c), which can be characterized as a post-treatment method the laser beam follows the plasma torch

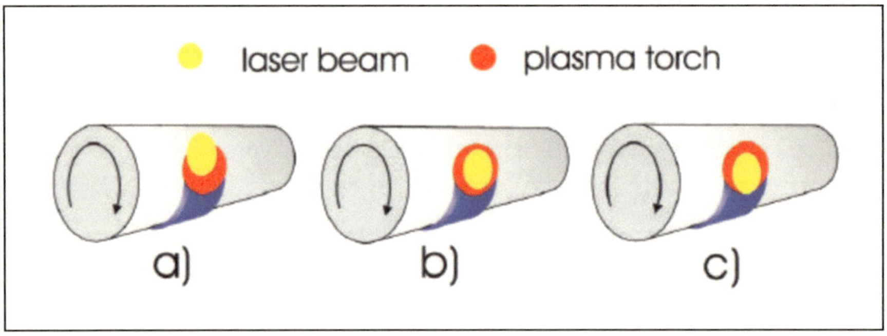

Fig. 3. Possible arrangements for laser thermal spray spots, after R. Zieris et al., 2003 (Zieris et al., 2003). The rotating cylinder is the substrate where the coating is deposited.

Two different LAAPS coatings were examined (Zieris et al., 2003); NiCrSiB onto mild steel substrates and Al_2O_3-3% TiO_2 coatings onto Inconel 718 substrate. Coatings were created on grit-blasted substrates as well as onto laser pre-treated substrates. For NiCrSiB sprayed coatings, the plasma spray gun was coupled with a laser diode (power of 2kW) and it was found that the adhesion strength of NiCrSiB was ~50 MPa onto grit-blasted substrates while it was increase at ~260 MPa in the case of laser ablated substrates. Similarly, in the case of Al_2O_3-3% TiO_2, the adhesion strength of the coating was increased by using LAAPS (~30 MPa onto as-received substrates) instead of conventional APS (~10 MPa onto grit-blasted substrates). For Al_2O_3-3% TiO_2, the plasma spray gun was coupled with Nd-YaG laser (power of 1 or 4 kW). Advanced coatings of AlSi30 (Zieris et al., 2004) and ZrO_2-8%Y_2O_3 (Dubourt et al., 2006) has been successfully deposited by employing LAAPS.

Formerly, the PROTAL® (French acronym for "PROjection Thermique Assisté par Laser", i.e: Laser Assisted Thermal Spray) technology comprises a method where pulsed laser (Q-switched Nd-YAG laser) irradiates the substrate surface prior to the deposition of the sprayed particles. It should be referred that PROTAL® is employed only as a pre-treatment method of the substrate. Thus, this case the laser treatment always precedes the spraying deposition. The purpose, of the laser irradiation is to eliminate the contaminations films and oxide layers, to generate a surface state enhancing deposit adhesion and to limit the recontamination of the deposited layer by condensed vapors (Costil et al., 2004a).

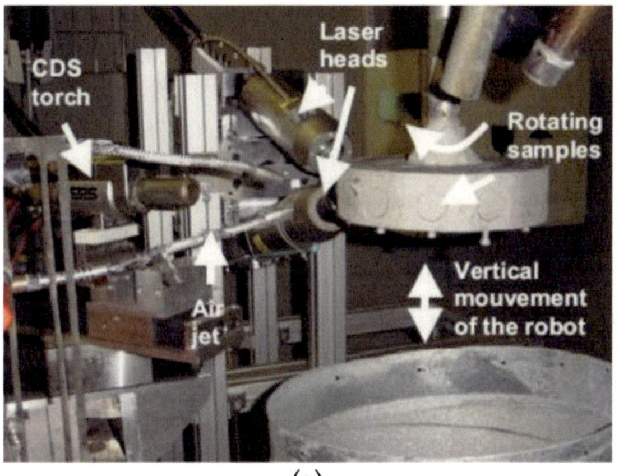

(a)

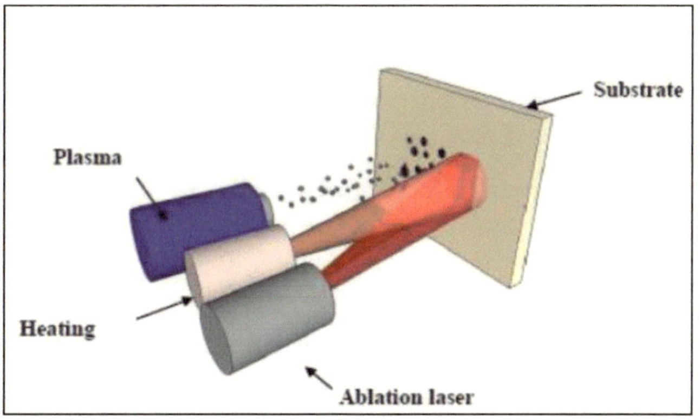

(b)

Fig. 4. (a) PROTAL® coupled to HVOF gun by Costil et al. (Costil et al., 2004a) (b) Two Nd-YAG lasers coupled with a plasma spray gun by Danlos et al. (Danlos et al., 2011)

The PROTAL® is equipped with two Q-switched Nd-YAG lasers operating at λ = 1.064 μm with an average power output of 40 W each (270 mJ per pulse with adjustable frequency up to 150Hz) and with a pulse duration (FWHM) of approximately 10 ns. The laser beam has a rectangular shape with a "top-hat" energy distribution thanks to a specific optical arrangement (Costil et al., 2004a).

The PROTAL® technology has been used already successfully in thermal spray technology (i.e. Plasma spraying, HVOF) (Barradas et al., 2004; Costil et al., 2004b) (figure 4a). Costil et al. (Costil et al., 2004b) have shown that in the case of plasma sprayed Ni-Al onto al-alloy substrate (AISI 2017), PROTAL® coatings can show higher adhesion strength compared to coatings formed onto substrates pretreated by conventional methods (i.e. grit-blasting).

Recently, Danlos et al. (Danlos et al., 2011) coupled two pulsed Nd-YAG lasers with a plasma spray gun (figure 4b). In this configuration the first Nd-YAG laser was used in order to heat the substrate and the second one (PROTAL® laser head) was used to ablate the substrate. The heating Nd-YAG laser was a pulsed millisecond laser with a wavelength of 1064 nm, while the ablation laser was a pulsed nanosecond laser with the same wavelength of 1064 nm. In order to impose both laser treatments simultaneously, the same frequency of 60 Hz was chosen for both heating and ablation laser. By using this set-up Ni-5Al coatings were deposited onto two different substrate; Al-alloy substrate (AISI-2017) and Ti-alloy substrate (Ti-6Al-4V). The substrates were pre-treated by conventional method (decreasing and grit-blasting) as well as by employing the set-up of figure 4a. For this combination of materials, it was found that the coatings adhesion was higher onto gritblasted substrates. It seems that the higher surface roughness which induced by the gritblasting, can explain the higher adhesion level onto the gritblasted substrates [47].

4. Adhesion determination of thermal sprayed coatings by employing laser techniques

The formation of thermal sprayed coatings can be divided into two stages: the creation of the first layer onto the substrate and the building-up of the coating itself onto as-sprayed layers. During the creation of the first layer, the adhesion of thermal sprayed particles onto the substrate influences strongly the bond strength of the cold sprayed coatings. The individual thermal sprayed particles (splat) is the basic structural building block in thermal sprayed coatings (Davis, 2004) and thus the study of these individual particles can give answers to fundamental queries about thermal sprayed coatings (Moreau et al., 1995; Christoulis et al., 2006; McDonald et al., 2006).

Furthermore, on macro-scale level, the characterization of the bond strength of thermal sprayed coatings is of high interest. Adhesion is a property of major concern for thermal spray coatings, because it is necessary for the coating to adhere to the substrate throughout the design life of the coating system (Davis, 2004).

Two laser techniques have been developed to determine the adhesion level, either of individual sprayed particles, or of thermal sprayed coatings. LASERFLEX which is presented in the first paragraph of this section shed light on the adhesion mechanism of thermal sprayed particle. At a macroscopic level characterization and quantitative determination of the bond strength of the cold sprayed coatings take place due to the development of a, one my say, unique test, i.e. Laser Shock Adhesion Test, namely LASAT®.

4.1 LASERFLEX (Laser Shock Flier Impact Experiments)

LASER shock Flier impact Experiments (LASERFLEX) (Barradas et al., 2007; Ichikawa et al., 2007a), primarily laser shock spallation can be used as an innovative technique to simulate experimentally the influence of "in-flight" material velocity and temperature on adhesion phenomena at the impact. LASERFLEX (figure 5) is a high-velocity impact simulation test that uses laser shock accelerated foil to impact on a substrate. Laser shock flier impact experiments consisted in high velocity cladding of a foil (the flier which simulates the cold-sprayed particle) onto bulk material (which acts as a substrate). The foils are accelerated by the laser shock at high velocity similar to the velocity of cold sprayed particle and then the foils impact on "substrates" at high velocity.

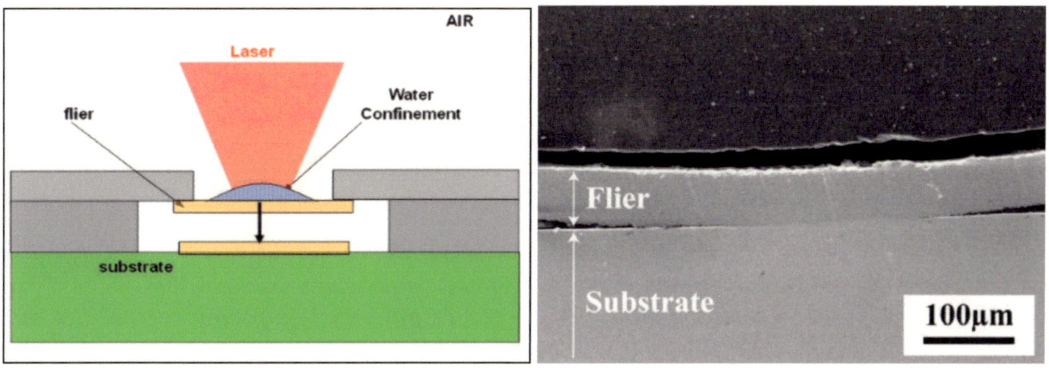

(a) (b)

Fig. 5. (a) Schematic illustration of the LASERFLEX (Barradas et al., 2007), (b) CoNiCrAlY flier deposited onto Inconel 615 substrate by employing LASERFLEX (Ichikawa et al., 2007a).

The flier is stuck to the steel sample holder horizontally due to a mere thin layer of grease, which left a distance to fly before reaching bulk substrate (Ichikawa et al., 2007a). For the shock, a one-shot laser beam from a Nd:YAG laser operating at 20 J during 20 ns focused on the foil in a spot of 4 mm in diameter. A transparent medium to the laser, i.e., water, was deposited at the surface of the sample. It led to a confining of the shock on the sample and increased the shock pressure and time so that the beam-matter interaction was extended to 50 ns.

The LASERFLEX technique has been used for the study of deposition of Cu flier onto Al substrate (Barradas et al., 2007) and CoNiCrAlY flier onto Inconel 615 substrate (Ichikawa et al., 2007a), revealing complicated phenomena which affect the coating-substrate adhesion. In the case of Cu flier onto Al substrate (Barradas et al., 2007) by using LASERFLEX evidence of local melting at the substrate-particle interface and the creation of intermetallics. The simulation of cold-sprayed particles by employing LASERFLEX, revealed that also in cold spraying, which is a solid state process, melting in micro level scale can be present in the interaction mechanism particle-substrate. On the other hand, in the case of the CoNiCrAlY flier onto Inconel 615, LASERFLEX revealed that the adhesion requires higher in-flight velocity and specific phase combination at the interface between flier and substrate (Ichikawa et al., 2007a). In this case, there is no evidence of melting of the flier or substrate.

4.2 LASAT (LAser Shock Adhesion Test)

LASAT is a laser innovative technique which has been developed for the determination of the adhesion and/or the cohesion of thermal sprayed coatings. LASAT has already been used to characterize the adhesion of thermal sprayed coatings (Barradas et al., 2004; Barradas et al., 2005; Boileau et al., 2005;Ichikawa et al., 2007b) and splat (isolated thermal sprayed particles) (Guetta et al., 2009). These experimental results have shown that LASATesting is a powerful tool for studying local adhesion of coatings and for determining the influence of metallurgical and morphological interface features on adhesion. Recently, LASAT was used also for other type of coatings and not only for thermal sprayed coatings. LASAT was applied successfully for the adhesion characterization of coatings which have been produced by EB-PVD (Fabre et al., 2011).

For the potted history, one may refer to shot peening as a process which was formerly applied to coatings (Al clads of aircrafts parts in this case) to characterize adhesion (Jeandin et al., 2003). The principle of this, one may say somewhat archaic, test was to shear the coating-substrate interface from coating plastic deformation due to peening effect. Above a certain peening level, coating spalling off can occur, which corresponds to a qualitative approximate adhesion limit.

In a similar way, the LASATesting was developed based on the use of a laser shock. The principle of the LASAT technology is presented in figure 6a. Irradiating the surface of a substrate with a high-power laser can generate a shock wave, the propagation of which leads to a tensile stress at the coating-substrate interface. This results from the crossing of the incident release wave with its reflected (at the reverse side, i.e. at the coating surface) release wave. To determine the adhesion strength, laser shocks are applied to samples at different levels of the incident laser power flux (figure 6b). The de-bonding limit, i.e. when exceeding the interface resistance, is inferred from coating surface velocity vs time profile obtained by Doppler laser interferometry using a VISAR ("Velocity Interferometer System for Any Reflector"). The adhesion strength is then given through simulation of the loading profile.

Below and above the laser shock adhesion threshold, two types of different velocity profiles of the coating surface are respectively achieved (figure 6b).

The two types of velocity signals show significant peaks in the velocity amplitude, which correspond to the interaction of the shock wave with the coating surface. In the curve typical of a laser flux below the threshold, the time between two subsequent peaks is that for the shock wave to propagate through the whole coated substrate material and go back. Therefore, the wave can go through the interface, which is the sign there is no interfacial damage.

The second typical shape of coating surface velocity profile is that recorded for a laser flux which is above the threshold (figure 6b). In this case, the distance between two velocity peaks is shorter and the time between the two peaks corresponds to the time necessary for the wave to cross two times the coating only. This signal is typical of de-bonding at the interface as the shock wave reflects on the void which was created at the first interaction with the interface.

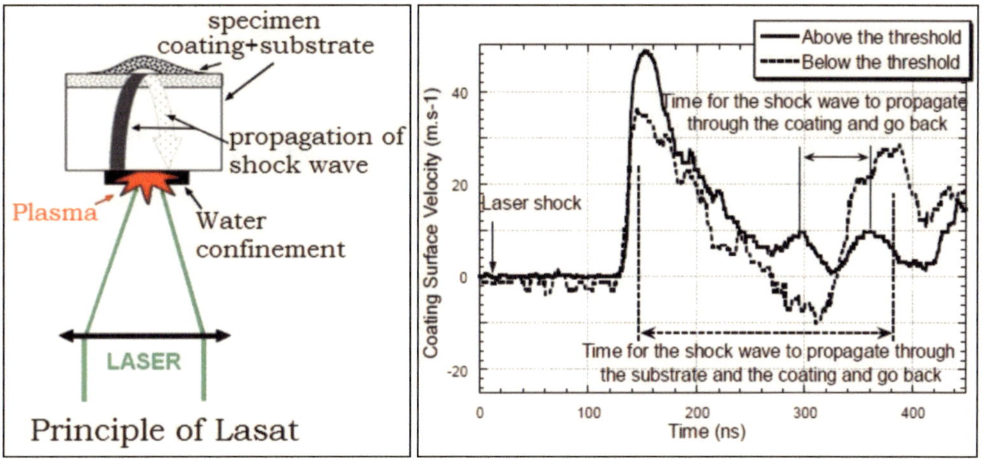

(a) (b)

Fig. 6. (a) Principle of LASAT, (b) Typical coating surface velocity vs time VISAR profiles, below (dotted) and above (plain) the laser shock adhesion threshold (Jeandin et al., 2003)

This interpretation of the velocity profiles, as for soundness criteria of the interface, was ascertained by metallographical studies of cross-sections of as-shocked specimens. Adhesion tests can therefore be interpreted using velocity signal diagnostics only.

Incidentally, LASATesting is based on a purely-mechanical process, which is different from (and more efficient than) other shock-based adhesion tests which, however, involve thermal phenomena and remain rather limited in their exploitation (Costil et al., 2001; Jeandin et al., 2003).

In contrast with conventional adhesion testing, e.g. "pull-off" testing, LASATesting can detect, in particular, the influence of well-defined microstructural features including those of a small size (below 1 µm typically) and/or of a given shape.

5. Laser-Assisted Cold Spray (LACS)

Laser Assisted Cold Spray (LACS) process combined laser pretreatment of the substrate and the coatings build-up in one step for the production of advanced protective or functional coatings of high adhesion and cohesion strength. Nowadays, to improve adhesion strength, conventional substrate pre-treatment are generally performed before applying the cold spray coating. This include grit-blasting that is performed prior to spraying. Often time, the grit blasting procedure requires masking of the part to protect the regions that are not intended to be coated. The grit blasting procedure results in the inclusion of small grit particles at the interface between the coating and substrate. This limits its use when fatigue properties are important. In addition, grit blasting produces waste and can be harmfull to the operator as discussed in section 3.1. The cohesive strength is improved by thermal post-treatments however, this adds another step in the production of a protective or functional coating but more importantly, in some cases, this procedure would hinder one of the advantages of the cold spray process which is to preserve the structure characteristics of the powder (i.e. nanostrcure, phases, etc.). The conventional techniques for the improvement of adhesion and cohesion techniques are two, three or even four step process while the coupling of cold spray system with a laser permits the creation of dense coatings with high bond strength and cohesion in a mere one-step process.

Bray et al. (Bray et al., 2009) coupled a laser diode of wavelength 980 nm with a cold spray gun (figure 7). In this case titanium powder (<45 μm) was sprayed onto mild steel substrates. The laser diode had a heating effect on the substrate, which was beneficial for the formation of the cold sprayed coatings. It was found that as the substrate temperature increased by the irradiation of the tlaser laser diode, the deposition efficiency of the titanium coatings increased, while the coatings porosity decreased.

In another configuration (figure 8), a pulsed Nd-YAG laser was coupled with the cold-spray gun to result in the laser beam passing milliseconds prior to the cold-spray jet for deposition, (Christoulis et al., 2009; Christoulis et al. 2010). Pulsed Nd-YAG laser were operated only during the first pass in order to clean the substrate surface. By using the experimental set-up of figure 8 two coating systems were studied: Aluminum and Ni-20Cr onto an Al-based

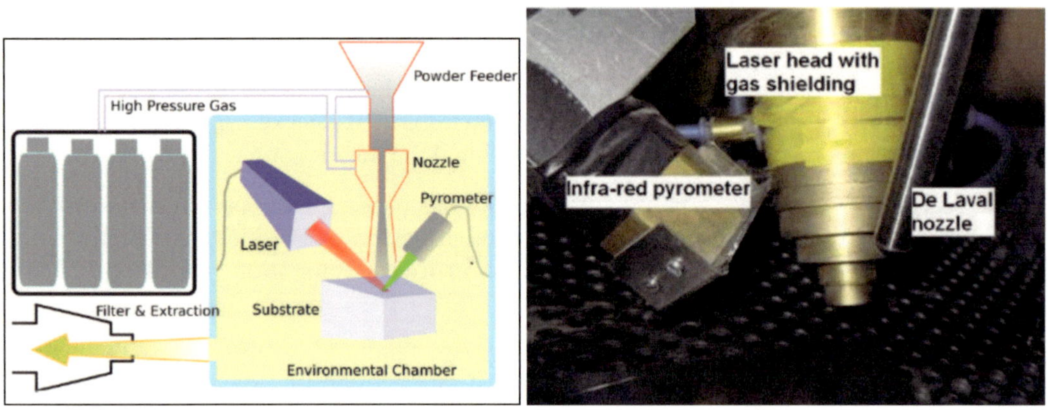

(a) (b)

Fig. 7. (a) Experimental set-up, (b) side view of experimental set-up (Bray et al., 2009).

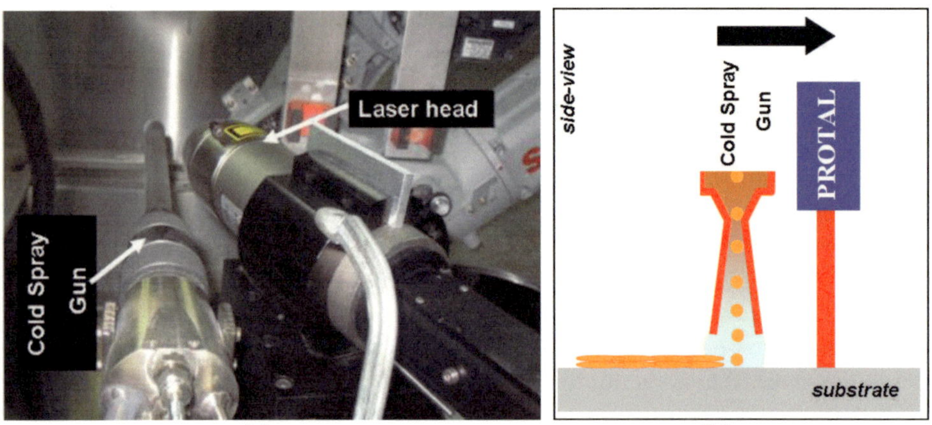

(a) (b)

Fig. 8. (a) Experimental set-up, (b) side view of experimental set-up (Christoulis et al., 2009)

alloy (AISI 2017) and a Ni-based (Inconel718) alloy respectively. The results of these experiments are presented in the next section. The surface preparations and sprayings experiments were carried out in the MAMADC cold spray laboratory located at the Industrial Materials Institute of the National Research Council of Canada.

6. Application of LACS

Al-based and Ni-based coatings were formed onto Nd-YAG laser pre-treated substrates as well as onto as-received substrates, and pre-treated by two conventional methods, namely polishing and grit-blasting. It is shown that the coating-substrate interface is significantly improved when pulsed laser ablation is performed at optimized parameters.

The spraying experiments were carried out by using nitrogen (N_2) as the propelling gas. The cold spraying conditions for both Al and Ni20Cr powders are presented in table 1. It should be mentioned that the spraying conditions were selected by measuring the particles mean velocity for various standoff distances. The particle velocity was measured by using the ColdSprayMeter® (Tecnar Automation Inc., St-Bruno, QC, Canada) (Christoulis et al., 2009; Christoulis et al., 2010; Jeandin et al. 2010).

Since the Nd-YAG laser was proceeding of the cold spray gun, the scanning of the substrate was realized in a specific way (figure 9a). The same movement geometry was chosen also for the conventionally prepared substrates since experiments indicate that substrate temperature increases during spraying due to the heated propelling gas (Irissou et al., 2008b). So, a unique movement geometry (figure 9b) was chosen to introduce uniform heat input to all the substrates.

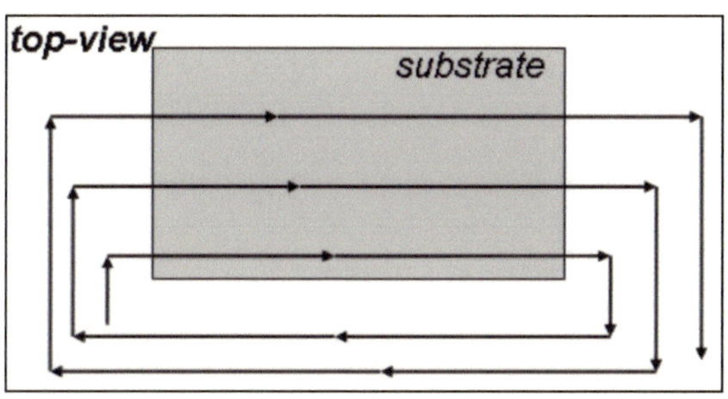

(a)

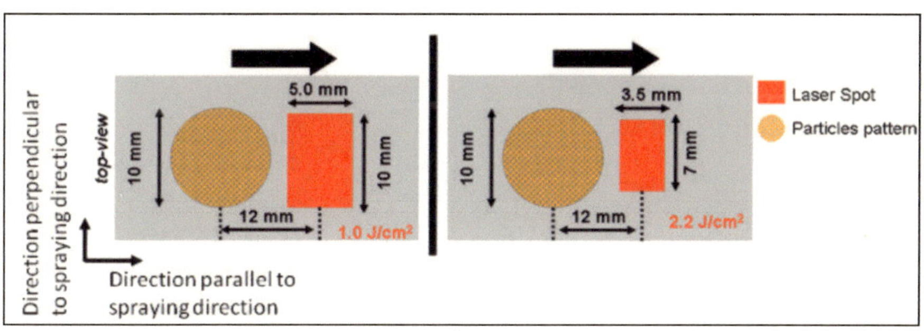

(b)

Fig. 9. (a) movement geometry of coupled cold-spray gun-laser head and (b) top view of the pattern of both cold-sprayed particles and laser spot

By taking into account, the frequency of the laser, the size of the spot and the movement of the gun (table 1), an arithmetic exercises suffices to calculate the overlapping percentage (table 2). In figure 10, it is shown the overlapped area for two sequential laser pulses for different laser energy and laser frequency. For both laser energies, under the lower laser frequency (18.75 Hz) no overlapping was created. In this case, inhomogeneous (treated and untreated) areas were created on the substrate (figure 10)

Powder	Al	Ni20Cr
Spraying Conditions		
Gas pressure (MPa)	3.0	3.0
Gas temperature (°C)	350	600
Standoff distance (mm)	20	40
Gun traverse speed (mm.s-1)	100	100
Nozzle Characteristics		
Type of the Nozzle	PBI-33	MOC
Exit diameter (mm)	10	6.6
Throat diameter	2.7	2.7
Expansion ratio	13.7	6.0
Total length (mm)	220	175

Table 1. Experimental Conditions

Substrate	Laser energy density ($J.cm^{-2}$)	Frequency (Hz)	Overlapping of 2 pulses (%)
As-received	No laser	No laser	No laser
Polished	No laser	No laser	No laser
Grit-blasted	No laser	No laser	No laser
As-received	1.0	18.75	No overlapping
As-received	1.0	37.5	24
As-received	1.0	150	82
As-received	2.2	18.75	No overlapping
As-received	2.2	37.5	47
As-received	2.2	150	87

Table 2. Pre-treatment of substrates by employing a Nd-YAG laser

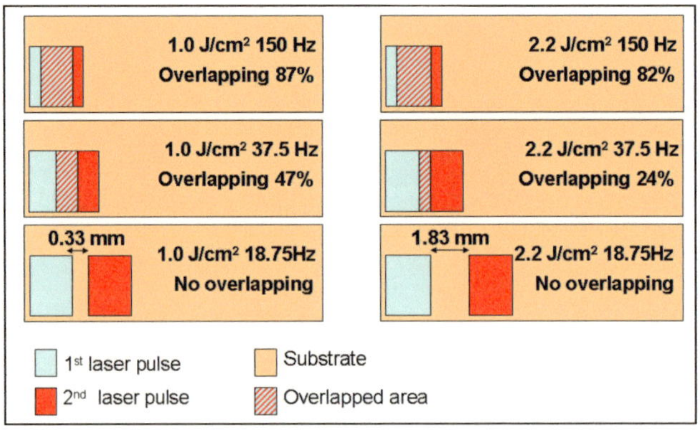

Fig. 10. Overlapping for 2 sequential Nd-YAG laser pulses

6.1 Al-based LACS Coatings

6.1.1 Materials

Fine gas-atomized aluminum powder (Alfa Aesar, Massachusetts, USA, 17-35 μm) of spherical morphology (figure 11a) was sprayed using a KINETICS® 3000-M System (CGT-GmbH, Ampfing, Germany). The particle size of the powder was measured by laser granulometry (figure 12b). The microhardness of the powder was measured at 28±1 HV0.01. The aluminum powder was sprayed onto AISI 2017 aluminum-based alloy.

6.1.2 Laser irradiation effects

Cracks and traces, which should have been induced during the production of the hot rolled Al alloy material, were found on the surface of as-received substrates (figure 12a). The surface of the substrates were observed by employing Scanning Electron Microscope (SEM (LEO 450VP, Germany)).

The observations of the ablated substrate revealed that for low energy density (1.0 J.cm^{-2}), the pulsed laser beam reacted only with the cracks provoking slight fusion at their borders (white arrows, Figures 3b and 3c). By comparing figures 3b and 3c, it seems that the increase of the laser frequency, provoked a more intensive fusion at the borders of the cracks.

Also, craters were formed after laser irradiation (white circles, figures 3b and 3c). The craters, are correlated either with the surface defects such as micro-inclusions and small scratches present prior to irradiation or, with the existence of precipitated phases (Al2Cu) of the used AISI 2017 alloy (Costil et al., 2004).

The increase of laser energy density to 2.2 J.cm^{-2} resulted in a strong change in surface morphology (figure 3c and 3d). Extensive substrate melting is observed. The pre-existing cracks disappeared and few craters (white circles, figure 3c) are still visible. Substrate surface seems to smoothen with the increase in the laser frequency (figure 3d), probably as a result of re-melting which is provoked by the high overlapping phenomenon.

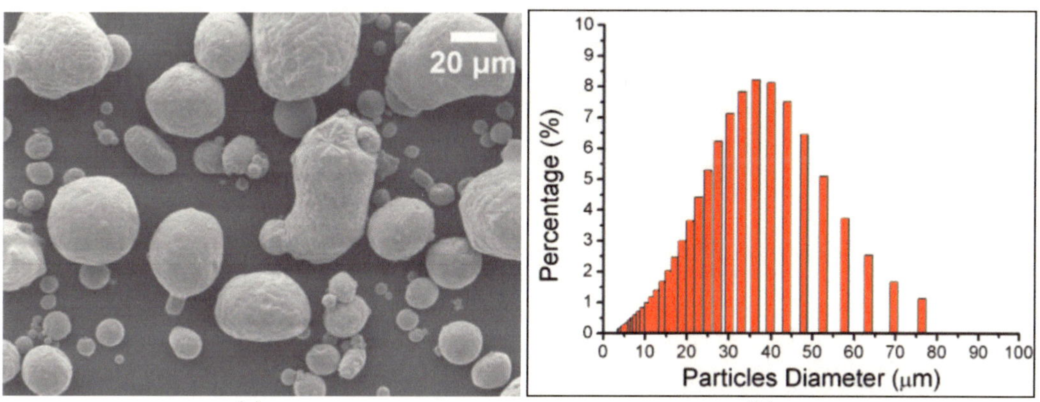

(a) (b)

Fig. 11. (a) SEM of Al powder and (b) Particle distribution by laser analysis.

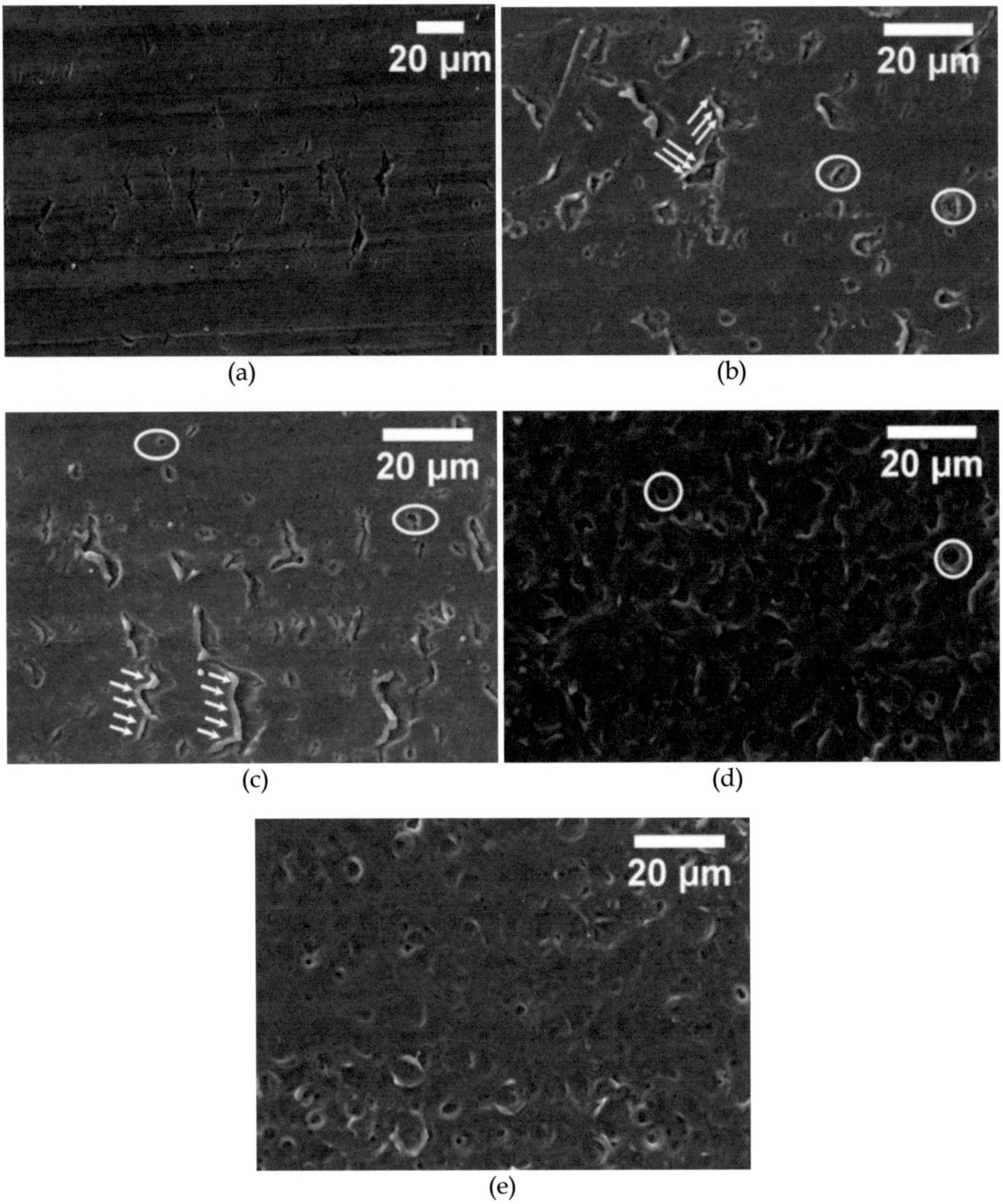

Fig. 12. SEM top view of AISI-2017 substrate surfaces: (a) un-treated and Nd-YAG laser treated (b) 1.0 J.cm-2, 37.5 Hz., (c) 1.0 J.cm-2, 150 Hz., (d) 2.2 J.cm-2, 37.5 Hz, (e) 2.2 J.cm-2, 150 Hz.

6.1.3 Formation of thick aluminum coatings

Thick coatings were deposited on the substrates (see figure 13). The mean thickness of the cold-sprayed coatings was also quantitatively calculated via image analysis (ImageJ®) of

SEM images. Coating mean thicknesses were almost the same for conventionally prepared substrates and for low energy (1.0 J.cm^{-2}) irradiated substrates (figure 14).

The mean thickness of the coatings was increased significantly as soon as the laser energy density was increased from 1.0 to 2.2 J.cm^{-2}. Interestingly, the difference in coating thickness

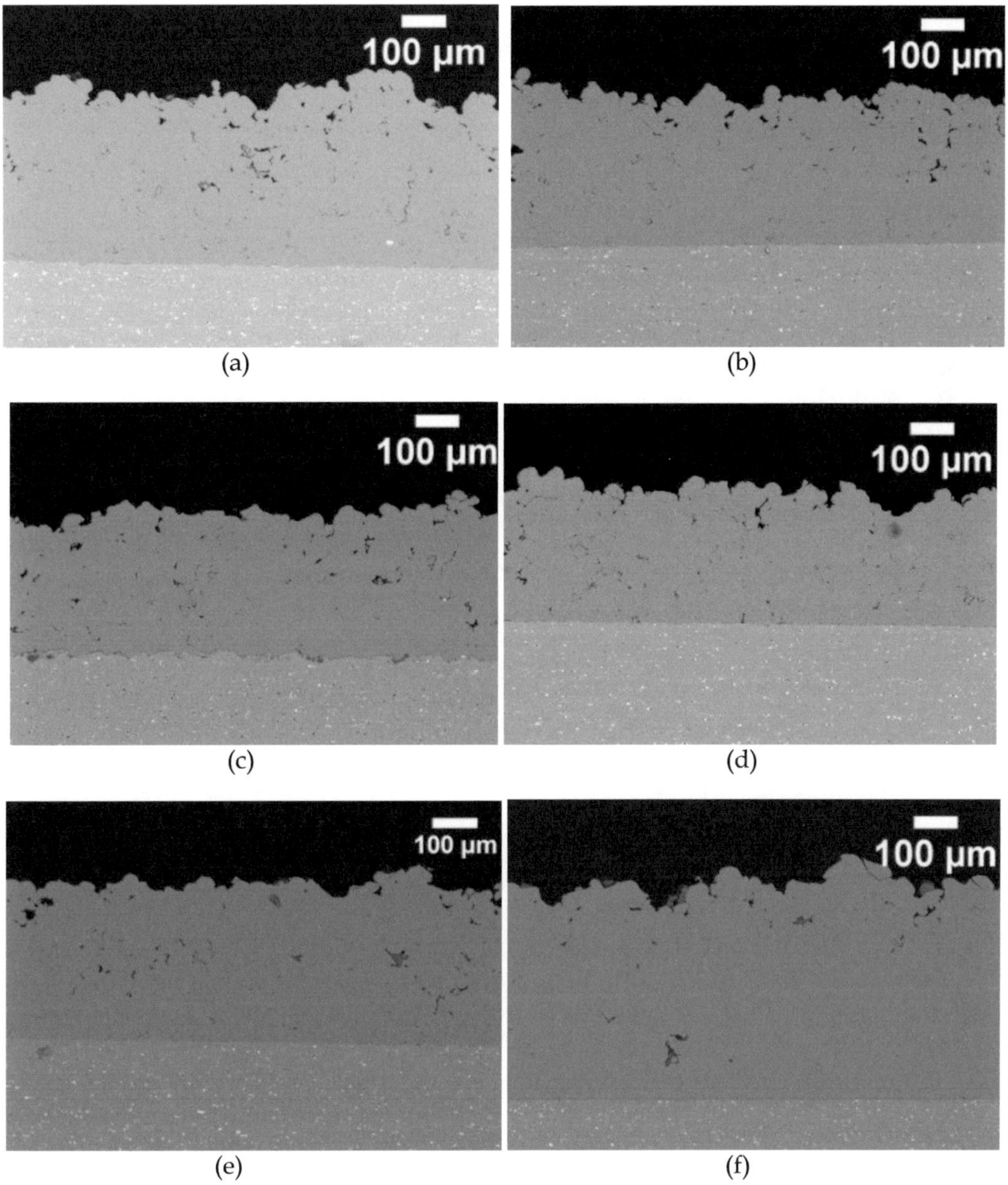

Fig. 13. SEM of cross-sections of cold-sprayed Al coatings formed onto (a) as-received, (b) mirror-polished substrate, (c) grit-blasted substrate, (d) laser irradiated substrate (1.0 J.cm^{-2}, 37.5 Hz), (e) laser irradiated substrate (2.2 J.cm^{-2}, 37.5 Hz), (f) laser irradiated substrate (2.2 J.cm^{-2}, 150 Hz)

of coatings deposited on untreated and on laser ablated (2.2 J.cm-2, 37.5 Hz) substrate is around 50 μm. One can conclude that this difference corresponds well to the difference of average coating thickness as shown in figure 13. The improved deposition efficiency with high fluence laser ablation is due essentially to the adhesion of a higher fraction of particles during the first pass as compare to other surface pre-treatment and laser ablation conditions. Nevertheless, it is clear that a large fraction of the impinging particles are not deposited during the first pass even at higher laser ablation fluence and so they should contribute to the surface modification as well.

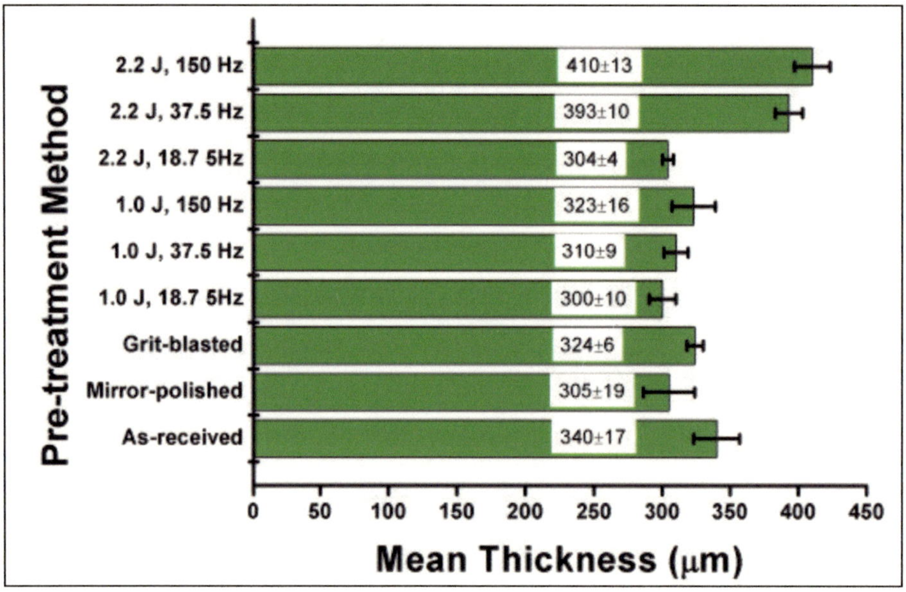

Fig. 14. Thickness of Al cold sprayed coatings for different pre-treatment methods

6.1.4 Coating-substrate interface

The percentage of cracked interfaces was determined by means of quantitative image analysis (ImageJ® (Ref 47)). About 10 mm of interface was examined parallel to the spraying direction. The examined 10 mm-region corresponds to about 60 SEM images of high magnification (×1000).

Representative interfaces between thick coating and the substrate are presented in figure 15. It was found that the grit blasting and polishing pre-treatment did not offer any reduction of cracks observed on the as-received substrates (figure 16). Surprisingly, the grit-blasting of the substrate contributed to the increase of the percentage of cracked interface (figure 15). Also, in the case of grit-blasted substrates, alumina particles entrapped onto the substrate were seen (figure 15c).

Conversely, Nd-YAG laser ablation promoted a better interface with much less interfacial cracks (figure 16). For the highest laser energy density (2.2 J.cm-2) and under the highest frequency (150 Hz), cracks could barely be found in direction perpendicular to spraying and did not appear at all in the parallel direction (Figure 16c).

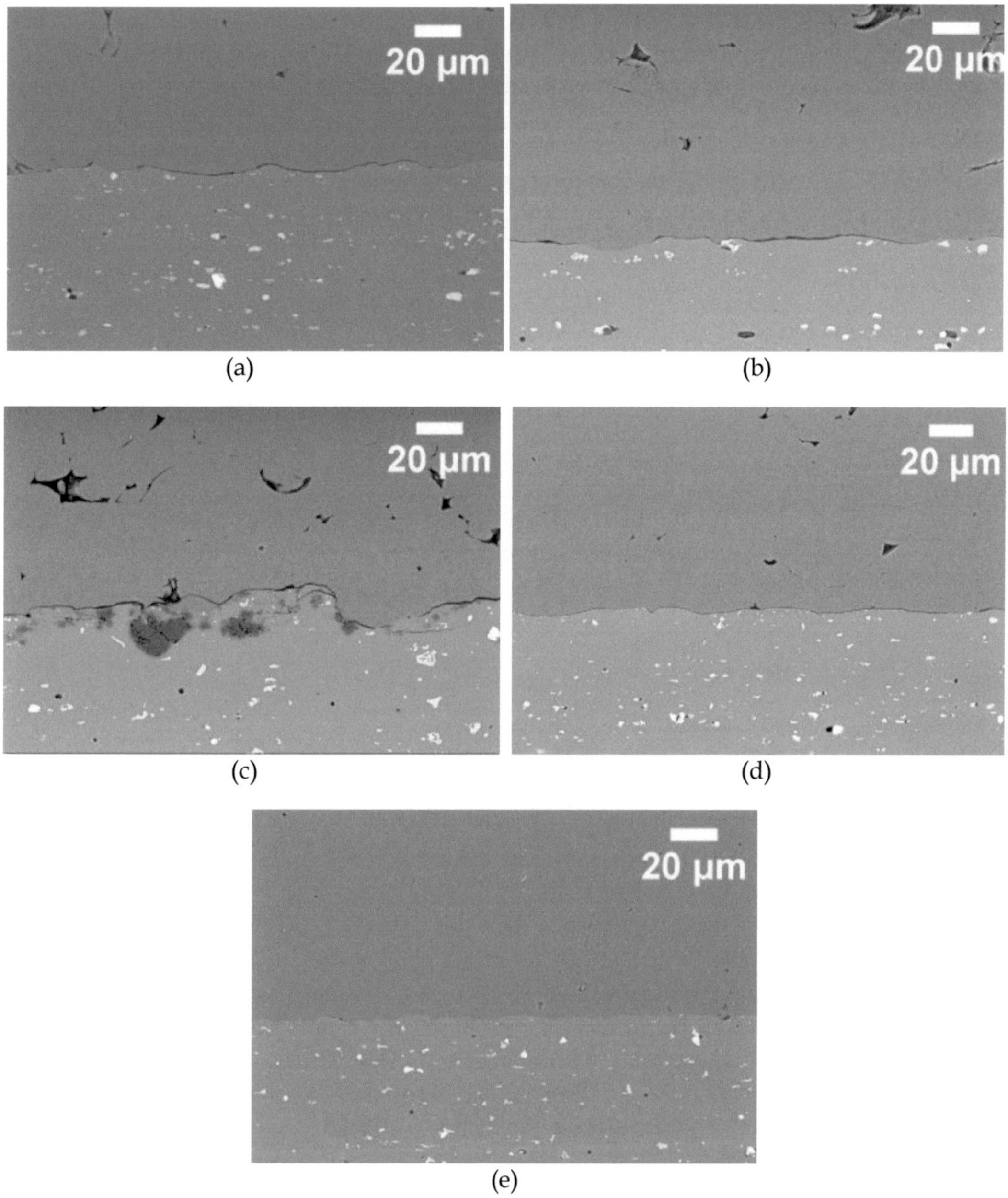

Fig. 15. SEM of cross-sections of cold-sprayed Al coatings formed onto (a) as-received substrate, (b) mirror-polished substrates, (c) grit-blasted substrate (d) laser irradiated substrate (1.0 $J.cm^{-2}$, 150 Hz) (e) laser irradiated substrate (2.2 $J.cm^{-2}$, 150 Hz).

The difference of the percentage of interfacial cracks between parallel and perpendicular is due to the difference in the size of the particle jet compared to the size of the laser spot (figure 11). In the perpendicular direction, fraction of sprayed particles impact on untreated areas and so these particles can present weaker adhesion (Christoulis et al., 2010). On the other hand, in the parallel direction, cold-sprayed particles always impact on laser pre-treated areas, if the cross section metallographic sample is prepared properly, and thus along this direction a reduced percentage of interfacial cracks is observed.

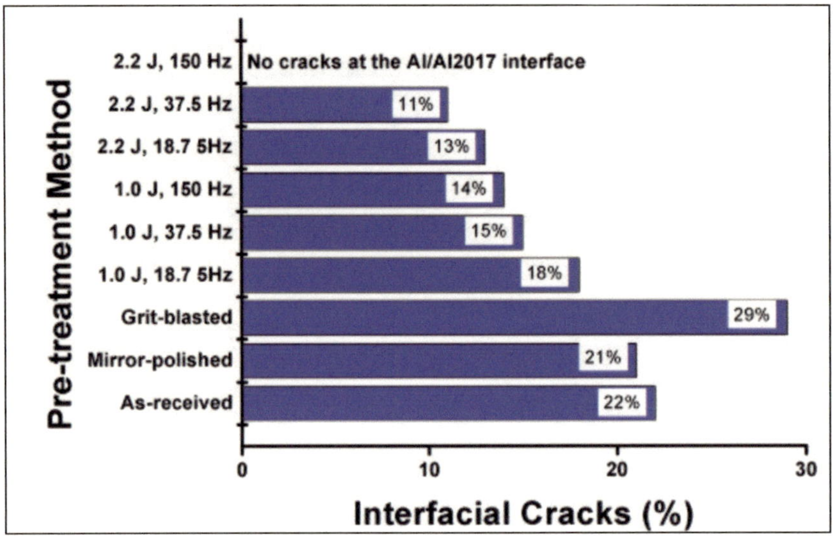

Fig. 16. Percentage of interfacial cracks between cold-sprayed Al coatings and substrates.

6.1.5 TEM observations

Thin foils of Al-coatings were prepared by FIB and they were examined with TEM.

A typical oxide layer of about 100 nm in thickness could be observed at the coating-substrate interface in cold-sprayed "as-received" Al 2017 (figure 17). The EDX indicates that this oxide layer is in fact divided in two regions of distinct Al/O ratio. (in gray in figure 17a). The layer at the substrate side has the stoichiometry of alumina indicating that it is the native oxide while the other one at the coating side is richer in oxygen (~35 %wt Al, 65 %wt O)

TEM analysis indicates that the oxide layer oxide at the coating side is amorphous since the grey contrast of this oxide layer was constant. Furthermore, EFTEM (Energy Filtered Transmission Electron Microscope) with GIF (Gatan Image Filter) image recorded on the low loss region (Plasmon) of the EELS spectrum highlights this interfacial layer, which is characteristic of amorphous structure (Gertsman et al., 2005).

The transformation of the native alumina oxide to an amorphous Al oxide could be the results of extensive peening by the impinging particles (similar to what is observed in mechanical milling for example (Gaffet et al., 1997; Li et al., 2009). Indeed, it was reported that amorphyzation can occur during cold spraying (Xiong et al., 2008). It is interesting to note that most of the cracks observed at the interface were propagated in between the two oxides.

In contrast, for laser-processed Al 2017, no oxygen could be detected at the interface (figure 18) neither on STEM images nor by EDX profile. This profile was obtained with a probe size of 1 nm (enlarged to 3 nm at the exit side of the thin foil), 10 nm between the consecutive analysis spots, and a limit of detection of 1 %wt for O. It can therefore be inferred that the native layer was removed by the laser treatment, and if an oxide thinner layer was formed prior to the particle reached the substrate, the thickness of this layer would not exceed a few nanometers

6.1.6 LASAtesting

Laser shock adhesion testing was performed both on coating formed onto as-received substrates and onto laser irradiated substrates (2.2 J.cm^{-2}, 150 Hz).

The lower adhesion of Al-coatings onto as-received substrates was confirmed by LASAT experiments. Coatings were totally de-bonded with the lower laser energy (1.7 GW.cm^{-2}) (figure 19a). These results are confirmed also by VISAR signals (see figure 20). The velocity peaks corresponded to shock wave interaction with the coating surface. For both LASAT energies, the absence of negative peaks corresponds to the damage of coating-substrate interface (Barradas et al., 2005).

Onto pre-treated substrates (2.2 J.cm^{-2}, 150 Hz), cold-sprayed aluminum coatings presented higher adhesion. When low laser energy (1.7 GW.cm^{-2}, figure 19b) was used, the coating remained adherent. The increase of LASAT energy (2.6 GW.cm^{-2}, Figure 19c) provoked the interlamelar cracking of the Al-coatings. This revealed that the cohesion of the coatings (bonding of cold-sprayed Al-Al) was weaker compared with the coating-substrate adhesion. Similar phenomenon has been found for cold-sprayed copper on AISI 2017 substrate (Barradas et al., 2005).

At higher LASAT laser energy (4.3 GW.cm^{-2}) the coating was de-bonded from the laser pre-treated substrate, (figure 19c). However, few particles still remained bonded reveling the high achieved adhesion strength.

For laser pre-treated substrates the VISAR results are well segregated (figure 20). For low LASAT energy (2.6 GW.cm^{-2}, black line) the aluminum surface velocity showed negative values after the first positive peak. The positive velocity peaks correspond to shock wave interaction with the coating surface while the negative peaks correspond to tensile stresses reaching the substrate surface (Barradas et al., 2005). The tensile stresses were those generated at the aluminum surface, after their propagation through the whole sample. On the other hand for high LASAT energy (4.6 GW.cm^{-2}) only a positive peak was detected since the tensile wave reflected on the thus-created cracks and the surface coating velocity did not show negative values. The tensile stress at the interface, $\sigma 22$, was calculated from modeling/numerical simulation of 1D and 2D shock wave effects within the coating-substrate system and averaged over the whole laser spot (Boustie et al., 2000). Bonding strength value could therefore be determined from "post-mortem" observation of interfacial cross-sections and study of VISAR velocity profiles (figure 20) during the test, since both show when the coating de-bonds (figure 20). The bond strength was found to be above 629 MPa but below 681 MPa for laser-ablated Al 2017 compared to below (one may assume much below) 562 MPa for as-received Al 2017. These bond strength values are significantly higher than what is typically reported for pull-off ASTM C633 testing which lies in the 10-80 MPa range (Price et al., 2006; Stoltenhoff et al., 2006; Shin et al., 2008; Triantou et al., 2008), while for some conditions of aluminum coating applied on Aluminum 7075 alloy substrate the failure mode was in the glue at over 60 MPa (Irissou et al., 2007).

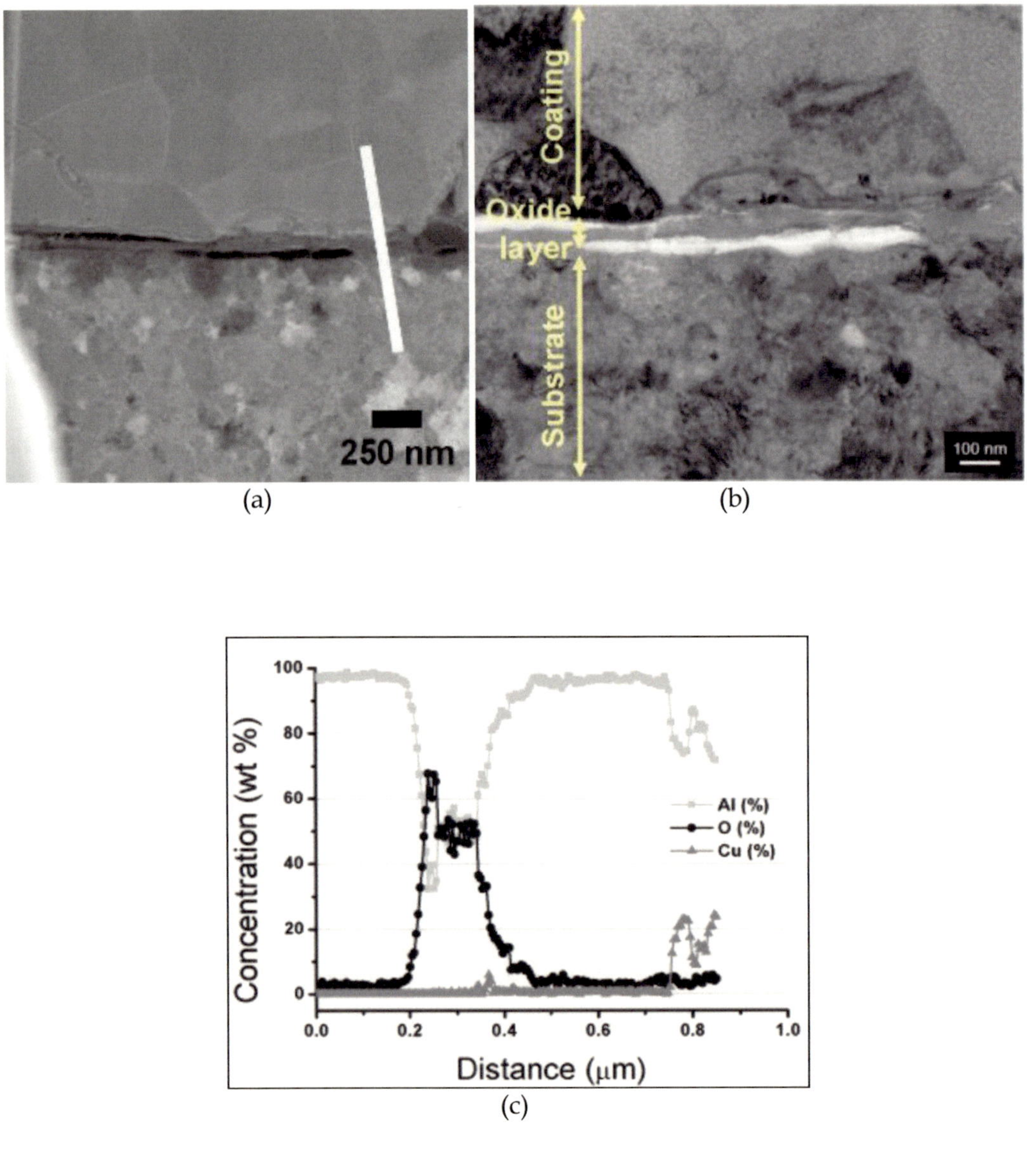

Fig. 17. TEM pictures of cold-sprayed Al on as-received AISI 2017 substrate (a) Bright Field (BF) image, (b) HAADF (High Angle Annular Dark Field) image, (c) EDX profiles along the white line.

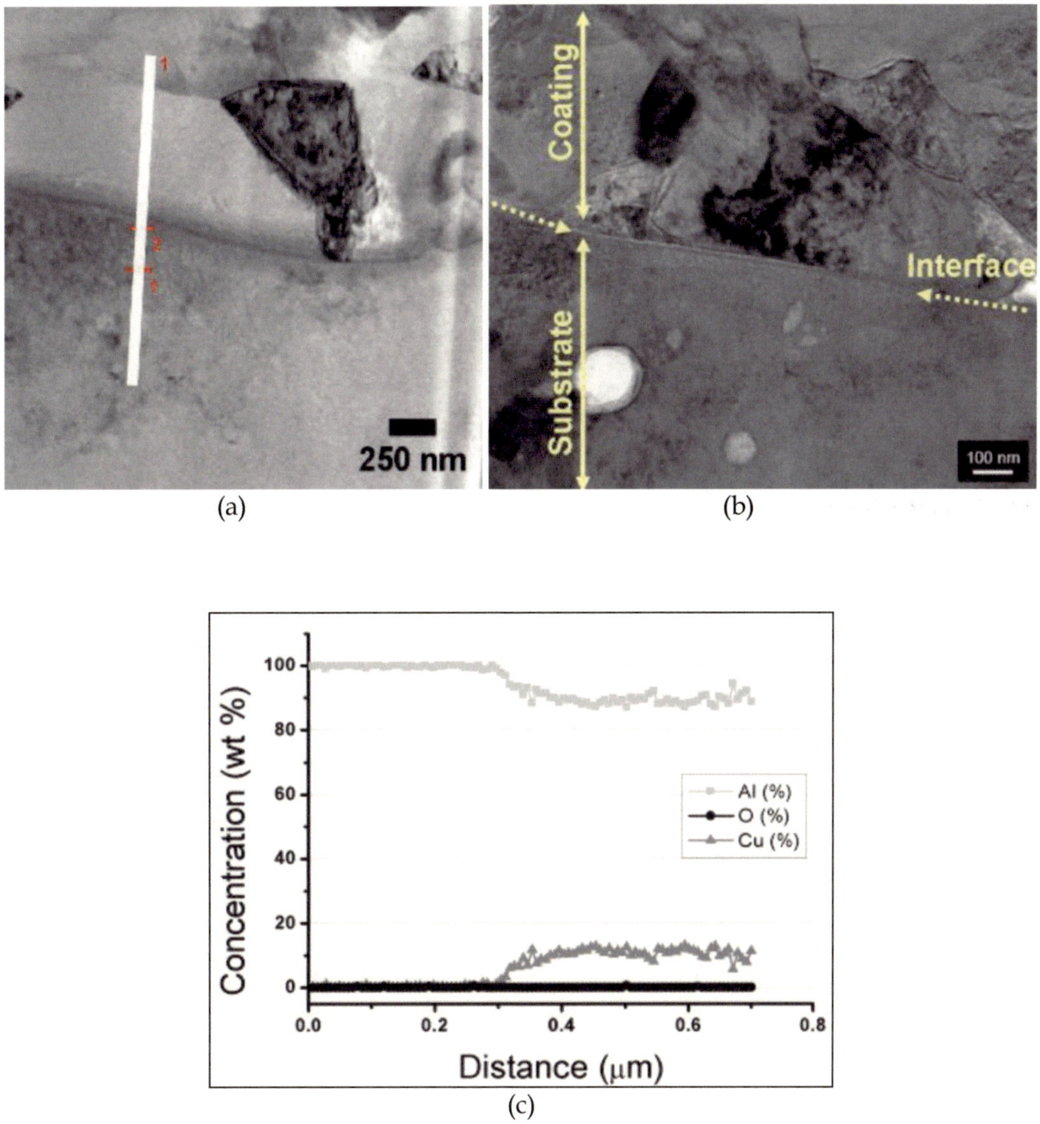

Fig. 18. TEM pictures of cold-sprayed Al on laser pre-treated (2.2 J.cm^{-2}, 150 Hz) AISI 2017 substrate (a)Bright Field (BF) image, (b) HAADF (High Angle Annular Dark Field) image, (c) EDX profiles along the white line

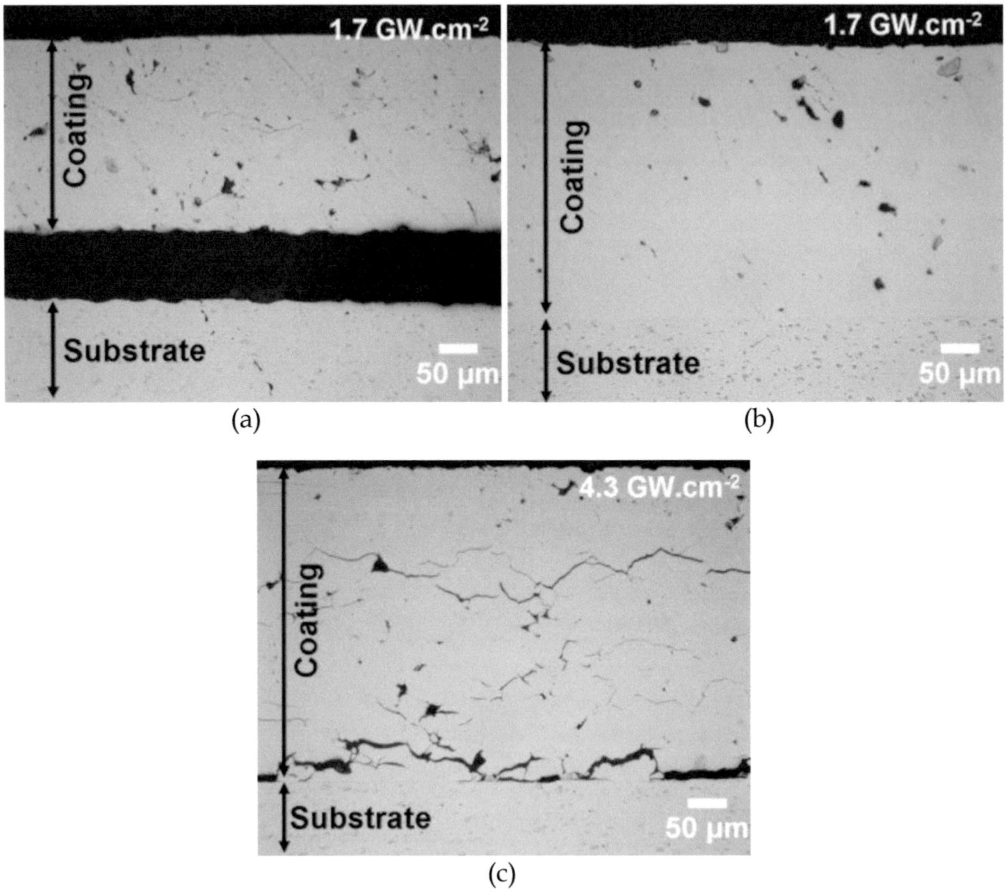

Fig. 19. Cross-section optical image of LASATested interface of cold-sprayed Al onto (a) as-received substrate when the applied LASAT energy was 1.7 GW.cm^{-2} (b) laser pre-treated (PROTAL 2.2 J.cm^{-2}) substrate when the applied LASAT energy was 1.7 GW.cm^{-2} (c) laser pre-treated (PROTAL 2.2 J.cm^{-2}) substrate when the applied LASAT energy was 4.7 GW.cm^{-2}.

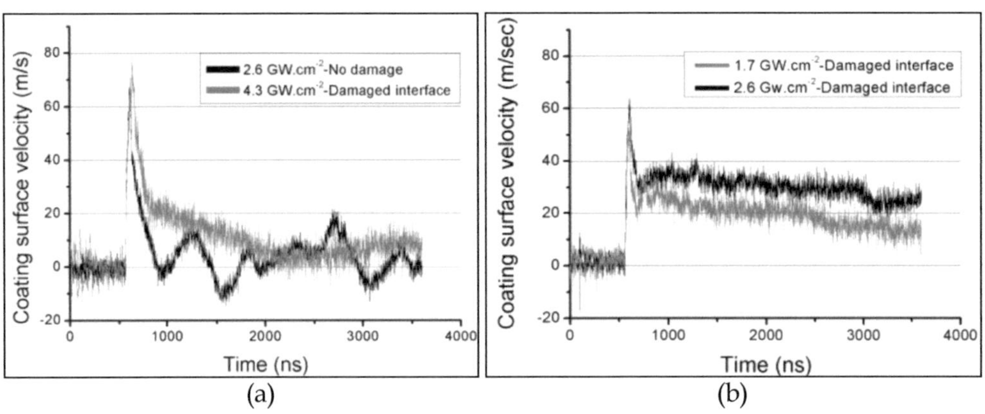

Fig. 20. Velocity profiles of cold-sprayed Al surface on (a) as- received AISI 2017 substrate (b) laser pre-treated (2.2 J.cm^{-2}) substrate AISI 2017

6.2 Ni-based LACS Coatings

6.2.1 Materials

The feedstock powder was commercial Ni-20Cr (Höganäs, 1616-09/PS) of spherical morphology (figure 21a) and with particle size ranging from 20-53 µm (figure 21b). The powders microhardness was measured at 192±15 $HV_{0.05}$. The powder was sprayed onto Inconel alloy 718 substrate.

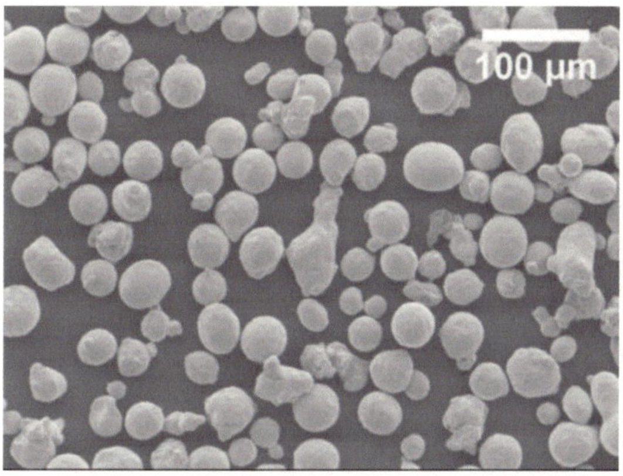

(a)

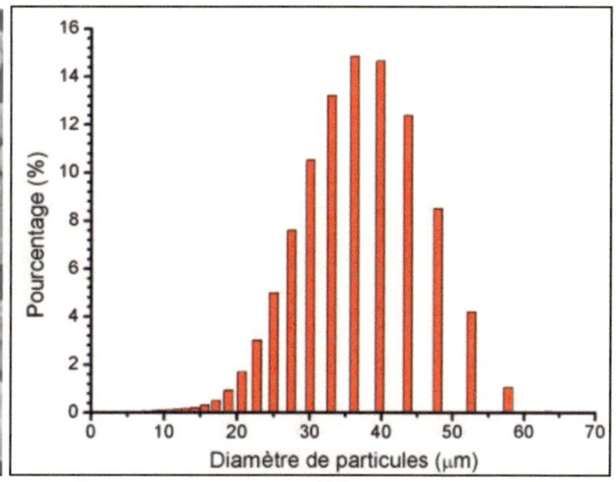

(b)

Fig. 21. SEM of Ni-20Cr powder (Höganäs, 1616-09/PS) and (b) Particle distribution by laser analysis.

6.2.2 Laser irradiation effects

Observations of the as-received substrates revealed a superficial "cellular" structure (Figure 22). The modifications induced on the substrates by the nano-pulsed Nd-YAG laser, are presented in Figure 22 as function of the laser energy and frequency. For low laser energy (1.0 $J.cm^{-2}$), at the lowest frequency (18.75 Hz) the laser provoked the partially removal of the "cellular" structure (some cells are still observed) and a new structure full of small craters was revealed. It seems that the laser impulses caused the local fusion of the substrate. The craters on the surface can be correlated either with the surface defects, such as micro-inclusions and small visible scratches prior to irradiation either with the existence of precipitated phases as it has been found in the case of al-alloy (Costil et al., 2004), but further studies should be done in order to verify this assumption. The gradual increase of the frequency from to 18.75 to 150 Hz resulted in the complete ablation of the initial "cellular" structure.

The increase of the laser energy to 2.2 $J.cm^{-2}$ induced more intensive melting of the substrate surface. The craters were reduced and it seems that the more intensive fusion led to their coverage. Increasing the frequency, which corresponds to the increase of overlapping, smoothened the substrate surface where neither the initial "cells" either their borders can be seen.

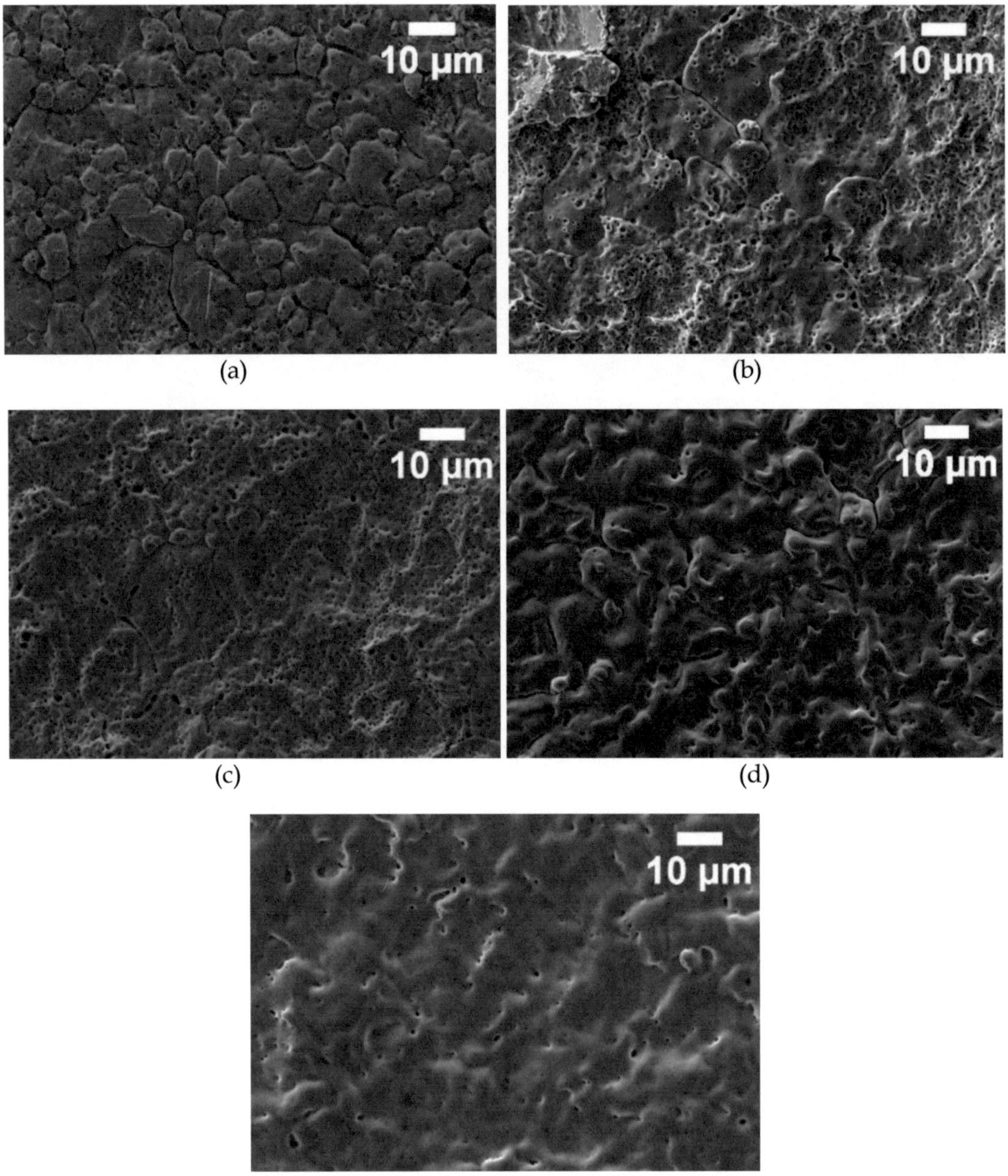

(a) (b) (c) (d) (e)

Fig. 22. SEM top view of Inconel 718 substrate surfaces: (a) un-treated and Nd-YAG laser treated (b) 1.0 $J.cm^{-2}$, 37.5 Hz., (c) 1.0 $J.cm^{-2}$, 150 Hz., (d) 2.2 $J.cm^{-2}$, 37.5 Hz, (e) 2.2 $J.cm^{-2}$, 150 Hz.

6.2.3 Formation of thick Ni-20Cr coatings

The cross-sections of the coatings were investigated by Scanning Electron Microscopy (SEM, LEO 450VP). Conventional image analysis software (ImageJ®) was used to calculate the mean thickness. The mean thickness of the coatings was calculated by observing 12 SEM images in standard magnification of x200 (figure 23). The 12 images correspond to about 6 mm length. The mean thickness of the coatings for the different pre-treatment methods of the substrate is presented in figure 24.

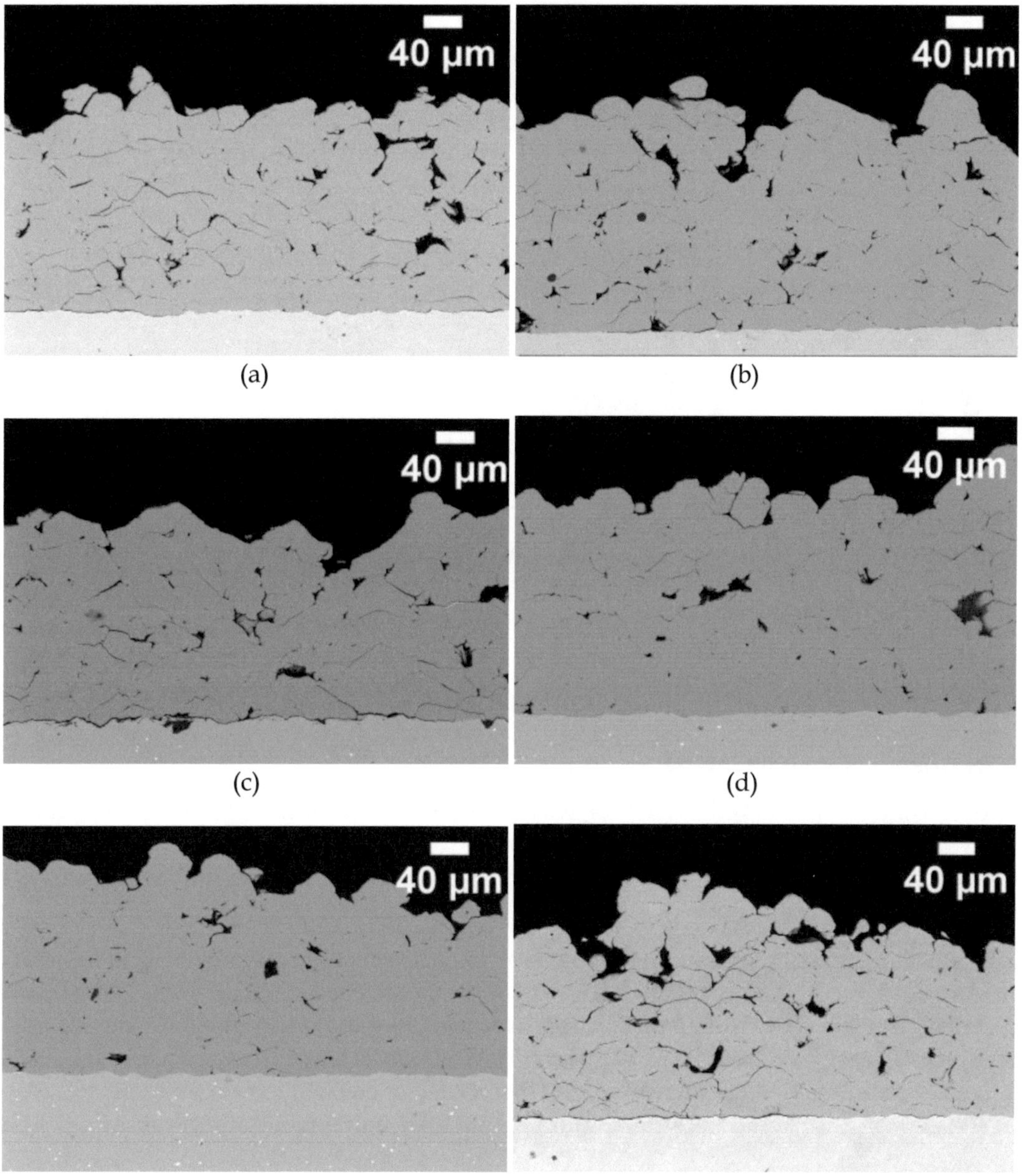

Fig. 23. SEM of cross-sections of Ni-20Cr cold-sprayed coatings formed onto (a) as-received substrate, (b) mirror-polished substrate, (c) grit-blasted substrate, (d) laser irradiated substrate (1.0 $J.cm^{-2}$, 37.5 Hz), (e) laser irradiated substrate (2.2 $J.cm^{-2}$, 37.5 Hz), (f) laser irradiated substrate (2.2 $J.cm^{-2}$, 150 Hz)

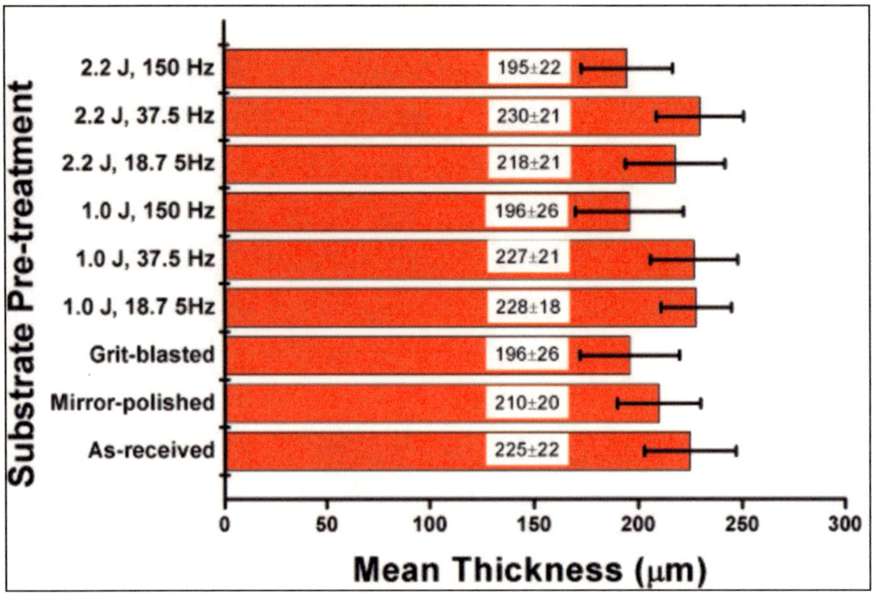

Fig. 24. Thickness of Ni-20Cr cold sprayed coatings for different pre-treatment methods

6.2.4 Coating-substrate interface

The percentage of cracked interfaces was determined by means of quantitative image analysis (ImageJ®). About 10 mm of interface was examined parallel to the spraying direction. The examined 10 mm-region corresponds to about 30 SEM images of high magnification (×500).

Representative interfaces between thick coating and the substrate are presented in figure 25. It was found that the grit blasting and polishing pre-treatment did not offer any reduction of cracks observed on the as-received substrates (figure 16). As it was found in the case of cold-sprayed Al coatings onto AISI 2017 substrates (figures 15 and 16) the grit-blasting of the substrate contributed to the increase of the percentage of cracked interface (figure 26). Also, in the case of grit-blasted substrates, alumina particles entrapped onto the substrate were seen (figure 25c).

On the other hand, Nd-YAG laser ablation promoted a better interface with much less interfacial cracks (figure 26). The optimum ablation conditions for LACSsprayed Ni-20Cr coatings were found to be: laser energy of 2.2 J.cm^{-2} and laser frequency of 37.5 Hz. Under these conditions the percentage of interfacial cracks between Ni-20Cr coatings and Inconel 718 substrate was decrease at the lowest value of 7.8% (figure 26).

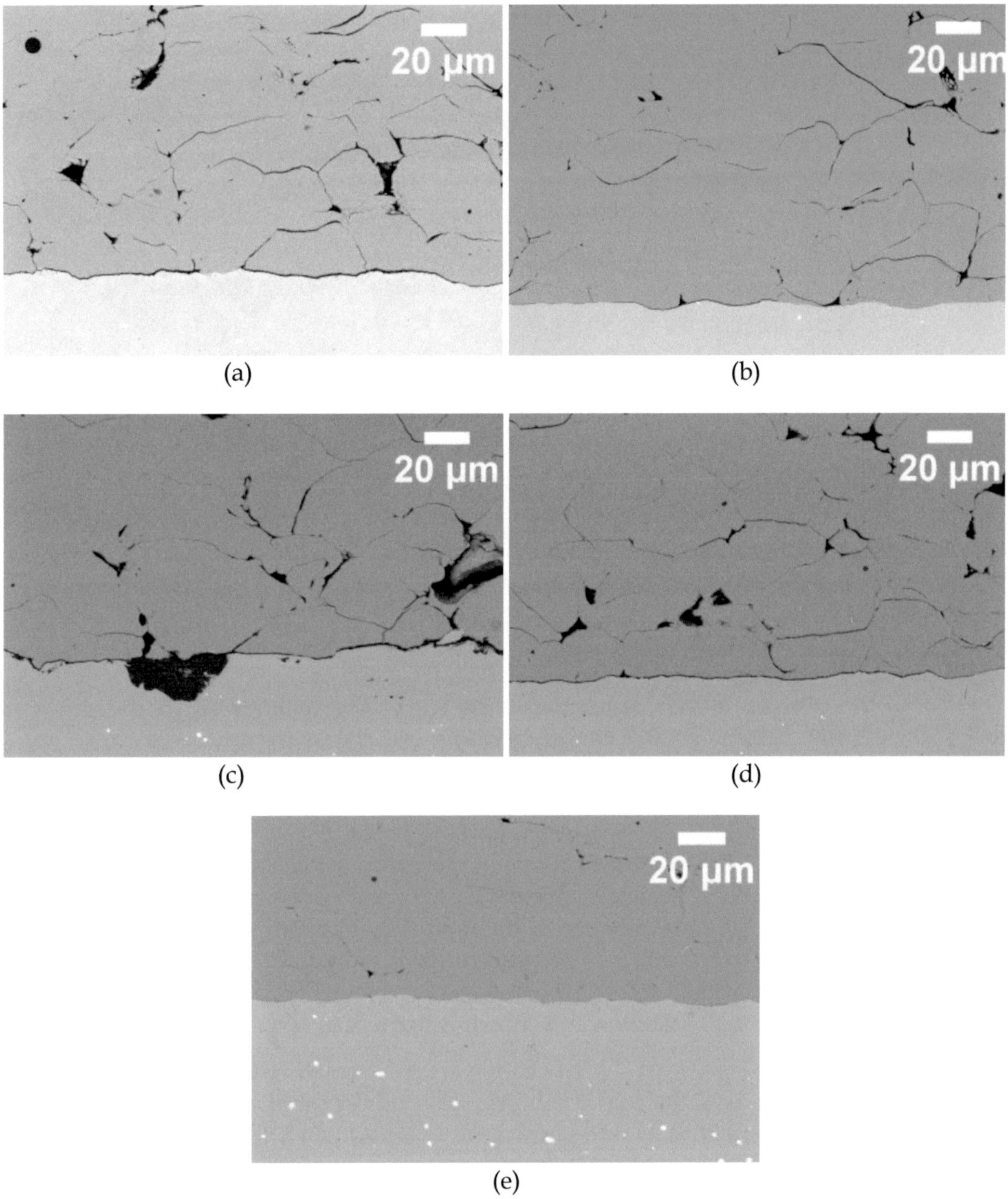

Fig. 25. SEM of cross-sections of cold-sprayed Al coatings formed onto (a) as-received substrate, (b) mirror-polished substrates, (c) grit-blasted substrate (d) laser irradiated substrate (1.0 $J.cm^{-2}$, 18.75 Hz) (e) laser irradiated substrate (2.2 $J.cm^{-2}$, 37.5 Hz).

The further increase of the laser frequency at 150 Hz for the highest laser energy (2.2 $J.cm^{-2}$) provoked a significant increase of the interfacial cracks (50%, figure 26). Based on SEM top view of Inconel 718 substrate surfaces (figure 22), it is assumed that the increase of the laser frequency resulted in extensive melting of the substrate, which in turn could increase the pores and the cracks due the change of the interaction between the sprayed particles and the substrate.

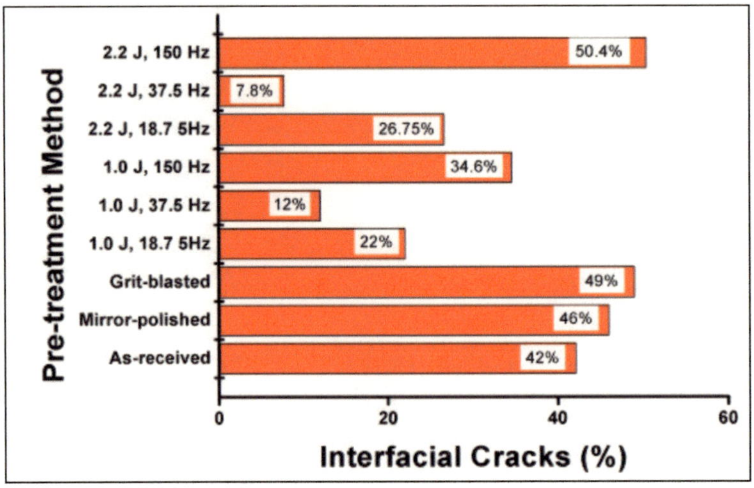

Fig. 26. Percentage of interfacial cracks between cold-sprayed Al coatings and substrates.

7. Conclusion

The use of the cold spray system for the formation of the coatings has the potential to overcome all the drawbacks which are induced by the conventional thermal spray technologies. Moreover, the Laser Assisted Cold Spray is a novel process, for the pretreatment of the substrate and the coatings build-up, which may lead to the production of advanced coatings of high adhesion and cohesion strength. Nowadays, the adhesion of cold sprayed coatings is improved by using environmental unfriendly pretreatment techniques (grit-blasting and chemical solvents) while the cohesion is improved by thermal post-treatments which can eliminate one of the main advantages of cold spray technology; the formation of oxide-free coatings. The conventional techniques for the improvement of adhesion and cohesion techniques are two-step process while LACS permits the creation of dense novel coatings of high adhesion and cohesion strength in one-step process.

Two different coating systems (Al-based coatings and Ni-based coatings) were studied by using a LACS system where the cold spray gun was coupled with a nano-pulsed Nd-YAG laser head. It was found that at optimized laser ablation conditions, coatings of higher thickness are formed. Also, under the optimized laser condition the interfacial cracks between the deposited coatings and the ablated substrates were reduced dramatically. For both Al-based and Ni-based coatings, cross section micrographs revealed that of a significant interface (coatings-substrate) length are cracked on samples prepared with conventional methods (grit blasting and polishing) polished or wrough substrates.

For Ni-20Cr coatings onto Inconel 718 substrate the optimized laser conditions found to be: 2.2 J.cm-2 Laser energy density and 37.5 Hz Laser frequency.

In the case of al-based coatings onto AISI 2017 substrates, the optimized laser conditions were found to be: 2.2 J.cm-2 Laser energy density and 37.5 Hz Laser frequency. TEM studies on the al-based coatings revealed that an amorphous aluminium oxide phase was created at the interface on the coating side while an alumina layer is still present on the substrate side. Most of the cracks observed at the interface were propagated in between the two oxides. The adhesive strength between the aluminium coating and the amorphous aluminium oxide layer thus seems higher than that between the two oxide layers. Pulsed laser ablation at optimized conditions (2 $J.cm^{-2}$, 37.5 Hz) eliminated the native oxide at the substrate surface, leading to the formation of al-coatings with the highest level of adhesion strength for the examined conditions.

The adhesion of LACSpraeyd al-based coatings was measured by employing a Nd-YAG laser based system (LASAT). The bond strength of Al cold-sprayed coating was numerically calculated from LASAT measurements and it was found to lie in between 629 and 681 MPa for samples prepared on laser pre-treated substrates (2.2 J.cm-2, 150 Hz) in between and below 562 MPa on as-received substrate. The experimental results shown that LASATesting is a powerful tool for studying local adhesion of coatings and for determining the influence of metallurgical and morphological interface features on adhesion.

8. Outlook

Laser Assisted Cold Spray process can shed light on an open query of cold spray technology. The presence of local transient melting at the interface of substrate and particles, and its contribution to the bonding of cold-sprayed particles has not been clarified yet. Experimental studies show evidence of local fusion at the substrate-particle interface (Barradas et al., 2007; Guetta et al., 2009) as well as interlamellar local melting (Li et al., 2007). In these cases, an increase of adhesion of cold sprayed coatings is observed due to the presence of molten phases (Barradas et al., 2007). On the other hand, other experimental results (Ref 12) and numerical simulations (Bae et al., 2009) indicate that, during the creation of cold-sprayed coating, melting cannot occur at the particle-substrate interface. While, in the case of cold sprayed aluminum onto tin substrate, the rebound of the particles is associated with substrate local fusion (Zhang et al., 2003). The presence or the absence of local melting seems to depend on the materials used and experimental conditions. It is therefore possible that local fusion can exist only under certain conditions (materials, gas and temperatures etc). The Laser Assisted Cold Spray can enlighten the role of the local melting at the interface between substrate and the sprayed particles. The laser ablation energy of the Nd-YAG laser should be adjusted to the appropriate level to provoke the melting of the substrate surface. These experiments will answer to an open fundamental query of cold spray technology: does the melting of the substrate enhance the adhesion strength of cold sprayed coating or not?

Furthermore, the Nd-YaG laser should be used, also, during coatings' build-up to enhance coatings' cohesion. The weak cohesion of cold-sprayed particles is one of the disadvantages of the cold spray process. The weak cohesion decreases coatings' properties and limits further industrial applications of cold spray technology. Presently, in order to improve the cohesion of cold-sprayed coatings, thermal post-treatment techniques are used (Hall et al., 2006; Makinen et al., 2007; Spencer et al., 2009). After the deposition of the cold sprayed coating, a thermal treatment follows. This two-step process can be replaced by a single step process: the coupling of cold-spraying with the Nd-YaG laser; thus cold-sprayed coatings of both high adhesion and cohesion strength will be created.

Laser-Assisted Cold Spray, should be coupled also with other type of laser which are suitable to create a particular structure on the substrate (Knapp et al., 2011). Among the various laser surface processes, laser structuring is currently one of the most developed and promising. Beyond the mere application to engraving, laser structuring of a substrate (whatever the type of material) is expected to improve bond-strength when coated with a layer. The latter can be obtained using cold spray in particular.

9. Acknowledgements

The authors would like to thank Mr. B. Harvey and Mr. M. Lamontagne both of Industrial Material Institute (Boucherville, Canada) for Cold-spray and PROTAL® process, respectively. Professor K. Ogawa, Associate Professor Y. Ichikawa, Mr. K. Sagakuchi and Dr. T. Miyazaki, of Tohoku University (Sendai, Japan) are acknowledged for FIB cuttings. The current project was carried out in the framework of the Cold-spray Club (www.mat.ensmp.fr/clubcoldspray).

10. References

Ajdelsztajn L., Zuniga A., Jodoin B., Lavernia E.J. (2006). Cold Gas Dynamic Spraying of a High Temperature Al Alloy, *Surface & Coatings Technology,* 2006, Vol. 201, pp. 2109-211.

Araujo P., Chicot D., Staia M., Lesage J. (2005). Residual stresses and adhesion of thermal spray coatings, *Surface Engineering*, Vol. 21, No. 1, pp. 35-40

Arrigoni M., Barradas S., Braccini M., Dupeux M., Jeandin M., Boustie M., Bolis C., Berthe L. (2006). A comparative study of three adhesion tests (EN 582, similar to ASTM C633, LASAT, bulge and blister test) performed on plasma sprayed copper deposited on aluminium 2017 substrates, Journal of Adhesion Science and Technology, Vol. 20, No. 5, pp. 471-487

Bae G., Kang K., Na H., Lee C., Kim H.J. (2009). Thermally Enhanced Kinetic Sprayed Titanium Coating: Microstructure and Property Improvement for Potential Applications, *Proceedings of the Expanding Thermal Spray Performance to New Markets and Applications*, ISBN-13: 978-1-61503-004-0, Las Vegas, United States of America, May, 2009

Bahbou M.F., Nylen P., Wigren J. (2004). Effect of Grit Blasting and Spraying Angle on the Adhesion Strength of a Plasma-Sprayed Coating, *Journal of Thermal Spray Technology*, Vol. 6, No. 2, pp. 508-514.

Bala N., Singh H., Prakash S. (2010). High Temperature Corrosion Behavior of Cold Spray Ni-20Cr Coating on Boiler Steel in Molten Salt Environment at 900°C, *Journal of Thermal Spray Technology*, Vol. 19, No. 1-2, pp. 110-118

Barbezat G. (2005). Advanced thermal spray technology and coating for lightweight engine blocks for the automotive industry, *Surface & Coatings Technology*, Vol. 200, pp.1990 - 1993

Barbezat G. (2006). Application of thermal spraying in the automobile industry, *Surface & Coatings Technology*, Vol. 201, pp. 2028–2031

Barradas S., Jeandin M., Bolis C., Berthe L., Arrigoni M., Boustie M., Barbezat G. (2004) Study of Adhesion of PROTAL® Copper Coating of Al 2017 Using the Laser Shock Adhesion Test (LASAT), *Journal of Materials Science*, Vol. 39, pp 2707-2716

Barradas S., Molins R., Jeandin M., Arrigoni M., Boustie M., Bolis C., Berthe L., Ducos M. (2005). Application of laser shock adhesion testing to the study of the interlamellar strength and coating substrate adhesion in cold sprayed copper coating of aluminum, *Surface & Coatings Technology*, Vol. 197,pp. 18-27

Barradas S., Guipont V., Molins R., Jeandin M., Arrigoni M., Boustie M., Bolis C., Berthe L., Ducos M. (2007). Laser Shock Flier Impact Simulation of Particle-Substrate Interactions in Cold Spray, *Journal of Thermal Spray Technology*, Vol. 16, No. 4, pp. 548-556

Bobzin K., Schlafer T., Warda T., Bruhl M. (2010). Thermally Sprayed Oxidation Protection Coatings for γ-TiAl Substrates, 2010, *Proceedings of Thermal Spray: Global Solutions for Future Application*, ISBN 978-3-87155-590-9, Singapore, May 2010

Boileau S., Guipont V., Jeandin M., Nivard M., Berthe L., Jerome J., Boustie M., Li H., Khor K.A. (2005). The Laser Shock Adhesion (LASAT) for Production Control of Thermally-Sprayed Ceramic Coatings, *Proceedings of Thermal Spray Connects: Explore its Surfacing Potential!*, ISBN 3-87155-793-5, Basel, Switzerland, May, 2005

Boustie M., Auroux E., Romain J.P. (2000). Application of the Laser Spallation Technique to the Measurement of the Adhesion Strength of Tungsten Carbide Coatings on Superalloy Substrates, *European Physical Journal-Applied Physics*, Vol. 12, pp. 47-53

Brandt O.C. (1995), Mechanical properties of HVOF coatings, *Journal of Thermal Spray Technology*, Vol. 4, No. 2, pp. 147-152

Bray M., Cockburn A., O'Neil W. (2009). The Laser-assisted Cold Spray process and deposit characterization, *Surface & Coatings Technology*, Vol. 203, pp. 2851–2857

Chavan N. M., Ramakrishna M., Sudharshan Phani P., Srinivasa Rao D., Sundararajan G. (2011). The influence of process parameters and heat treatment on the properties of cold sprayed silver coatings, *Surface & Coatings Technology*, Vol. 205, No. 20, pp. 4798-4807

Christoulis D.K., Pantelis D.I., Borit F., Guipont V., Jeandin M. (2006). Effect of Substrate Preparation on the Flattening of Plasma Sprayed Aluminium Bronze Powders., *Surface Engineering*, Vol. 6, pp. 420-431

Christoulis D.K., Guetta S., Guipont V., Berger M.H., Jeandin M., Boustie M., Costil S., Ichikawa Y., Ogawa K., Irissou E., Legoux J.-G., Moreau C. (2009). Cold Spraying Combined to Laser Surface Pre-treatment using PROTAL®, *Proceedings of the Expanding Thermal Spray Performance to New Markets and Applications*, ISBN-13: 978-1-61503-004-0, Las Vegas, United States of America, May, 2009

Christoulis D.K., Guetta S., Irissou E., Guipont V., Berger M.H., Jeandin M., Legoux J.-G., Moreau C., Ichikawa Y., Ogawa K., Costil S., Boustie M. (2010). Cold-Spraying Coupled to Nano-Pulsed Nd-YaG Laser Surface Pre-treatment, *Journal of Thermal Spray Technology*, Vol. 19, No. 5, pp.1062-1073.

Christoulis D.K., Guetta S., Guipont V., Jeandin M. (2011). The Influence of the Substrate on the Deposition of Cold-Sprayed Titanium: An Experimental and Numerical Study, *Journal of Thermal Spray Technology*, Vol. 20, No. 3, pp. 523-533

Costil S., Coddet C., Rosa G., Psyllaki P., Oltra R. (2001). Non Destructive testing for the Estimation of the Adhesion of Plasma Sprayed Coatings, *Proceedings of News Surfaces for the New Millenium*, ISBN 0-87170-737-3, Singapore, 2001

Costil S., Li H., Coddet C., Barnier V., Oltra R. (2004a). Role of Laser Surface Activation During Plasma Spray Coating of Metallic Materials, *Proceedings of 18th International*

Conference on Surface Modification Technologies, ISBN 0871708337, Dijon, France, November, 2004

Costil S., Li H., Coddet C. (2004b). New Developments in the PROTAL® Process, *Proceedings of Thermal Spray 2004: Advances in Technology and Application*, ISBN 0-87170-809-4, Osaka, Japan, May, 2004

Danlos Y., Costil S., Liao H., Coddet C. (2011). Influence of Ti-6Al-4V and Al 2017 substrate morphology on Ni-Al coating adhesion - impacts of laser treatments, *Surface & Coatings Technology*, Vol. 205, No. 8-9, pp. 2702-2708.

Davis J. (2004). *Handbook of Thermal Spray Technology*, 2004, ASM International., ISBN 0-87170-795-0, United States of America

Day J., Huang X., RicFhards N.L. (2005), Examination of a Grit-Blasting Process for Thermal Spraying Using Statistical Methods, *Journal of Thermal Spray Technology*, Vol. 14, No. 4, pp 471-479.

Dorfman M.R., Nonni M., Mallon J., Woodard W., P. Meyer. (2004). Thermal Spray Technology Growth in Gas Turbine Coatings, *Proceedings of Thermal Spray 2004: Advances in Technology and Application*, ISBN 0-87170-809-4, Osaka, Japan, May, 2004

Dubourg L., Moreau C. (2006). Microstructure and Mechanical Properties of Yttria Partially Stabilized Zirconia Coatings Deposited by Laser-Assisted Air Plasma Spraying, *Proceedings of Building on 100 Years of Success: Proceedings of the 2006 International Thermal Spray Conference*, Seattle, United States of America, May, 2006

Evans B., Panza-Giosa R., Cochien Brikaras E., Maitland S. (2006). HVOF-Applied WC-CO-Cr as a Hard Chrome Replacement for Landing Gear, *Proceedings of Building on 100 Years of Success: Proceedings of the 2006 International Thermal Spray Conference*, Seattle, United States of America, May, 2006

Fabre G., Guipont V., Jeandin M., Boustie M., Cuq-Lelandais J.P., Berthe L., Pasquet A., Guedou J-Y. (2011). LAser Shock Adhesion Test (LASAT) of Electron Beam Physical Vapor Deposited Thermal Barrier Coatings (EB-PVD TBCs), *Advanced Materials Research*, Vol. 278, pp. 509-514

Fukanuma H., Ohno N. (2003). Influences of Substrate Roughness and Temperature on Adhesive Strength in Thermal Spray Coatings, *Proceedings of Thermal Spray 2003: Advancing the Science & Applying the Technology*, ISBN 0-87170-785-3, Orlando, United States of America, May 2003.

Gaffet E., Michel D., Mazerolles L., Berthet P., Effects of High Energy Ball Milling on Ceramic Oxides, *Materials Science Forum*, Vol. 235-238, pp. 103-108

Garcia-Alonso D., Serres N., Demian C., Costil S., Langlade C., Coddet C. (2011). Pre-/During-/Post-Laser Processes to Enhance the Adhesion and Mechanical Properties of Thermal-Sprayed Coatings with a Reduced Environmental Impact, *Journal of Thermal Spray Technology*, 2011, Vol. 20, No. 11, pp. 719-735

Gartner F., Schmidt T., Stoltenhoff T., Kreye H. (2006). Recent developments and potential applications of cold spraying, *Advanced Engineering Materials*, Vol. 8, No. 7, pp. 611-618

Gertsman V.Y,. Kwok Q.S.M. (2005). TEM Investigation of Nanophase Aluminum Powder, *Microscopy and Microanalysis*, Vol. 11, No. 5, pp. 410-420

Golesich B.T., Anderson K. (2008). Effects of Surface Preparation on the Performance of Cold Spray Coatings, *Proceeding of 19th AeroMat Conference & Exposition*, 23-26 June 2008, Austin, Texas, USA

Gonzalez-Hermosilla W.A., Chicot D., Lesage J., La Barbera-Sosa J.G., Gruescu I.C., Staia M.H., Puchi-Cabrera E.S. (2010). Effect of substrate roughness on the fatigue behavior of a SAE 1045 steel coated with a WC–10Co–4Cr cermet deposited by HVOF thermal spray, *Materials Science and Engineering A*,Vol. 527, pp. 6551–6561

Greving D.J., Shadley J.R., Rybicki E.F., Greving D.J., Shadley J.R., Rybicki E.F. (1994), Effects of coating thickness and residual stresses on the bond strength of ASTM C633-79 thermal spray coating test specimens, *Journal of Thermal Spray Technology*, Vol. 3, No. 4, pp. 371-378

Guetta S., Berger M.H., Borit F., Guipont V., Jeandin M., Boustie M., Ichikawa Y., Sakaguchi K., Ogawa K. (2009). Influence of Particle Velocity on Adhesion of Cold-Sprayed Splats, *Journal of Thermal Spray Technology*, Vol. 18, No. 3, 2009, pp. 331-342

Guipont V., Espanol M., Borit F., Llorca-Isern N., Jeandin M., Khor K.A., Cheang P. (2002). High-Pressure Plasma Spraying of Hydroxyapatite Powders, *Materials Science and Engineering A*, Vol. 25, No. 1-2, pp. 9-18

Hall A.C., Cook D.J., Neiser R.A., Roemer T.J., Hirschfeld D.A.. (2006). The effect of a simple annealing heat treatment on the mechanical properties of cold-sprayed aluminium, Journal of Thermal Spray Technology Vol. 15, No. 2, pp. 233-238

Hartmann S. (2010). New industrial applications for cold spraying, *Proceedings of Thermal Spray: Global Solutions for Future Application,* ISBN 978-3-87155-590-9, Singapore, May 2010

Ichikawa Y., Sakaguchi K., Ogawa K., Shoji T., Barradas S., Jeandin M., Boustie M. (2007a). Depostion mechanisms of cold gas dynamic sprayed MCrAlY Coatings, *Proceedings of Thermal Spray 2007: Global Coating Solutions*, ISBN 0-87170-809-4, Beijing, China, May 2007

Ichikawa Y., Barradas S., Borit F., Guipont V., Jeandin M., Nivard M., Berthe L., Ogawa K., Shoji T. (2007b). Evaluation of Adhesive Strength of Thermal-Sprayed Hydroxyapatite Coatings using the Laser Shock Adhesion Test (LASAT), *Materials Transactions*, Vol. 48, No. 4, pp. 793-798

Ichikawa Y., Miyazaki T., Ogawa K., Shoji T., Jeandin M., (2008). Deposition Mechanism of Cold Sprayed MCrAlY Coatings Focused on Nanostructure, *Proceedings of Thermal Spray 2008: Thermal Spray Crossing Borders*, ISBN: 978-3-87155-979-2, Maastricht, The Netherlands, June 2008

Irissou E., Arsenault B. (2007). Corrosion study of cold sprayed aluminum coatings onto Al 7075 alloy substrates, *Proceedings of Thermal Spray 2007: Global Coating Solutions*, ISBN 0-87170-809-4, Beijing, China, May 2007

Irissou E., Legoux J.G., Ryabinin A.N., Jodoin B., Moreau C. (2008a). Review on cold spray process and technology: Part I - Intellectual property, *Journal of Thermal Spray Technology*, Vol. 17, No. 4, pp. 495-516

Irissou E., Legoux J.-G., Moreau C., Ryabinin N. (2008b). How Cold is Cold Spray? An Experimental Study of the Heat Transfer to the Substrate in Cold Gas Dynamic Spraying, *Proceedings of Thermal Spray 2008: Thermal Spray Crossing Borders*, ISBN: 978-3-87155-979-2, Maastricht, The Netherlands, June, 2008

Jeandin M., Barradas S., Arrigoni M., He H.L., Boustie M., Bolis C., Berthe L. (2003). Thermal Spray and Lasers, Proceeding of the 2nd International Conference on Materials Processing for Properties and Performance (MP3), Yokohama, Japan, October, 2003.

Jeandin M., Christoulis D., Borit F., Berger M.H., Guetta S., Rolland G., Guipont V., Irissou E., Legoux J.G., Moreau C., Nivard M., Berthe L., Boustie M., Ludwig W., Sakaguchi K., Ichikawa Y., Ogawa K. Costil S. (2010). Lasers and Thermal Spray, *Materials Science Forum*, Vol.. 638-642, pp.171-184

Knapp W., Jeandin M., Behm V., Giraud D., Descurninges L.L., Guinard C., Zeralli Y., Christoulis D. (2011). Laser Structuring to improve cold-sprayed coating-substrate bond strength. *Presented at 1st EUropean COLd Symposium (EUCOS 2011),* Paris, France, May, 2011

Koivuluoto H., Lagerbom J., Vuoristo P. (2007). Microstructural Studies of Cold Sprayed Copper, Nickel, and Nickel-30% Copper Coatings, *Journal of Thermal Spray Technology*, Vol. 16, No. 4, pp. 488-497

Koivuluoto H., Vuoristo P. (2010). Structural Analysis of Cold-Sprayed Nickel-Based Metallic and Metallic-Ceramic Coatings, *Journal of Thermal Spray Technology*, Vol. 19, No. 5, pp. 975-989

Kroemmer W., Heinrich P. (2006). Cold spraying-Potential and new application ideas, *Proceedings of Building on 100 Years of Success: Proceedings of the 2006 International Thermal Spray Conference*, Seattle, United States of America, May, 2006

Li W.-Y., Zhang C.,. Guo X. P, Dembinski L., Liao H.L., Coddet C., Li C.-J. (2007) Impact fusion of particle interfaces in cold spraying and its effect on coating microstructure, *Proceedings of Thermal Spray 2007: Global Coating Solutions*, ISBN 0-87170-809-4, Beijing, China, May 2007

Li W.-Y., Zhang C., Liao H., Li J., Coddet C. (2008). Characterizations of cold-sprayed Nickel–Alumina composite coating with relatively large Nickel-coated Alumina powder, *Surface & Coatings Technology*, Vol. 202, pp. 4855–4860

Li P., Xi S., Zhou J. (2009). Phase Transformation and Gas-Solid Reaction of Al_2O_3 During High-Energy Ball Milling in N2 Atmosphere, Ceramics International, Vol. 35, No. 1, pp. 247-251

Liu X.-Y., Ding C.-X. Microstructure and Biological Properties of Plasma Sprayed Novel Bioactive Coatings, *Proceedings of the Expanding Thermal Spray Performance to New Markets and Applications*, ISBN-13: 978-1-61503-004-0, Las Vegas, United States of America, May, 2009

Lu S.P., Kwon O.Y. (2002). Microstructure and bonding strength of WC reinforced Ni-base alloy brazed composite coating, *Surface & Coatings Technology*, Vol. 153, No. 1, pp. 40-48

Makinen H., Lagerbom J., Vuoristo P. (2007). Adhesion of cold sprayed coatings: Effect of powder, substrate and heat treatment, *Proceedings of Thermal Spray 2007: Global Coating Solutions*, ISBN 0-87170-809-4, Beijing, China, May 2007

Marrocco T., McCartney D.G., Shipway P.H., Sturgeon A.J. (2006). Production of titanium deposits by cold gas dynamic spray: Numerical modelling and experimental characterization, *Journal of Thermal Spray Technology*, Vol. 15, No. 2 pp. 263-272.

McDonald A., Lamontagne M., Moreau C., Chandra S. (2006). Impact of plasma-sprayed metal particles on hot and cold glass surfaces, *Thin Solid Films*, Vol. 514, pp. 212-22

Moreau C, Gougeon P., Lamontagne M., (1995). Influence of Substrate Preparation on the Flattening and Cooling of Plasma-Sprayed Particles, *Journal of Thermal Spray Technology*, Vol. 4 No. 1, 1995, pp. 25-33.

Padture N.P, Gell M., Jordan E.H. (2002). Thermal barrier coatings for gas-turbine engine applications, *Science*, Vol. 296, No. 5566, pp. 280-284

Papyrin A., Kosarev V., Klinkov S., Alkhimov A., Fomin V. (2007), *Cold Spray Technology*, Elsevier Ltd, ISBN: 978-0-08-045155-8, The Netherlands

Paredes R.S.C., Amico S.C., d'Oliveira A.S.C.M. (2006). The effect of roughness and pre-heating of the substrate on the morphology of aluminium coatings deposited by thermal spraying, *Surface & Coatings Technology*, Vol. 200, pp. 3049 - 3055

Petavratzi E., Kingman S., Lowndes I. (2005). Particulates from Mining Operations: A Review of Sources, Effects and Regulations, *Minerals Engineering*, Vol. 18, No. 12, pp. 1183-1199.

Price T.S., Shipway P.H., McCartney D.G. (2006). Effect of Cold Spray Deposition of a Titanium Coating on Fatigue Behavior of a Titanium Alloy, *Journal of Thermal Spray Technology*, Vol. 15, pp. 507-512

Richer P., Jodoin B., Ajdelsztajn L., Lavernia E.J. (2006). Substrate Roughness and Thickness Effects on Cold Spray Nanocrystalline Al-Mg Coatings, *Journal of Thermal Spray Technology*, Vol. 15, No. 6, pp. 246-254

Sarafoglou Ch.I., Pantelis D.I., Beauvais S., Jeandi M. (2007). Study of Al2O3 coatings on AISI 316 stainless steel obtained by controlled atmosphere plasma spraying (CAPS), *Surface & Coatings Technology*, Vol. 202, No. 1, pp. 155-161

Saral U., Toplan N. (2009) Thermal cycle properties of plasma sprayed YSZ/Al2O3 thermal barrier coatings, Surface Engineering, Vol. 25, No. 7, pp. 541-547

Schlaefer T., Bobzin K., Ernst F., Richardt K., Verpoort C., Schreiber A., Schwenk A., Flores G. (2008). Plasma transferred wire arc spraying of novel wire feedstock onto cylinder bore walls of AlSi engine blocks, *Proceedings of Thermal Spray 2008: Thermal Spray Crossing Borders*, ISBN: 978-3-87155-979-2, Maastricht, The Netherlands, June 2008

Schmidt T., Assadi H., Gartner F., Richter H., Stoltenhoff T., Kreye H., Klassen T. (2009). From Particle Acceleration to Impact and Bonding in Cold Spraying, *Journal of Thermal Spray Technology*, Vol. 18, No. 5-6, pp. 794-808

Shin S., Xiong Y., Ji Y., Kim H.J., Lee C. (2008). The influence of process parameters on deposition characteristics of a soft/hard composite coating in kinetic spray process, *Applied Surface Science*, Vol. 254, No. 8, pp. 2269-2275

Stoltenhoff T., Borchers C., Gartner F., Kreye H. (2006). Microstructures and key properties of cold-sprayed and thermally sprayed copper coatings, *Surface & Coatings Technology*, Vol. 200, No. 16-17, pp. 4947-4960

Spencer K., Zhang M.-X. (2009). Heat treatment of cold spray coatings to form protective intermetallic layers, *Scripta Materialia*, Vol. 61, pp. 44–47

Sundararajan G., Chavan N. M., Sivakumar G., Sudharshan Phani P. (2010). Evaluation of Parameters for Assessment of Inter-Splat Bond Strength in Cold-Sprayed Coatings, *Journal of Thermal Spray Technology*, Vol. 19, No. 6, pp. 1255-1266

Taylor, T.A. (1995). Surface roughening of metallic substrates by high pressure pure waterjet, *Surface & Coatings Technology*, Vol. 76–77, pp. 95–100.

Triantou K. I., Sarafoglou Ch. I., Pantelis D.I., Christoulis D.K., Guipont V., Jeandin M., Zaroulias A., Vardavoulias M. (2008) A Microstructural Study of Cold Sprayed Cu Coatings on 2017 Al Alloy, *Proceedings of Thermal Spray 2008: Thermal Spray Crossing Borders*, ISBN: 978-3-87155-979-2, Maastricht, The Netherlands, June 2008

Tollier L., Fabbro R., Bartnicki E. (1998a). Study of the laser driven spallation process by the velocity interferometer system for any reflector interferometry technique. I. laser shock characterization, *Journal of Applied Physics,* Vol. 83, No. 3, pp. 1224-1230

Tollier L., Fabbro R., Bartnicki E. (1998b). Study of the laser driven spallation process by the velocity interferometer system for any reflector interferometry technique. II. Experiment and simulation of the spallation process, *Journal of Applied Physics,* Vol. 83, No. 3, pp. 1231-1237

Van-Steenkiste T.H., Smith J.R., Teets R.E., Moleski J.J., Gorkiewicz D.W., Tison R.P., Marantz D.R., Kowalsky K.A., Riggs W.L., Zajchowski P.H., Pilsner B., McCune R.C., Barnett K.J. (1999). Kinetic spray coatings, *Surface & Coatings Technology,* Vol. 111, pp. 62-71.

Van-Steenkiste T.H., Smith J.R., Teets R.E. (2002). Aluminum coatings via kinetic spray with relatively large powder particles, *Surface & Coatings Technology,* Vol.154, pp. 237-252.

Wong W., Irissou E., Ryabinin A. N., Legoux J.-G., Yue S. (2011), Influence of Helium and Nitrogen Gases on the Properties of Cold Gas Dynamic Sprayed Pure Titanium Coatings, 2011, *Journal of Thermal Spray Technology*, Vol. 20, No. 1-2, pp. 213-226

Xiong Y., Kang K., Bae G., Yoon S., Lee C. (2008). Dynamic Amorphization and Recrystallization of Metals in Kinetic Spray Process, *Applied Physics Letters*, Vol. 92, No. 19, pp. 19401-19404

Zhang D., Shipway P.H., McCartney D.G. (2003). Particle-substrate interactions in cold gas dynamic spraying *Proceedings of Thermal Spray 2003: Advancing the Science & Applying the Technology*, ISBN 0-87170-785-3, Orlando, United States of America, May 2003.

Zhang J., Xu N., Hou W.L., Wang J.Q., Quan M.X., Chang X.C. (2009), Fabrication and Property of MCrAlY/Nanostructured YSZ Thermal Barrier Coating, *Proceedings of the Expanding Thermal Spray Performance to New Markets and Applications*, ISBN-13: 978-1-61503-004-0, Las Vegas, United States of America, May, 2009

Zecchino M. (2003). Why Average Roughness is Not Enough, *Advance Materials & Processes*, Vol. 161, No. 3, pp.25-28

Zieris R., Nowotny S., Berger L.-M., Haubold L., Beyer E. (2003). Characterization of Coatings Deposited by Laser-Assisted Atmospheric Plasma Spraying, , *Proceedings of Thermal Spray 2003: Advancing the Science & Applying the Technology*, ISBN 0-87170-785-3, Orlando, United States of America, May 2003.

Zieris R, Langner G., Berger L.-M., Nowotny S., Beyer E. (2004). Investigation of AlSi coatings prepared by laser-assisted atmospheric plasma spraying of internal surfaces of tubes, *Proceedings of Thermal Spray 2004: Advances in Technology and Application*, ISBN 0-87170-809-4, Osaka, Japan, May, 2004

6

Diode-Pumped Nd:YAG Green Laser with Q-Switch and Mode Locking

V.I. Donin, D.V. Yakovin and A.V. Gribanov
Institute of Automation and Electrometry, Siberian Branch of RAS, Novosibirsk
Russia

1. Introduction

Gaining of a high peak power in the visible light range from a solid state laser with continuous diode pumping is a challenging task in several applications (high-precision material processing, nonlinear optics and Raman-spectroscopy, medicine, etc.). The technique of modulating the Q-factor of laser cavity (Q-switch) enables growth of the peak power approximately as τ_{sp} / τ_{ph} (here τ_{sp} is the upper laser level lifetime and τ_{ph} is the photon lifetime in the cavity). When we take a typical Nd:YAG laser, this gain in peak power is about 10^3- 10^4 times. The further growth of the peak power is possible through methods of mode locking of the laser (ML). However, realization of mode-locking together with Q-switch (unlike the case of continuous operation mode) is a technically challenging task: we face a high amplification, almost uncontrollable nonlinear effects, damage of optical elements in the laser, etc. In prior art, the steady mode of generation for Q-switch coupled with mode locking (so called QML) is accomplished with using of two acousto-optic modulators (AOM) in a cavity; one modulator operates in the traveling acoustic wave, and the other modulator has the standing wave (see, e.g., Kuizenga, 1981). Actually, the QML mode is possible when we introduce absorbing elements within the cavity (Herrmann & Wilhelmi, 1987; He et al., 1996; Chen et al., 2001; Agnesi et al., 2001; Pan et al., 2007), but in this case the higher pumping level means a higher pulse repetition rate. Therefore the practical levels of peak power are very moderate.

In this chapter we inform about a new technique for obtaining a steady QLM lasing mode in a diode-pumped green Nd:YAG laser; this type of operation is possible with only one travelling-wave AOM. As for the further shortening of lasing pulse duration ($\Delta\tau$), it is achieved through formation of a Kerr lens in a doubling-frequency crystal. Since the process of make-up a Kerr lens is very rapid, this trick gives us $\Delta\tau$ as small as the theoretical limit $\approx 1/\Delta\nu$ (here $\Delta\nu$ is the spectral bandwidth of generation).

2. The general scheme of experiment and key parameters of Nd:YAG laser

Now the class of solid-state lasers demonstrate a rebirth due to replacement of pumping with arc lamps by diode-laser pumping; this ensures a higher efficiency of the device (higher than 10%) and a longer service life (more than 10,000 hrs). Typically, diode laser (DL) can be used in different schemes of pumping – longitudinal or transversal. Solid-state lasers with longitudinal pumping are more efficient and the emission beam has better quality ($M^2 \approx 1$) (Tidwell et al.,1992; Tsunekane et al., 1998). However, it is very difficult to gain a high output power at this geometry of pumping. When the required output power is above 100 W, the good choice is lasers with transverse pumping of the active element (e.g., in Ostermeyer et al., 2002; Fujikawa et al., 2001; Lee et al., 2002). But the efficiency of this scheme is lower and lasing with $M^2 \approx 1$ usually faces serious difficulties.

In this paragraph we describe the experimental scheme of a Nd:YAG laser with transverse pumping by DL arrays and with time-averaged output power of 15 W (in TEM_{00}-mode). When the effective doubling of laser emission frequency is achieved, this scheme enables lasing with the output power of 12 W (here the wavelength is λ = 532 nm). The laser scheme with a Z-shaped and 4-mirrored cavity (Donin et al., 2004) is depicted in Fig.1. The reflection coefficient (r) for mirrors M1-M4 at the wavelength 1064 nm was better than 99.5%. Mirror M4 was a dichroic mirror with r > 99.5 % (at λ=532 nm); at the latter wavelength the transparency coefficient (T) for mirror M3 was 92 %. To gain the light generation at the second harmonics between spherical mirrors *M3* and *M4* (they have curvature radii R =200 and 150 mm, correspondingly), a nonlinear crystal (BBO, KTP or LBO) was placed into the zone of cavity waist . The beam diameter in this zone was about 100 -150 μm. To gain a maximum power of emission at the wavelength λ=1064 nm, the mirror M1 was replaced with a mirror possessing an optimal transmission coefficient (but there was no need in the nonlinear crystal within the resonator).

The active element was a Nd^{+3} :YAG crystal (mass concentration for Nd was 1 %) with the diameter of 2 mm and the length of 63 mm; this element was illuminated from three sides with bars of LD arrays emitting at the wavelength λ_p =808 nm (Fig. 1b). The laser's active element together with DLs is cooled by distilled water flow. This cooling is accomplished by a circulation-type close-loop cooler; this device ensures temperature stability within 0.1 ^{0}C.

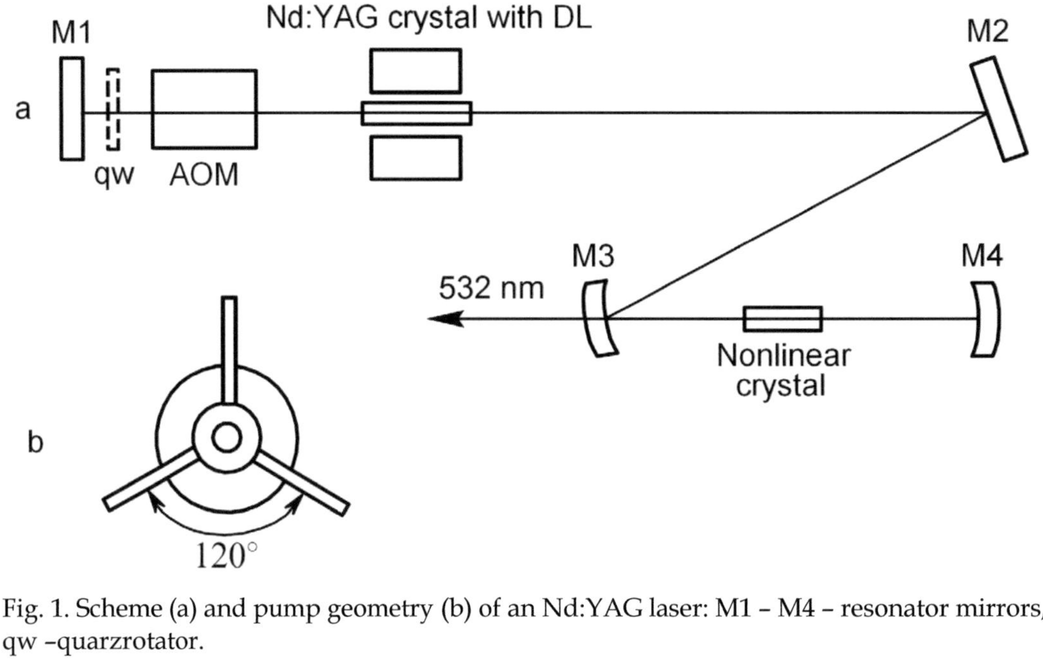

Fig. 1. Scheme (a) and pump geometry (b) of an Nd:YAG laser: M1 – M4 – resonator mirrors, qw –quarzrotator.

The operation temperature of laser was close to 28 ^{0}C and it can be regulated in a wide range. The dependency of laser output power on the coolant water temperature is plotted in Fig.2. The Q-factor of laser was modulated with the MZ-305 type AOM with the travelling acoustic wave. The AOM with the driving frequency of 50 MHz was manufactured from crystalline quartz and equipped with water cooling facility. The quarzrotator 5 could be used for compensation of induced birefringence.

The laser resonator was calculated with the matrix method with account for a heat lens in the active element and dispersion of the air gap between the nonlinear crystal and mirror 4. We have also estimated the diffraction losses of TEM_{oo}-mode (≈ 10%) and higher-order modes, e.g., for TEM_{01} mode (≈ 50%). All these losses ensured lasing in the regime close to TEM_{oo}-mode.

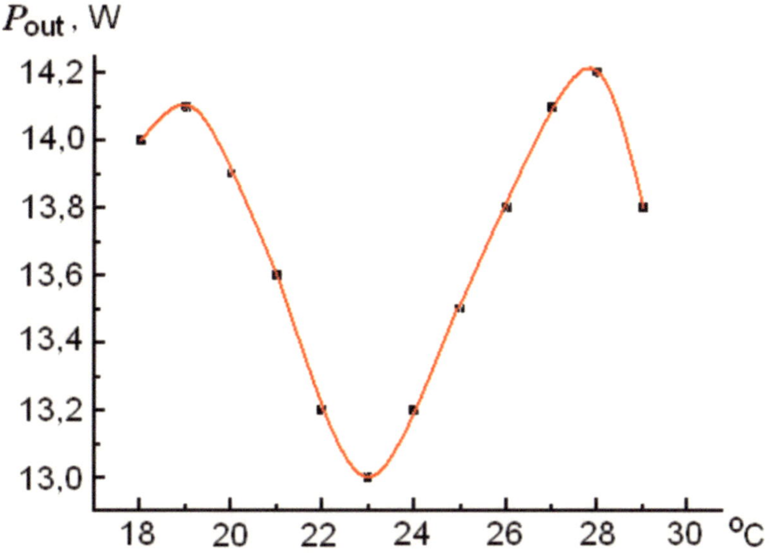

Fig. 2. Output laser power vs. temperature.

When the dimensions of the active element were chosen in the view of most effective usage of this element, we took into consideration the following things. During of active element pumping, the birefringence effect takes place: this causes depolarization of beam passing the crystal and makes the losses higher. The depolarization-caused losses are proportional to ω^4, and the output power has the law ω^2 (here ω is the radius of TEM_{00}-mode in active element). This means that we can find the optimal value for the resonator's fundemental mode. The estimate for optimal ω for our conditions was about ≈ 1 mm; this coincides with results from (Matthew et al., 1996). Therefore the optimal diameter of rod should be take ≈ 2 mm.

The dependencies of output power (P_{out}) of the laser at λ=1064 nm (at the AOM switching frequency or the pulse repetition rate F = 20 kHz) on the current (I) through DL, as well as function of emitted light power (P_{light}) and consumed electric power (P_{electr}) of DLs are plotted in Fig.3. The saturation of P_{out} along with current growth is caused by the resonator mismatch due to thermo-optic effects. The thermal lens in active element has the focal distance (at saturation current) close to 25 cm. The efficiency coefficient on the total electric power was 5%, and the light efficiency was 12.5 % (the differential coefficients of efficiency were 10% and 21

%, correspondingly). These curves were obtained for a configuration with an exit mirror M1 (see Fig.1) which has the optimal transmission coefficient T = 20 %. Under described conditions, generation took place in the TEM_{oo}-mode. Fig.4 shows the plotting of the output power and the lasing pulse duration Δτ vs. the pulse repetition rate (F). The pulse length was measured with an avalanche photodiode type LFD-2 with the time resolution better than 1 ns.

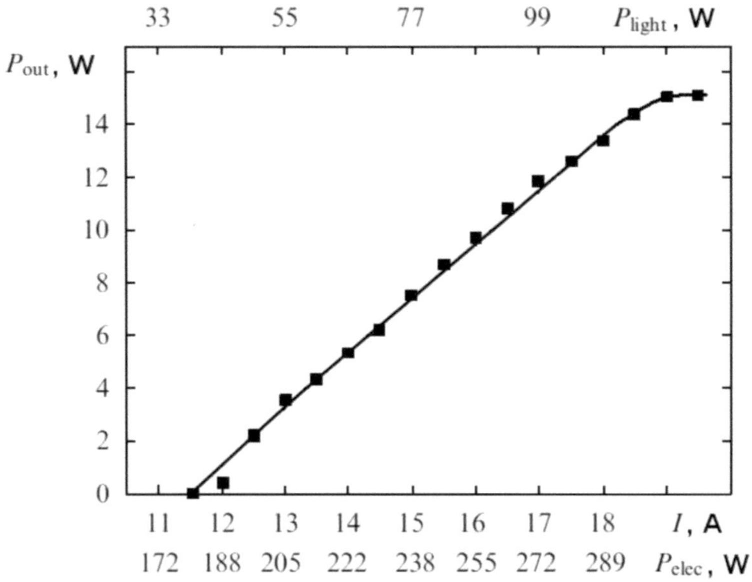

Fig. 3. Dependence of the 1064 nm output power on the current, light output power and the consumed electric power of the DL.

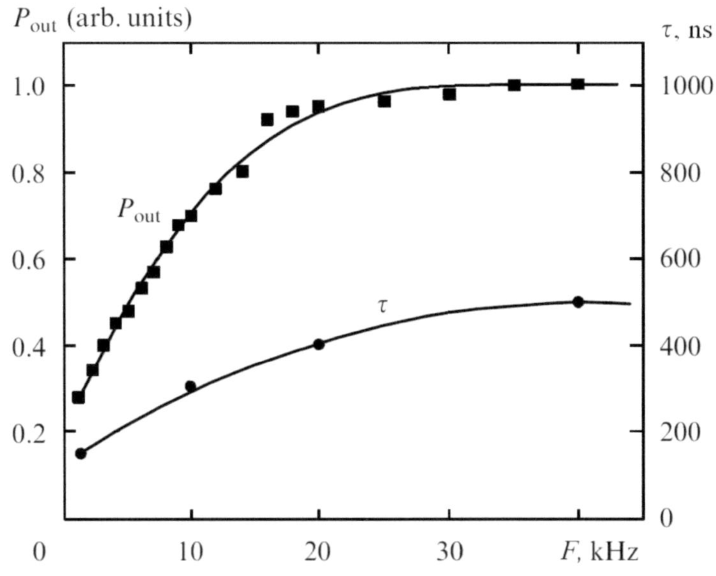

Fig. 4. Dependence of the output power at 1064 nm and the laser pulse duration on the pulse repetition rate.

3. Frequency doubling for laser emission

Let us focus on the problem of frequency doubling (second harmonic generation) for laser emission. It is known (Smith, 1970) that while intra-cavity frequency doubling the losses brought by a nonlinear crystal must be balanced to the losses brought by an optimal mirror for a laser without nonlinear crystal. This condition enables a maximum output power at the doubled frequency.

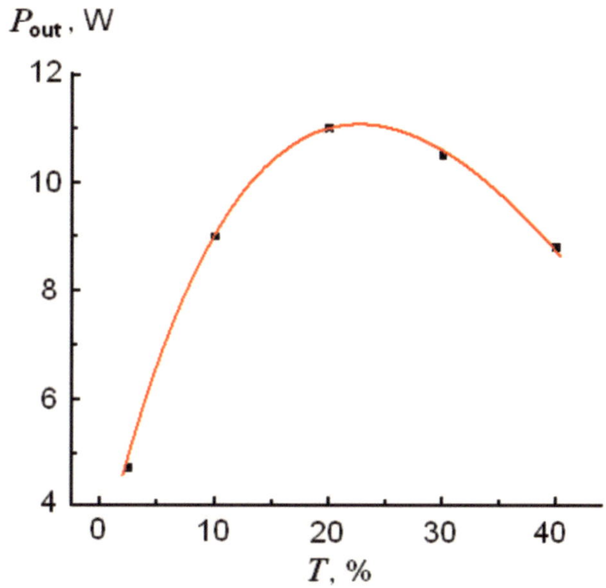

Fig. 5. Laser output power (λ=1064 nm) vs. transmission of output mirror T.

Fig.5 presents an experimental plotting of laser output power as a function of transmission coefficient for the output Mirror (DL current I = 17A).

In choosing the type of nonlinear crystals for the second harmonic generation and picking up optimal parameters (including the crystal length d_{cr}), we made computer simulations for the double-frequency output power. These simulations started from a system of truncated equations for the real parts of electric field amplitudes; this system of equations was derived from the wave equation by the method of slowly varying amplitude in low-absorption and low-nonlinearity medium within the plain wave approximation. These equations were solved by using software specially developed for determining the all required parameters of nonlinear crystals (even in the case of deviation from precise phase matching).

We applied a simple differential scheme for a particular case of second harmonics generation when effects with normalized phase ($\Delta kz + \pi/2$) accumulates most rapidly. At the initial conditions z=0, $a_2(0)=0$ we solved numerically this equation system (Dmitriev & Tarasov, 1982):

$$\partial a_1/\partial z = -\sigma_1 a_1 a_2 \cos(\Delta kz)$$

$$\partial a_2/\partial z = \sigma_2 a_1^2 \cos(\Delta kz).$$

For the case of precise phase matching (wave detuning $\Delta k=2k_1-k_2=0$) this equation system has an exact analytical solution:

$$a_2(z)=a_{10}(\sigma_2/\sigma_1)^{1/2}\tanh[(\sigma_1\sigma_2)^{1/2}a_{10}z]).$$

Here $\sigma_1=4\pi^2 d_{eff}/\lambda_1 n_1$, $\sigma_2=2\pi^2 d_{eff}/\lambda_2 n_2$ are the nonlinear coupling coefficients, d_{eff} is the effective nonlinearity, λ_i is the wavelength, n_i is the refraction index, k_i is the wave vector, a_i is the electric field amplitude, a_{10} is the initial amplitude of the electric field for the fundemental emission, z is the running coordinate.

In these calculations we took the formulae and input data for calculating the effective nonlinearity d_{eff} and Sellmeyer's equations for determining the refraction index n_i as in (Lin et al., 1990; Gurzadyan et al., 1991). The phase-matching angles for the wavelengths of fundemental radiation (1.07-0.75 μm) were calculated by well-known formulae (see Dmitriev & Tarasov, 1982). These calculations were performed for nonlinear crystals LBO, KTP, BBO. Fig. 6 presents calculated the second-harmonic power density S_2 as a function of the crystal length. We took the power density of fundamental input radiation $S_1(0)$ equal to 150 MW/cm², which corresponds to the Q-switch frequency F = 10-15 kHz. The power density was defined as $S_i = c\, n_i\, a_i^2/8\pi$, here c is the velocity of light in vacuum. The approximate lengths of crystals which are optimal for generating output power at the second harmonics are the following: ≈ 10 mm (LBO, phase-matched type I), ≈ 2 mm (KTP, phase-matched type II), and ≈ 5 mm (BBO, phase-matched type I).

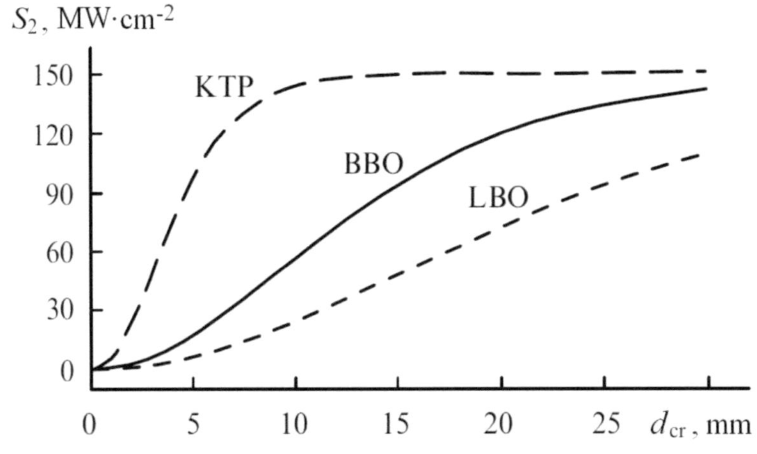

Fig. 6. Calculated dependences of the second-harmonic power densities S_2 on the crystal length.

The walk-off angle was calculated by formula

$$\tan\beta_i = \frac{\left[1-(n_0/n_e)^2\right]\tan\theta}{1-(n_0/n_e)^2\tan^2\theta}$$

where θ is the phase-matching angle, n_o and n_e - ordinary and extraordinary refractive index. The same software was used for estimating of the angular and spectral phase-matching width.

In our experiment on study of the second harmonics generation, we used the BBO crystals (phase-matched type I: θ=22.8^0, φ=90^0; d_{cr} =5 mm), KTP (phase-matched type II: θ=90^0, φ=23.5^0; d_{cr} =5 mm), and LBO crystal (phase-matched type I: θ=90^0, φ=11.6^0, d_{cr} =12 mm). All the tested crystals had two-hump antireflection coatings at the operating wavelength with $r < 0.5\%$. This allowed us to obtain the maximal power of generation: BBO - 5 W (F =15 kHz), KTP - 12 W (F >20 kHz), and LBO - 8.3 W (F = 10 kHz). We should note that for the BBO crystal the maximal power was restricted by coating destruction. A typical dependence of second-harmonic output power on the DL current for a LBO crystal is shown in Fig.7. One can see that at the initial interval, this dependency has the second power law, but later it goes slower due to resonator mismatch caused by thermo-optic effects. The curve of output power vs. the pulse repetition rate frequency for an LBO crystal had a maximum at the F = 10 kHz.

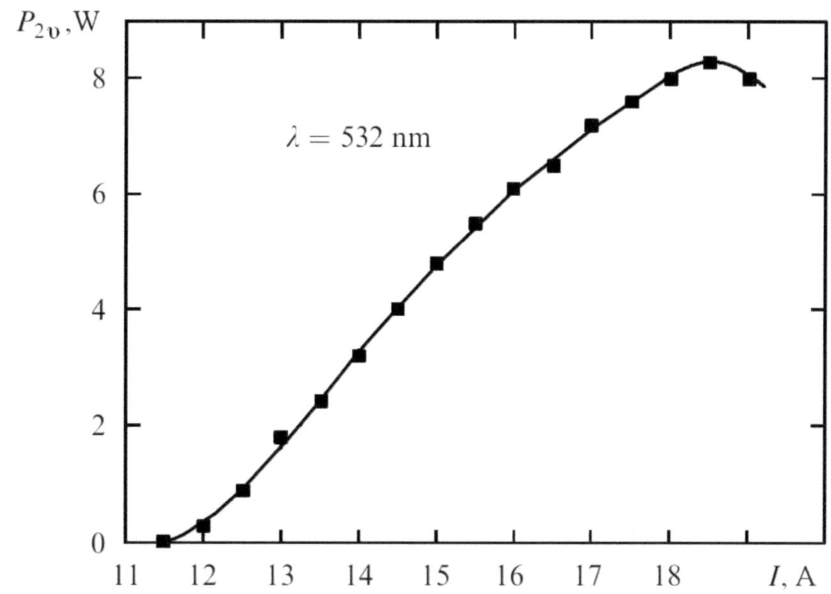

Fig. 7. Dependence of the second-harmonic output power P_{2v} on the DL current I for an LBO crystal at F = 10 kHz.

For a KTP crystal, the generated power is growing steadily up to F=20 kHz, and becomes constant in the interval from 20 to 40 kHz. For a KTP crystal, the coefficient of conversion into second harmonic was 80 %. For LBO the maximal power of second harmonic was obtained at driving frequency of 10 kHz, and this corresponds to 10.5 W generation at wavelength λ=1064 nm (Fig.7). Therefore, the coefficient of conversion into second harmonic for this type of crystal was 79%.

Fig.8 presents the dependency of pulse energy and peak power at the wavelength 532 nm as a function of Q-switch frequency (the case of LBO crystal and a cavity on Fig. 9).

Our results demonstrated the efficient conversion of the fundamental frequency into the second harmonic for the case of Q-switched pulse energy ≈ 1 mJ; this gives hope for developing an effective laser with output at wavelength λ = 532 nm and with the average output power for TEM_{00}-mode as high as 100 W and even higher.

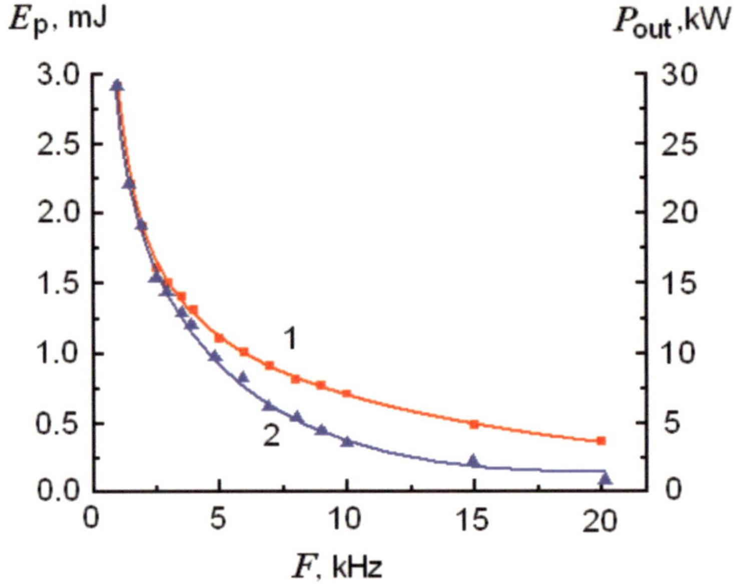

Fig. 8. Dependence of pulse energy E_p (1) and power P_{out} (2) at λ = 532 nm on the pulse repetition rate F.

4. Q-switch and mode locking achieved with a travelling-wave AOM

Here we develop a new method (Donin et al., 2011) of QLM lasing; the design includes a end-spherical mirror (SM) in the cavity and one travelling-wave AOM (abbreviation – SMAOM). The operation principle for SMAOM configuration and laser scheme is illustrated in Fig. 9. The Nd:YAG green laser was assembled by the key diagram shown in Fig.1. The mirror curvature radii were 200, -900, 200 and 150 mm accordingly for the mirrors M1, M2, M3 and M4. The cavity optic length was L =1.5 m. The AOM was placed at the Bragg angle (θ_B) to the optic axis of cavity near the terminal spherical mirror M1. The AOM's center is distanced from the mirror's reflecting surface by distance R1, which is the mirror curvature radius. When the driving frequency is f = 50 MHz, which equals a half of laser's intermode interval c/2L =2f , a travelling acoustic wave is created in the quartz block of AOM (it is shown by small bold-symbol arrow in Fig. 9); this travelling wave is a source for Bragg's diffraction for laser emission. Since the light beam (with frequency ν_0) passes through the AOM from right to left, two beams enter the mirror (1 and 2). Beam 1 goes along the cavity axis and being reflected from the mirror backward without any change in initial frequency ν_0. The beam 2 feels the Bragg diffraction and travels to the same mirror with frequency (ν_0 + f). The reflects from the spherical surface of the mirror and returns to the AOM, where it is split into a beam with the unchanged frequency (ν_0+ f), exiting the cavity in the backward direction at the angle $2\theta_B$, and another beam after repeated diffraction (on the AOM's quartz block). The latter kind of beam has a frequency (ν_0 + 2f) and travels backward along the cavity axis. This beam produces the effect of mode locking.

The beam with frequency (ν_0 + f) exiting the cavity at angle $2\theta_B$ provides modulating of losses in the cavity; so the laser operates in the mode of Q-switch with the pulse repetition rate which is given by the AOM switching frequency (F~1÷100 kHz). The good feature of this scheme is that when the driving frequency is off, the acoustic wave inside then AOM dies during a time $t = d_b/V_{snd} = 0.2\text{cm}/5\cdot10^5\,\text{cm/s} \approx 0.4\,\mu\text{s}$ (here d_b is the laser beam diameter in the quartz block, V_{snd} is the sound velocity in the material). The duration of the lasing pulse in the Q-switch mode was ≈ 100 ns, i.e., during time t due to the beam of repeated diffraction with frequency (ν_0+ 2f) mode locking is achieved in the generation pulse.

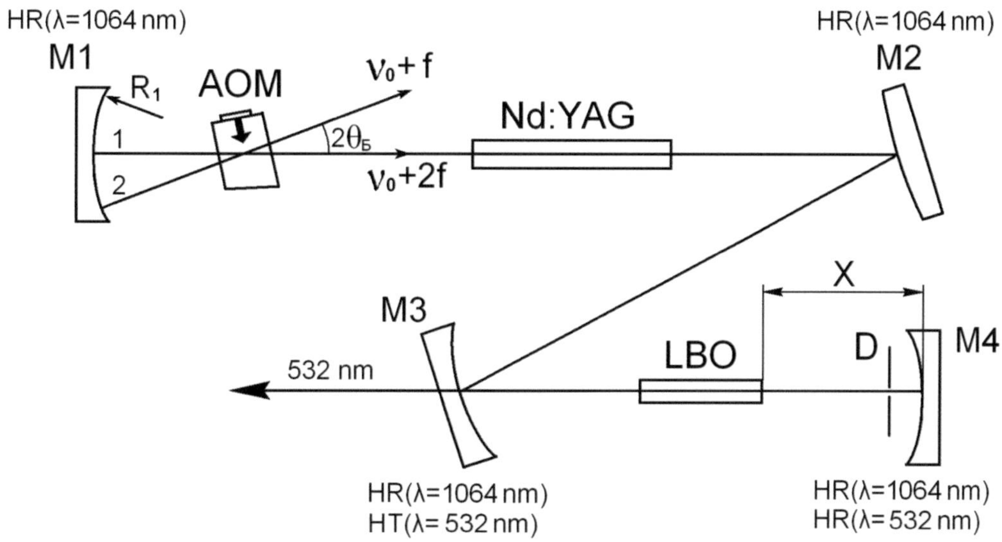

Fig. 9. Laser diagram and operation principle for SMAOM. M1-M4 – cavity mirrors, AOM - acoustic-optic modulator, Nd: YAG – active element, LBO – nonlinear crystal, D – diaphragm.

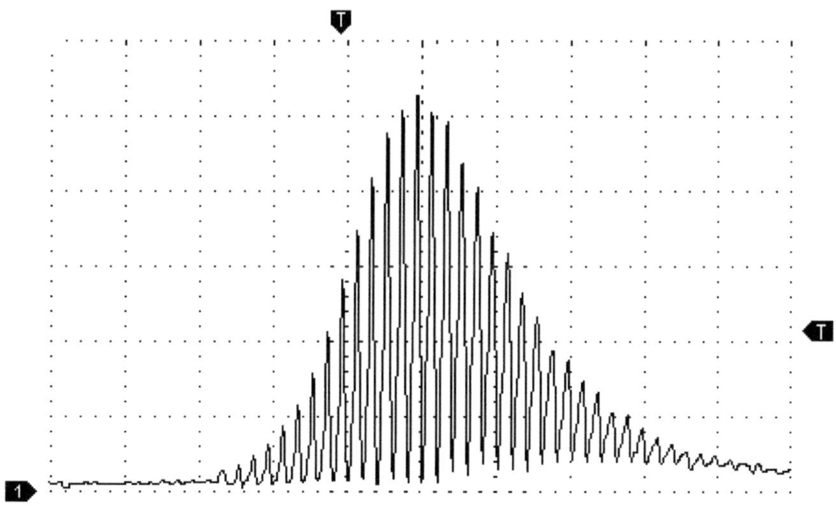

Fig. 10. Oscillogram of generated pulse at the wavelength λ = 1064 nm produced in QML mode. The division value for abscissa axis is 50 ns.

On the first stage, we performed measurements without nonlinear crystal and diaphragm (this means no frequency doubling and no Kerr lens formation). In this case we replaced the mirror M1 with another one with similar curvature radius, but possessing the transparency of T = 11% at λ = 1064 nm. The oscillogram of a Q-switch pulse with mode locking is depicted in Fig.10. The average power of laser was 2 W (at the Q-switch frequency equal to 2 kHz). The registration system resolution time (photodiode and oscillograph) ≈ 2 ns did not allow us to determine the actual duration of pulses inside a "train", so this task had required assembling an optical correlator for pulse registration on the second harmonic in the KTP crystal (collinear scheme). This optical correlator gave the pulse duration for mode locking generation about 40 ps (see Fig. 14 a), i.e., a single peak power was ≈ 2 MW.

5. The Kerr lens based on nonlinear crystal

The further gain in reducing a single pulse duration and growth of its peak power has been achieved using a Kerr lens and a diaphragm (laser generation at the line λ = 532 nm as shown in Fig. 9). Kerr lens is formed in a nonlinear crystal for 2nd harmonic generation (LBO, phase-matched type I, the length of d_{cr} = 20 mm). For the first time, the phenomena of mode locking with a Kerr lens (or self-locking of mode) has been studied in (Spence et al., 1991). The foundation of this method is a radiation self-focusing in a media; this opens the way to creating an affect similar to absorbing elements in a laser cavity. Self-focusing develops due to dependency of material refraction index on radiation intensity: $n = n_0 + n_2 I$. This effect creates a lens with focus depending on light intensity. This type of nonlinear lens matched with a diaphragm works as a kind of saturable absorber. Another design is too possible: the same lens without a diaphragm, and the cavity elements work as a diaphragm. In the case when self-focusing is driven by electron polarization under impact of the field of light wave, this would create an almost zero-inertia "saturable absorber" - the response time is ~10^{-15}s (see, e.g., Shapiro, 1977). The calculating of cavity's parameters was done with the matrix method. The following matrix was used for describing the beam passing through the Kerr element (Magni et al., 1993):

$$\mathrm{M}=\sqrt{1-\gamma}\begin{pmatrix} 1 & d_e \\ -\gamma/[(1-\gamma)d_e] & 1 \end{pmatrix},$$

where $d_e = d_{cr}/n_0$ is the effective length of media at the inter-resonator power P = 0, and here

$$\gamma = p\left[1+\frac{1}{4}\left(\frac{2\pi\omega_c^2}{\lambda d_e}-\frac{\lambda d_e}{2\pi\omega_0^2}\right)^2\right]^{-1},$$

where $p = P/P_c$ ($P_c = c\varepsilon_0\lambda^2/(2\pi n_2)$ is the critical power for self-focusing), ω_c is the beam size in the middle of medium, and ω_0 is the beam size in the waist, calculated at p = 0. The effect of "saturable absorber" requires a shrinking of the beam size with its intensity growing in the plane of the diaphragm. Quantitatively, this effect is described by a parameter (Magni et al., 1993)

$$\delta = \frac{1}{\omega}\frac{d\omega}{dp}\bigg|_{p=0},$$

where ω is the radius of Gaussian beam in the specific plane within the cavity. To obtain the shorter pulse, parameters δ should be negative and high in modulus. The diaphragm was placed in the plane at the end-mirror M4 (see Fig. 9). The variable parameter in our simulations was the distance between the end-mirror M4 and nonlinear crystal: this distance is marked as X in our diagrams. Computations gave us a plotting for parameter δ - the graph is shown in Fig. 11. One can see from this plotting that δ takes a maximal negative value at the boundaries of stability zone. Therefore our choice for the distance was X ≈14.06 cm.

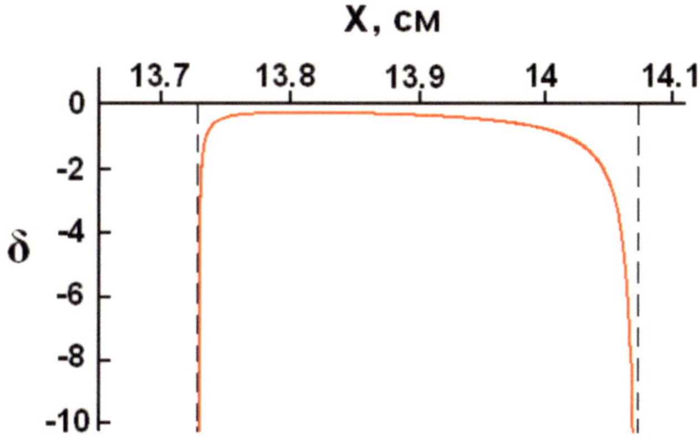

Fig. 11. Parameter δ vs. distance X. The vertical dash lines depict the boundaries of stability zone.

Fig. 12 shows the cavity's stability zone in coordinates distance X and power p. One can see that at low power (the beginning of forming a Q-switch pulse), the laser operates at the boundary of stability zone, but when the Kerr lens has been launched, operation moves to a more stable mode.

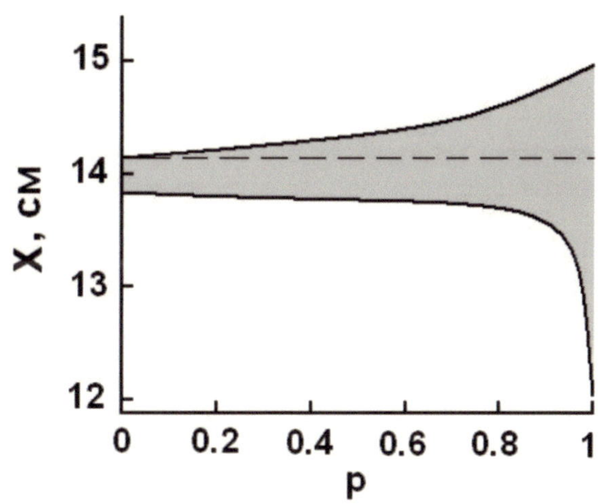

Fig. 12. Stability zone for a resonator (grey shading). The horizontal line indicates the working distance X.

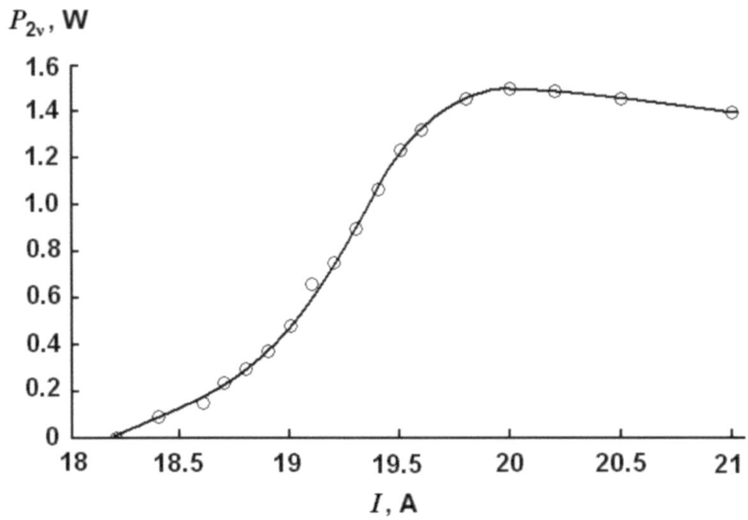

Fig. 13. Dependence of the second-harmonic output power P_{2v} on the DL current I for an LBO crystal at F = 2 kHz

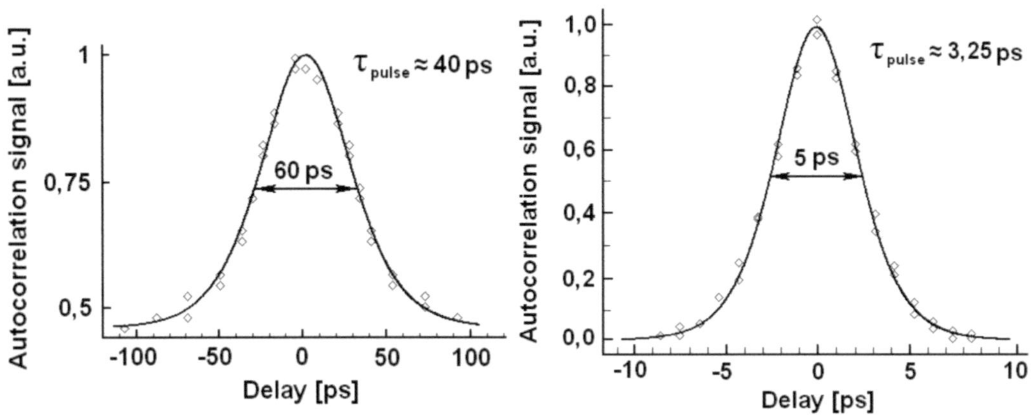

(a) (b)

Fig. 14. Measured autocorrelation traces of locking pulses and their sech[2] fits.

When the optical correlatator for pulse registration on the two-photon-induced photocurrent in a GaAsP photodiode (type G1116, Hamamatsu) was applied for measurements, the duration of a single pulse from the Nd:YAG laser was 3.25 ps; this was at the average output power 1.5 W and the pulse repetition rate of Q-switch equal 2 kHz (the dependence of the second-harmonic output power P_{2v} on the DL current I for an LBO crystal at F = 2 kHz is shown on Fig. 13). Fig. 14b demonstrates the measured autocorrelation function for a pulse. The scanning Fabry-Perot interferometer (parameters: free spectral range – 1500 GHz, reflection coefficient of mirrors – 0.96) was used for measuring the spectral bandwidth of generation $\Delta v \approx 200$ GHz (see Fig. 15, where this spectral width was slightly increased due to contribution from instrumental broadening ≈ 10%). Therefore, we obtain $\Delta v \cdot \Delta \tau \approx 0.65$, which is with accuracy to factor 2 close to the case of unchirped sech^2-shaped pulses.

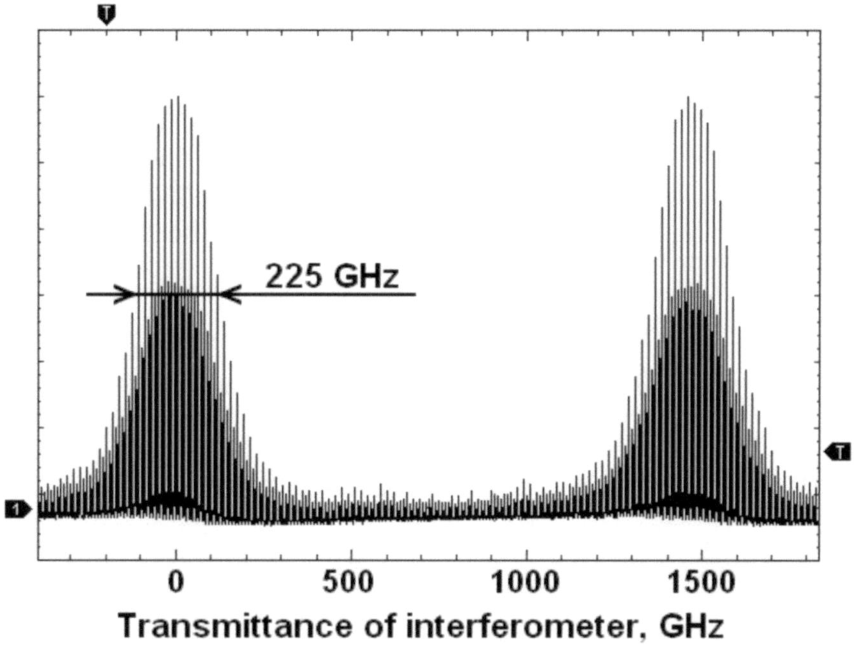

Fig. 15. The optical spectrum, measured with a scanning Fabry–Perot interferometer.

The peak power of a separate pulse near the maximum of envelope for Q-switch pulses was ≈ 50 MW. In this connection, it should be noted that $\Delta\tau$ was measured from the autocorrelation function at wavelength λ = 1064 nm. When we made measurements of $\Delta\tau$ for the Q-switch mode, it was shown that at λ = 532 nm the pulse duration was approximately twice shorter. One might expect that this proportion in the pulse lengths will be almost the same for operating in the QLM mode, so the actual peak power of lasing may be about ≈ 100 MW.

6. Conclusion

In summary, we should note that ML mode of generation for a continuous laser using a travelling-wave AOM was obtained in papers (Kornienko et al., 1981; Kravtsov et al., 1983; Nadtocheev & Nanii, 1989). The authors had noted that the band of mode locking was increased by ≥ 10 times in comparison with the case of standing-wave AOM. However, in these researches the feedback was ensured using additional mirrors within the laser cavity; this made the entire design more complicated and the Q-switch mode was absent. Our solution SMAOM, when a single AOM is enough for obtaining a stable QML mode, in combination with forming a Kerr lens, will provide high levels of pulse energy ≈ 1 mJ and peak power ≈ 50 MW at least. This design of laser does not require any additional "starting" for the making of Kerr lens and it exhibits good (long-run and short-run) stability in output characteristics and there is no need in auto-adjustment schemes.

7. References

Agnesi A., Guandalini A.,Reali G., Jabczynski J.K., Kopczynski K., Mierczyk Z. Diode pumped Nd:YVO_4 laser at 1.34 μm Q-swithed and mode locked by a V^{3+}:YAG saturable absorber. *Opt. Commun.*, 194, 429-433 (2001).

Chen Y.F., Huang K.F., Tsai S.W., Lan Y.P.,Wang S.C., Chen J. Simultaneous mode locking in a diode-pumped passively Q-swithed Nd:YVO_4 laser with a GaAs saturable absorber. *Appl. Opt.* 40, 6038-6041 (2001).

Dmitriev V.G., Tarasov L.V., *Applied Nonlinear Optics: Second Harmonic Generators and Parametric Light Generators*, (Moscow: Radio i Svyaz, 1982), pp. 45-57.

Donin V. I., Nikonov A. V., Yakovin D. V. Efficient frequency doubling in a transversely diode-pumped Nd:YAG laser. *Quantum Electronics* 34, 930-932 (2004).

Donin V. I., Yakovin D. V., Gribanov., A. V. A laser with Q-swithing and mode-locking. Patent pending No 2011123043/28, (7.06.2011).

Fujikawa S., Furuta K., Yasui K. 28% electrical-efficiency operation of a diode-side-pumped Nd:YAG rod laser, *Opt. Lett.* 26, 602-604, (2001).

Gurzadyan G.G., Dmitriev V.G., Nikogosyan D.N.: *Nonlinear Optical Crystals. Properties and Applications in Quantum Electronics. Handbook* (Radio i Svyaz, Moscow, 1991), pp. 1–160.

Herrmann J., Wilhelmi B. *Lasers for ultrashort light pulses* (Akademie-Verlag Berlin, 1987).

He G.S., Cui Y., Xu G.C., Prasad P.N. Multiple mode-locking of Q-switched Nd:YAG laser with a coupled resonant cavity. *Opt. Commun.* 96, 321-329 (1996).

Kornienko L.S., Kravtsov N.V., Nanii O.E., Shelaev A.N. Solid-state ring laser with diffraction acoustooptic mode feedback. *Sov J Quantum Electron.* 11, 1557-1559 (1981).

Kravtsov N.V., Magdich L.N., Shelaev A.N., Shniser P.I. Mode-locking in a laser using the modulator of a traveling acoustic wave. *Pis'ma Zh. Tekh. Fiz.* 9, 440-443 (1983).

Kuizenga D.J. Short-pulse oscillator development for the Nd:Glass laser-fusion systems. *IEEE J Quantum Electron.* 17, 1694-1708 (1981).

Lee S., Yun M., Cha B.H., Kim C.J., Suk S., Kim H.S. Stability analysis of a diode-pumped, thermal birefringence-compensated two-rod Nd:YAG laser with 770-W output power. *Appl. Opt.* 41, 5625-5631 (2002).

Lin S., Sun Z., Wu B., Chen C. The nonlinear optical characteristics of a LiB_30_5 crystal. *J. Appl. Phys.* 67, 634-638 (1990).

Magni V., Cerullo G., De Silvestri S. ABCD matrix analysis of propagation of Gaussian beams through Kerr media. *Opt. Commun.* 96, 348-355 (1993).

Magni V., Cerullo G., De Silvestri S. Closed form analysis of resonators containing a Kerr medium for femtosecond lasers. *Opt. Commun.* 101, 365-370 (1993).

Matthew P. Murdough and Craig A. Denman. Mode-volume and pump-power limitations in injection-locked TEM_{00} Nd:YAG rod lasers. *Appl. Opt.* 35, 5925-5936 (1996).

Nadtocheev V.E., Nanii O.E. Use of traveling acoustic waves for mode locking in lasers. *Sov J Quantum Electron.* 19, 1435-1437 (1989).

Ostermeyer M., Klemz G., Kubina P., Menzel R. Quasi-continuous-wave birefringence-compensated single- and double-rod Nd:YAG lasers. *Appl. Opt.* 41, 7573-7582 (2002).

Pan S., Xue L., Fan X., Huang H., He J. Diode-pumped passively Q-swithed mode-locked Nd:YLF laser with uncoated GaAs saturable absorber. *Opt. Commun.* 272, 178-181 (2007).

Shapiro S. L. (editor). *Ultrashort Light Pulses. Picosecond Techniques and Applications.* (Springer-Verlag, Berlin Heidelberg New York, 1977).

Smith R.G. Theory of intracavity optical second-harmonic generation. *IEEE J. Quantum Electron.* 6, 215-223 (1970).

Spence D.E., Kean P.N., Sibbett W. 60-fsec pulse generation from a self-mode-locked Ti:sapphire laser. *Opt. Lett.* 16, 42-44 (1991).

Tidwell S.C., Seamans J.F., Bowers M.S., Cousins A.K. Scaling CW Diode-End-Pumped Nd : YAG Lasers to High Average Powers. *IEEE J. Quantum Eectron.* 28, 997-1009 (1992).

Tsunekane M., Taguchi N., Inaba H. Reduction of thermal effects in a diode-end-pumped, composite Nd:YAG rod with a sapphire end. *Appl. Opt.* 37, 3290-3294 (1998).

7

Study on Polishing DF2 (AISI O1) Steel by Nd:YAG Laser

Kelvii Wei Guo
MBE, City University of Hong Kong
Hong Kong

1. Introduction

Dies and moulds used in the fabrication of metal, glass and plastic products require a high quality surface finish. Some surface improvement, of easily accessible areas, can be carried out using small rotary abrasive wheels or vibrating tools. However, the majority of the surface finishing operations applied to these dies and moulds are manual. The finishing or polishing is based on the use of abrasive cloths or papers of successively decreasing abrasive particle size culminating in the buffing of the mould with impregnated cloth.

Operator shortcomings cause these manual polishing methods to have a number of limitations. The operator has to be trained in suitable techniques and the effective application of these techniques and for polishing flat small radii requires considerable experience. Consistency and repeatability are also required. For example, a skilled worker can be required to spend ten days polishing ten complex dies and needs to ensure that, on completion, all the dies are identical. Finally, it is important that, in the achievement of the polished surface, there is no loss of dimensional accuracy. Consequently, a trained and experienced worker, skilled in polishing, becomes an expensive but essential part of the manufacturing process.

The process is also extremely time-consuming. It has been reported that the polishing time of a mould can represent up to 37% of the total production time of the entire mould (Steen, 2003). Improvement in the finishing or polishing of the mould or die to lessen the input of a skilled operator or to reduce the processing time has the potential to dramatically reduce the cost of the finished item.

Whilst automated processes, suitable for the polishing of closed dies, they are limited in their application. For example, precision machining using a single point diamond tool is slow, requires conditions not readily available in an industrial environment and is limited to flat surfaces (Kurt et al., 1997; Ryk et al., 2002; Steen, 2003). Chemical polishing and electrochemical polishing are limited in their application and can be difficult to control (Ryk et al., 2002). Some research has been carried out into the use of robot controlled polishing tools (Duely, 1983; Lugomer, 1990; Ryk et al., 2002; Steen, 2003). However, the use of a rotary wheel or ultrasonic chisel requires that the wheel axis or angle of the chisel is almost parallel to the surface being polished. This limits the application to almost flat workpieces.

Controlled material removal can be achieved by the use of ultrasonic machining (Ryk et al., 2002; Steen, 2003). In this process, the workpiece material is removed by the high frequency hammering of abrasive particles, in the form of the water-based slurry, into the workpiece surface. By the use of suitably formed tools, complex shapes can be created in the workpiece. As a potential method of polishing closed dies, this machining method has the disadvantage that it is normally used for fixed location drilling or workpiece indentation.

Abrasive flow machining is currently used in industrial applications to successfully and efficiently polish open dies. It uses a mix of abrasive particles suspended in a pliable polymer base. This mix is hydraulically powered over the surface to be polished. The process is however normally restricted to the polishing of open dies which are tubular or hollow in form with an entry and exit for the flowing mix (Kurt et al., 1997; Steen, 2003).

It is well known that high power lasers are presently used for a variety of machining application especially to process exotic materials. These materials range from metals to non-metals including ceramics. Basically, laser machining is a highly localized thermal processing.

Now, Laser is widely used as a machine tool to modify the surface of the engineering materials, such as laser surface alloying, laser cladding, surface texturing, laser physical vapor deposition *etc* (Duley, 1983; Lugomer, 1990; Steen, 2003). In recent years, laser polishing is becoming an attractive new technique to polish diamond, lens and so on (Duley, 1983; Kurt et al., 1997; Lugomer, 1990; Ryk et al., 2002).

By considering the unique characteristics of the laser radiation and excellent non-contact micromachining possibility with the laser, an attempt has been made to polish DF2 steel employing the concepts of micromachining (Guo, 2009; Guo, 2007). As a powerful surface modification technique, it is imperative to investigate on the laser polishing of metals systematically.

2. Experimental material and procedures

2.1 Experimental material

The chemical composition of DF2 cold work steel was shown as Table 1.

Element	C	Si	Mn	Cr	W	V	Fe
(Wt.%)	0.9	0.3	1.2	0.5	0.5	0.1	Bal.

Table 1. Chemical composition of DF2 cold work steel

2.2 Experimental procedures

The materials were machined into 10 mm×10 mm×25 mm, which had surface roughness of 0.2, 0.4, 0.6 μm, and its corresponding initial 3D surface morphologies were measured by Surface Texture Tester/Form Talysurf PGI.

The specimens were carefully cleaned by acetone and pure ethyl alcohol to remove any contaminants on its surface. The initial morphology of 0.4 μm and its corresponding SEM surface image were shown in Fig. 1. Figure 2 illustrated the microstructure of DF2 steel (AISI

O1). A GSI Lumonics Model JK702H Nd:YAG TEM_{00} mode laser system, with wavelength of 1.06 μm, defocused distance of 15 mm, and focus spot diameter of approximately 1.26 mm, was used to irradiate the DF2 steel.

The different laser polishing conditions employed were as follows:

Polishing feedrate: 100, 200, 300, 400 mm/min
Pulse energy: 1, 2, 3 J
Pulse duration: 2, 3, 4, 5, 6 ms
Pulse frequency: 15, 20, 25, 30 Hz

After polishing, the surface morphology was observed by Taylor Hobson/Form Talysurf PGI. In order to understand mechanism of laser polishing, the generated morphology was observed by Scanning Electron Microscope (SEM) JEOL/JSM-5600.

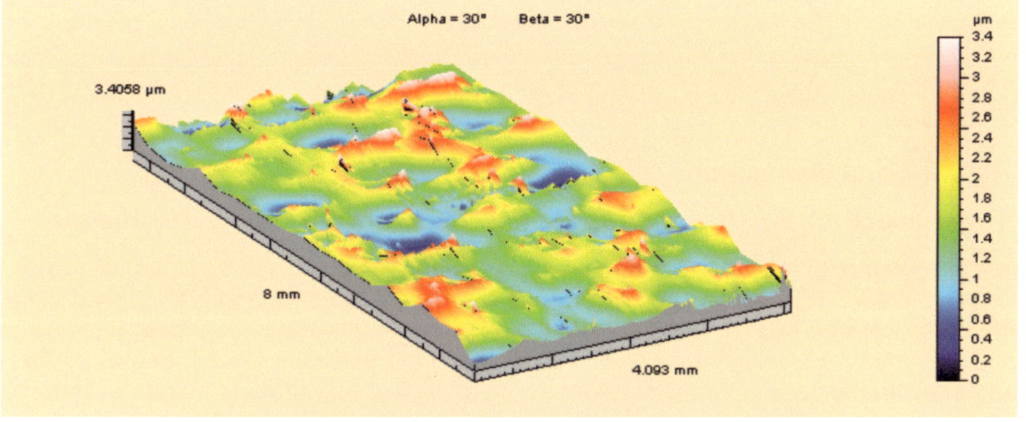

(a) Initial morphology of specimen

(b) SEM of the initial surface

Fig. 1. Morphology and its corresponding SEM of the initial surface (Ra=0.4 μm)

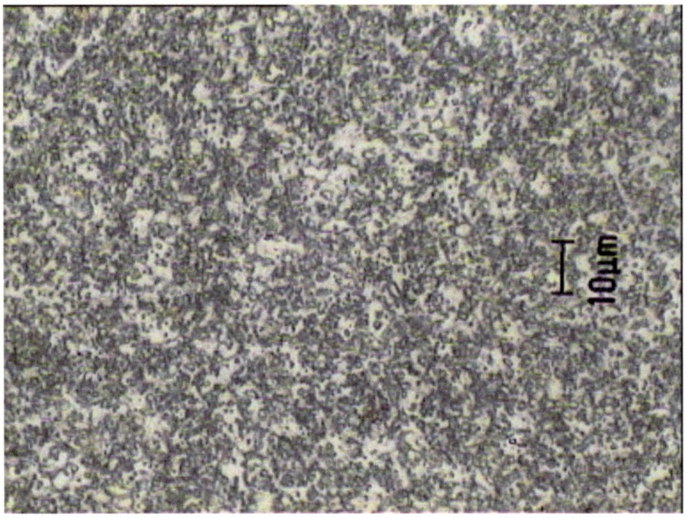

Fig. 2. Microstructure of DF2 steel (AISI-O1)

3. Results and discussion

3.1 Influence of relative irradiating speed (feedrate) between laser and workpiece

Although laser irradiated surface removed most high plateaus from the original surface of the specimens (cf. Fig. 1(a) and Fig. 3), improper setting of irradiation might roughen surface rather than smoothening it. Laser irradiation with original surface roughness *R*a = 0.4 μm under different feedrates with operational conditions of pulse energy 1 J, frequency 25 Hz and pulse duration 3 ms (Fig. 3) suggested that there was an optimal feedrate in achieving smoothest surface 3D morphology, which was at 300 mm/min. As the variation of morphology for this particular study (Fig. 3) illustrated that the smoother surface was obtained when feedrate increased from 100 mm/min to 300 mm/min, and it became coarser once again when it was above 300 mm/min. When the feedrate was above 300 mm/min, increasing the magnitude of feedrate led to unsmooth surface (Figs. 3(a) and (b)). When feedrate was below 300 mm/min, increasing in irradiation feedrate improved the surface topography (Figs. 3(c) and (d)).

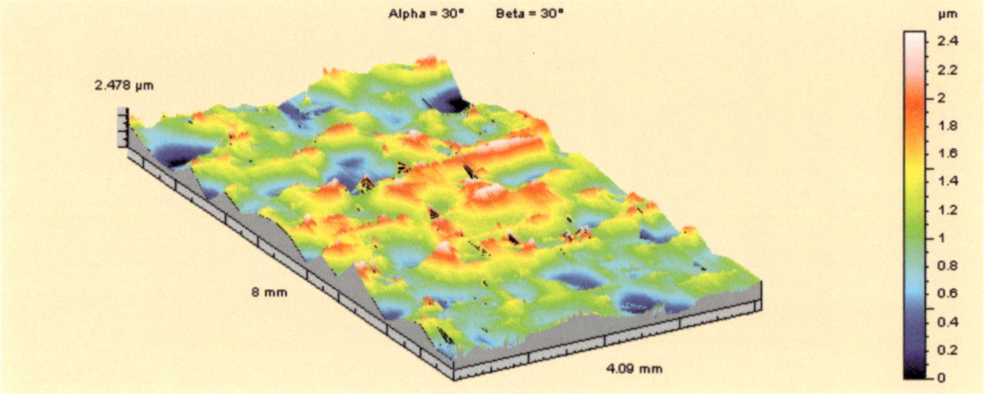

(a) 400 mm/min

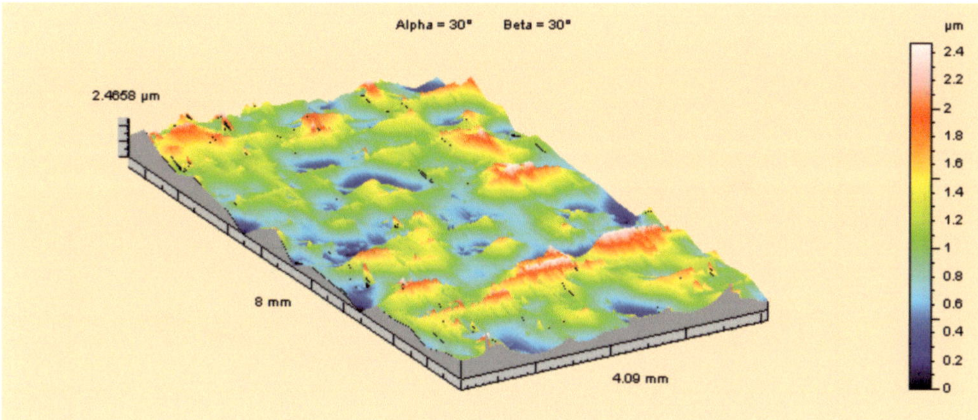

(b) 300 mm/min

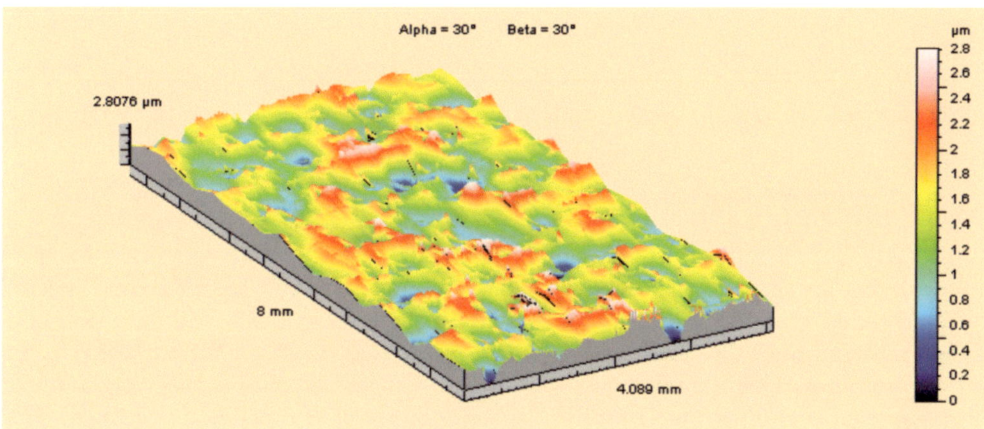

(c) 200 mm/min

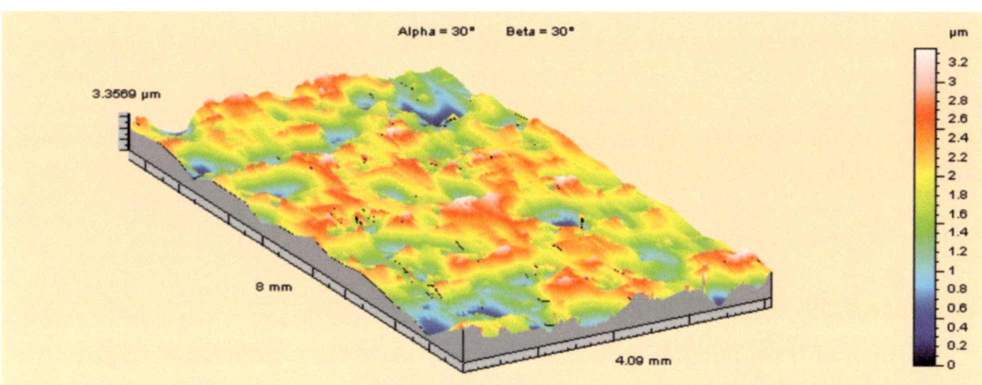

(d) 100 mm/min

*R*a=0.4 μm, Pulse energy= 1 J, *f*=25 Hz, PD=3 ms

Fig. 3. Surface Textures of laser polishing at various feedrates

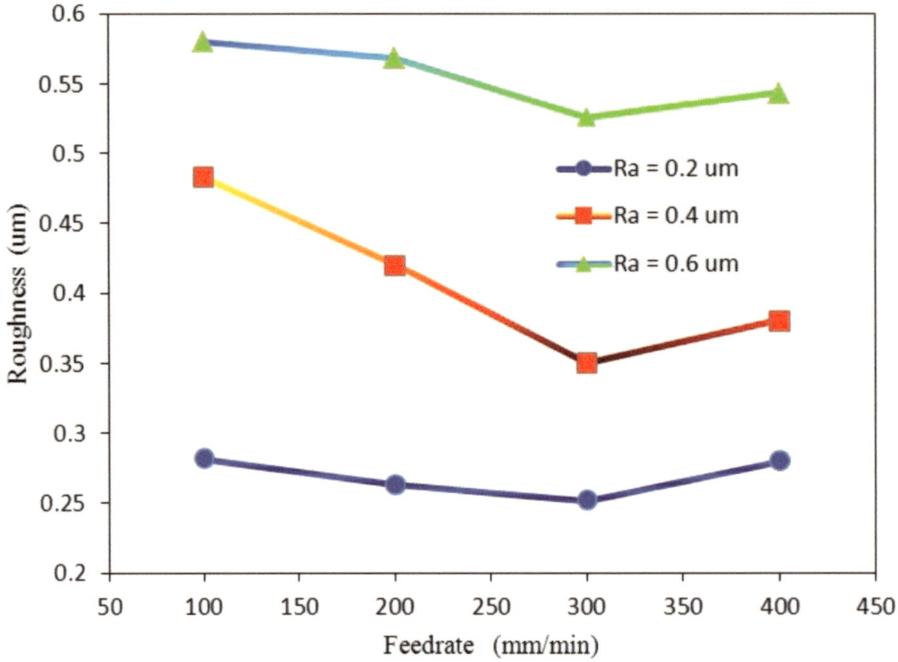

Fig. 4. Relationship between *Ra* and feedrates

Figure 4 showed the relationship between the arithmetic mean surface roughness *Ra* and the relative moving speed (feedrate) between the irradiating laser and the irradiated specimen. Results indicated that the initial increase in feedrate accompanied with the decrease in surface roughness until the feedrate reached 300 mm/min at which the surface roughness was the minimum. Then, further increase in the feedrate increased the roughness once again. Measurements illustrated that the surface roughness of irradiated specimens was to certain extend getting smoother along the irradiation path except with those specimens having surface roughness *Ra* = 0.2 μm. For those DF2 steel specimens with *Ra*=0.6 μm, the roughness of the irradiated surfaces was consistently and evidently deceased inspire of whether the relative feedrate being at or exceeding 300 mm/min. For those specimens with *Ra*= 0.4 μm, the irradiated surface roughness was generally higher than their initial condition when the relative feedrate was below 300 mm/min. However, it decreased with the increase in the relative feedrate. When the relative feedrate was at 400 mm/min, the irradiated roughness was increased to some level. For those DF2 tool steel specimens with *Ra*=0.2 μm, the irradiated roughness was consistently higher than their initial value within the various feedrates in the experimental studies.

3.2 Influence of laser irradiating pulse energy

3.2.1 Influence of laser irradiating pulse energy on surface topography

As studies in Sec. 3.1 giving optimal surface at the feedrate of 300 mm/min, subsequent studies were performed with this feedrate. Figure 5 showed the effect of laser irradiating energy on surface topography. Generally, increase in pulse energy simultaneously raised up the heat input to the surface substrate. Too high of the pulse energy was likely to melt the surface of substrate which subsequently changed the surface topography and mechanical properties of the DF2 steel specimens. When compared with the originally machined surface of Ra = 0.4 μm (Fig. 1(b)), its treated surface (Fig. 5(a)) irradiated under a feedrate=300 mm/min, f=25 Hz, PD=3 ms and pulse energy = 2 J gave the coarser surface. And, its further magnified counterpart (Fig. 5(b)) showed sign of some micro-pits and round agglomerates loosely scattering over both troughs and crests on the surface. Those round agglomerates would be the nucleus for re-solidifying the melt and subsequent shrinkage led to some locations slightly sinking below their surrounding material. Such surface topography with scattering of micro-pits and micro ball-like features (Fig. 5(b)) implied that there was some level of change of properties of DF2 surface. This change was not really anticipated since it was initially expected, from Fig. 5(a), that the properties of the original surface topography at pulse energy of 2 J would be the same as its as-received condition. At higher pulse energy (i.e. 3 J), except those originally machining grooves ridges (Fig. 5(c)) that scattering on the irradiated surface with shallower depth, more severe melting and re-solidification were observed. A large amount of microholes and weblike cracks were distributed on the polished surface (Figs. 5(c) and 5(d)). Such melting and re-solidification (Fig. 5(d)) changed the properties of the initial surface drastically resulted in the failure of polishing.

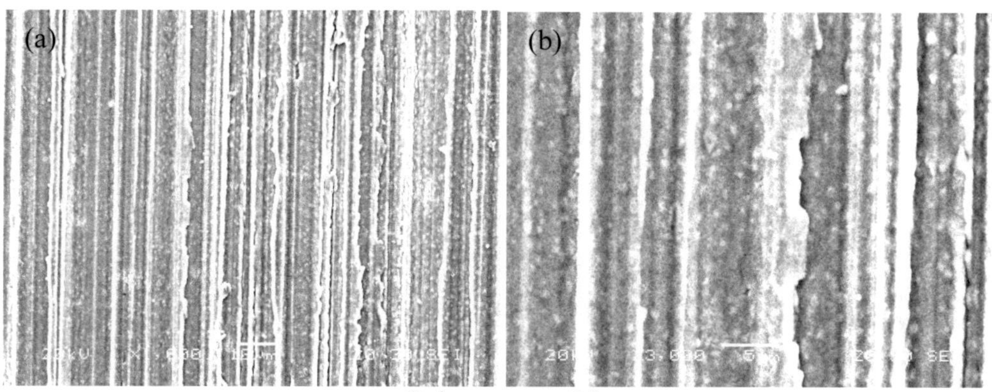

Pulse energy=2 J

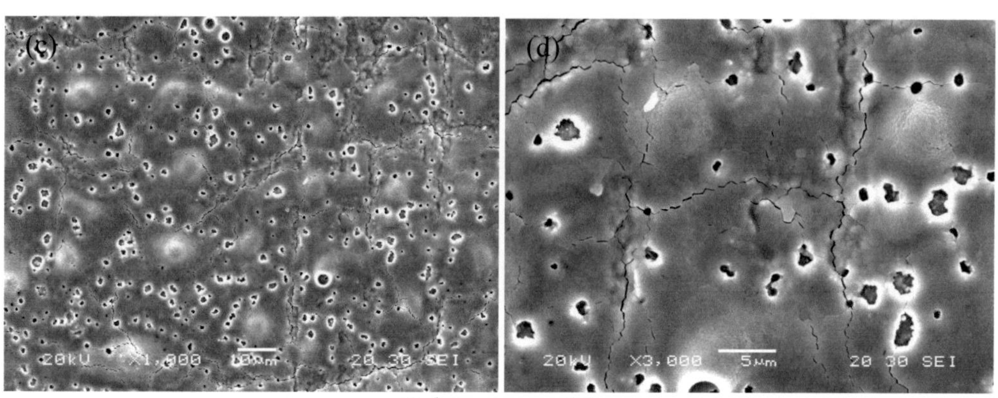

Pulse energy=3 J

Ra=0.4 μm, feedrate=300 mm/min, f=25 Hz, PD=3 ms

Fig. 5. SEM of laser polishing surface topographies at various pulse energies

3.2.2 Influence of laser pulse energy on surface roughness

Figure 6 showed the textures of a surface irradiated at various pulse energies. Basically, the surface gradually became coarser when pulse energy was increasing. Figure 7 plotted the experimental data of the laser pulse energy and the irradiated surface roughness. It showed the surface roughness increased with the irradiated pulse energy. At pulse energy of 2 J, the measured surface roughness was 0.49 μm as a result of the appearance of those protrusions on the surface (see Fig. 6(a)). The surface roughness approximately increased to 0.7 μm when the pulse energy was increasing further because of the occurrence of more prominent protrusions on the irradiated surface. Furthermore, the width of the laser irradiated band became larger as observed in Fig. 6(b), implying that laser polishing at pulse energy at and above 2 J under the setting conditions was generally not achievable.

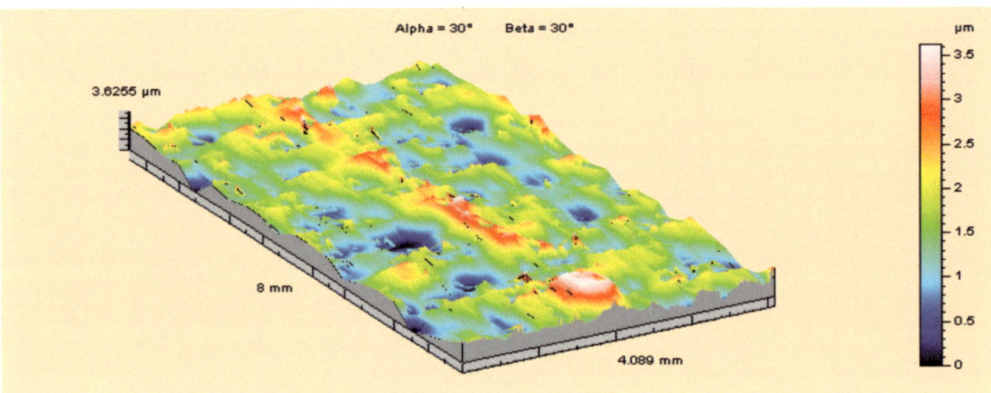

(a) Pulse energy=2 J

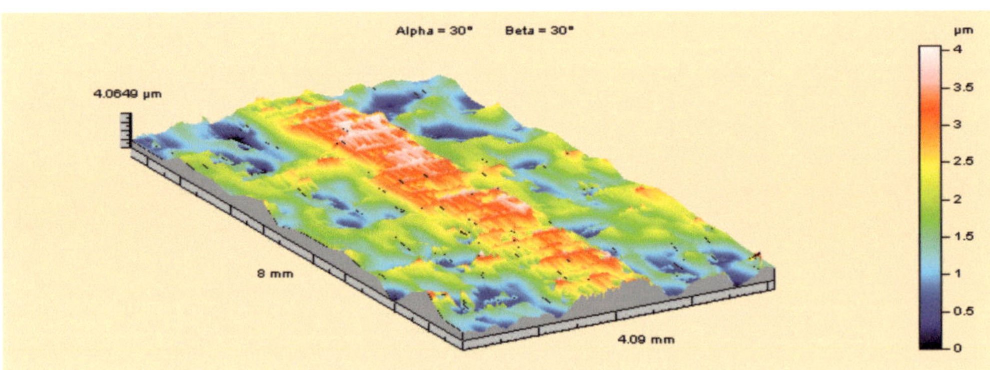

(b) Pulse energy=3 J

Ra=0.4 μm, feedrate=300 mm/min, f=25 Hz, PD=3 ms

Fig. 6. Surface Textures of laser polishing at various pulse energies

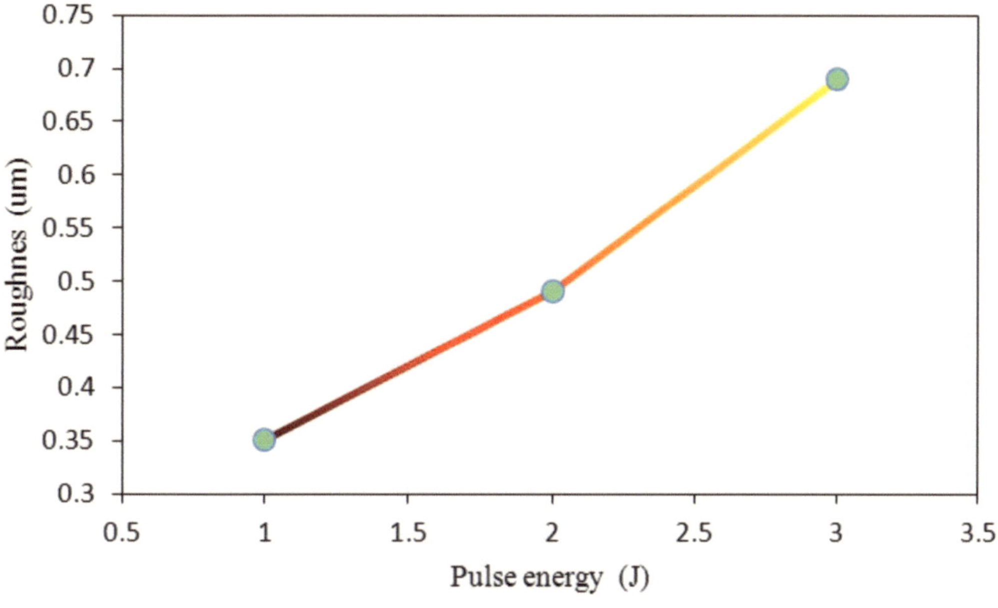

Fig. 7. Relationship between *R*a and pulse energy

3.3 Influence of laser irradiation frequency (Repetition rate)

3.3.1 Influence on surface topography

Result (Fig. 8) showed that the increase in pulse frequency generally led the topography of the irradiated surface to become smoother and smoother. But, the irradiated surface became coarser when the pulse frequency was over 25 Hz, as shown in Fig. 8.

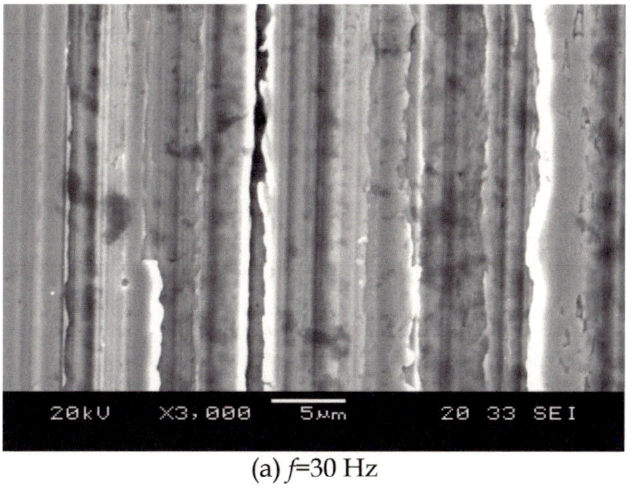

(a) *f*=30 Hz

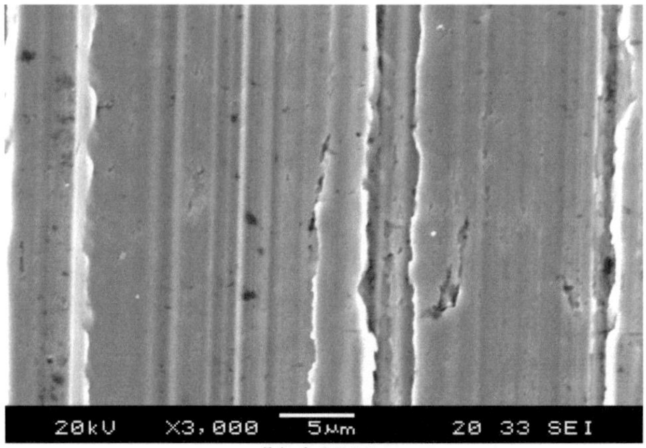

(b) f=25 Hz

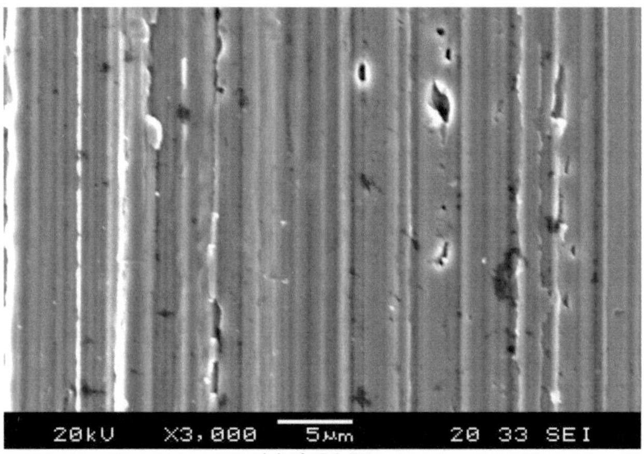

(c) f=20 Hz

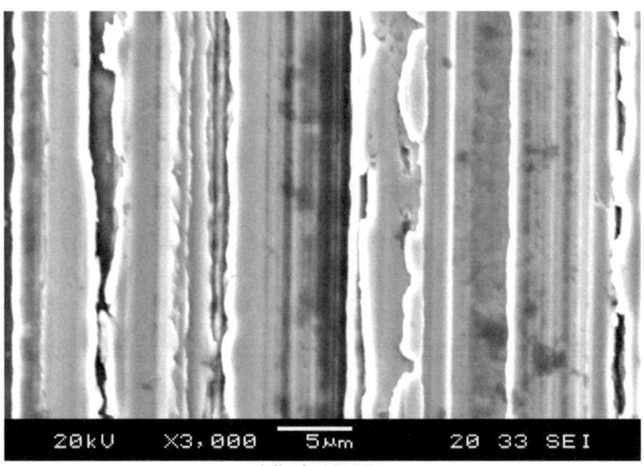

(d) f=15 Hz

Ra=0.4 μm, Pulse energy=1 J, feedrate=300 mm/min, PD=3 ms

Fig. 8. SEM of laser polishing surface topographies at various pulse frequencies

3.3.2 Influence on surface roughness

Figure 9 showed the laser irradiated surface scanned at various pulse frequencies. It showed that the roughness of the irradiated surface decreased with the pulse frequency. At a pulse

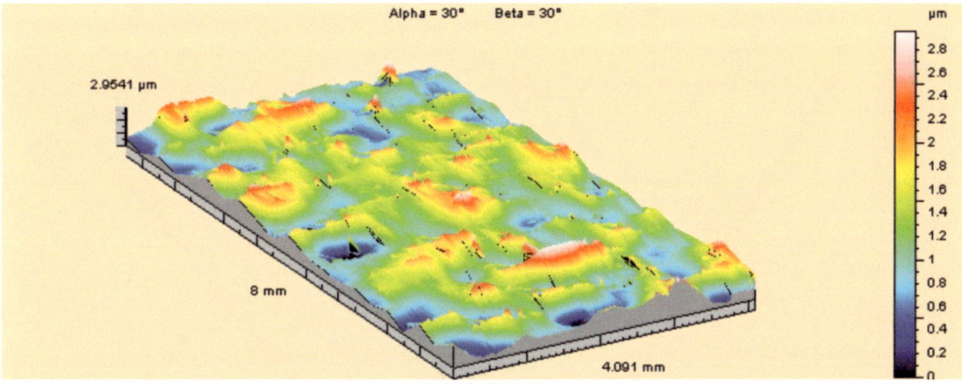

(a) f=30 Hz

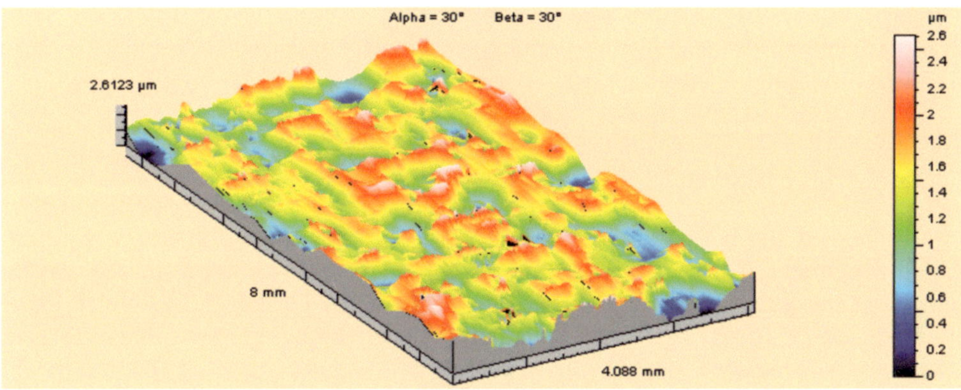

(b) f=20 Hz

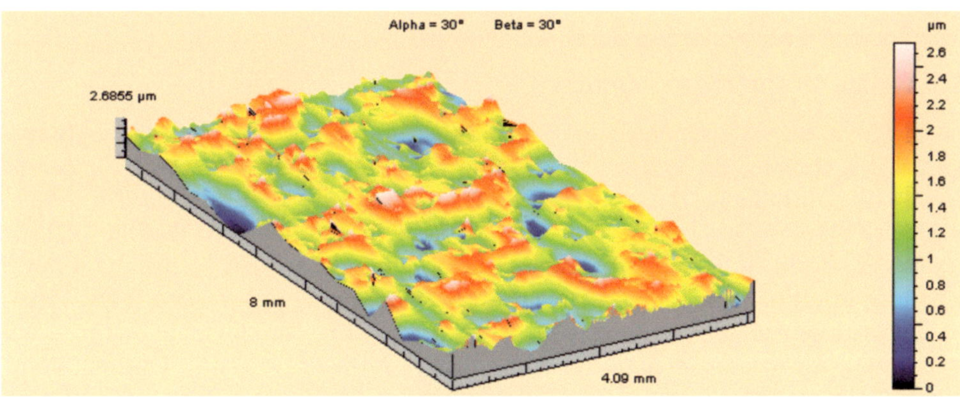

(c) f=15 Hz

Ra=0.4 μm, Pulse energy=1 J, feedrate=300 mm/min, PD=3 ms

Fig. 9. Surface textures of laser polishing at various pulse frequencies

frequency above 25 Hz, the increase in the roughness of the irradiated surface resulted in a coarser irradiated surface.

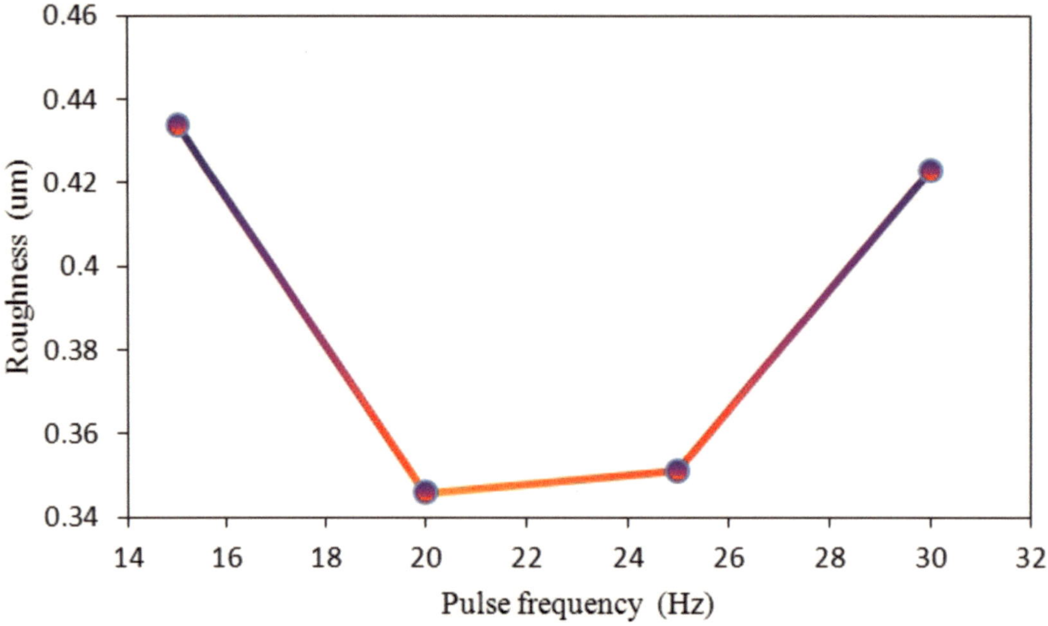

Fig. 10. Relationship between *Ra* and pulse frequencies

Figure 10 showed the corresponding relationship between the pulse frequency and the irradiated roughness. It illustrated that the roughness of the irradiated surface was below 0.4 μm when the pulse frequency was in the range of 20 Hz to 25 Hz, which gave the minimum value of roughness value. When the pulse frequency was above 25 Hz, the roughness of the irradiated surface was increasing with again, implying that the irradiated surface became coarser when the pulse frequency was either below 20 Hz or above 25 Hz.

3.4 Influence of laser polishing pulse duration (PD)

3.4.1 Influence on surface topography

Figure 11 illustrated the influence of pulse duration on the polished surface roughness and topography. It showed the non-uniform changes of the surface topography irradiated at different pulse durations. The variation of the topography of the surface irradiated at shorter pulse duration was usually detected markedly. It was observed that the DF2 surface was polished effectively when the pulse duration was in the range of 3 ms and 4 ms. With longer pulse duration, some level of micro-scale melted ridges was seen (Figs. 11(d) and 11(e)). In the present study, the Nd:YAG laser was irradiating the substrate that was moving relatively with a constant feedrate. The inputting laser energy at any location along the irradiated path on the substrate was the summation of (i) the energy per pulse at the irradiation point and (ii) the accumulated energy due to the effect of overlapping of the successive focusing spots along irradiation path and the heat transfer from its neighboring spots within a specific pulse frequency prior to the reaching of quasi-steady.

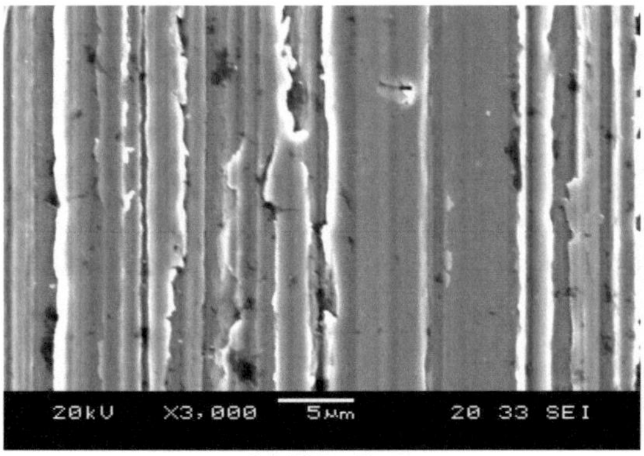

(a) PD=2 ms

(b) PD=3 ms

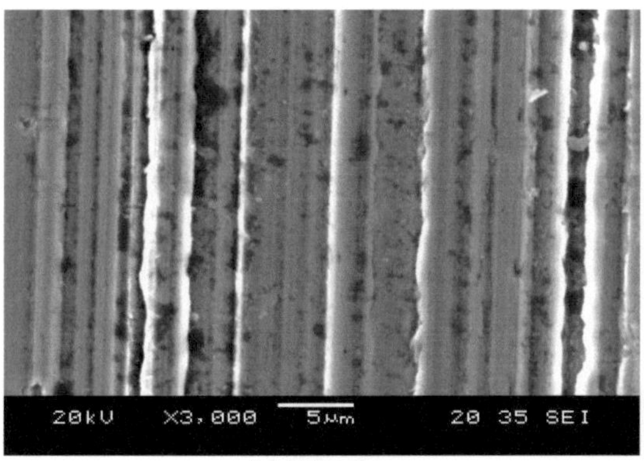

(c) PD=4 ms

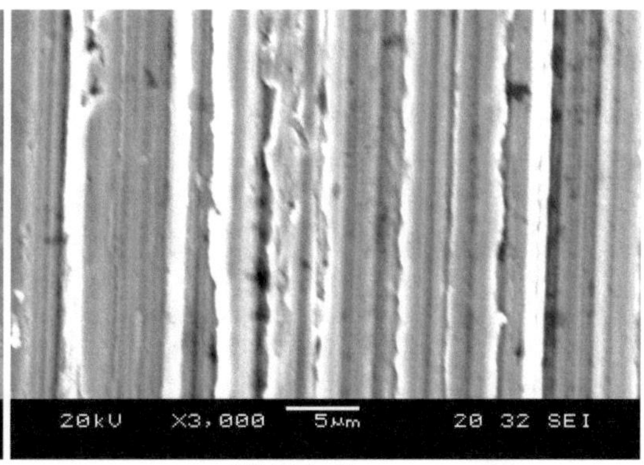

(d) PD=5 ms

(e) PD=6 ms

Ra=0.4 μm, Pulse energy=1 J, feedrate=300 mm/min, f=20 Hz

Fig. 11. SEM of laser polishing surface topographies at various pulse durations

3.4.2 Influence on surface roughness

Figure 12 showed the relationship of the measured roughness and pulse duration. It showed that the decrease in pulse duration reduced the roughness of the irradiated surface correspondingly. At the pulse duration of 3 ms, the irradiated surface roughness among the tests reached its minimum value. When the pulse duration was lower than 3 ms, the roughness of irradiated surface started to increase again to some respective level. It was observed that the irradiated surface became rougher at longer pulse duration of 4 ms, which typically illustrated the effect of longer pulse duration on the irradiated surface.

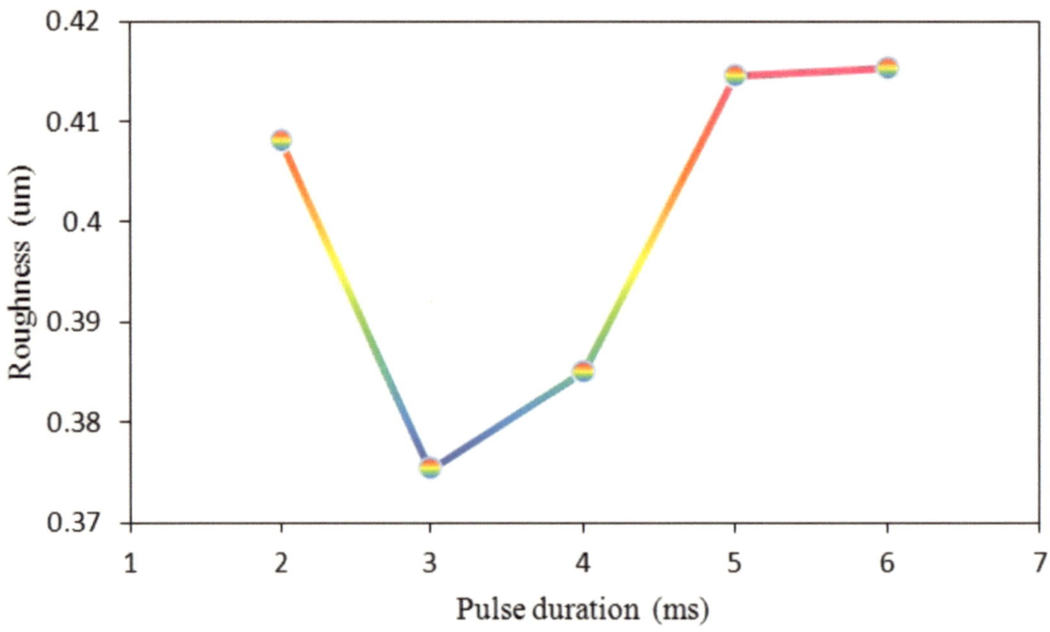

Fig. 12. Relationship between *Ra* and pulse duration

3.5 Surface morphology

Figure 13(a) showed the morphology of polished surface, and its corresponding SEM was shown in Fig. 13(b). It indicated that the polished surface was smoother than its initial surface. According to Fig. 13(a), it demonstrated that there were some small ridges on the polished surface. Compared Fig. 13(b) with Fig. 1(b), the grooves on the surface became shallower, together with thinner ridges. The big ridges could not be detected and the bulky chippings disappeared, resulted in smoother polished surface shown as Fig. 13(a). Moreover, the initial surface morphology had been properly maintained and no cracks can be found on the polished surface.

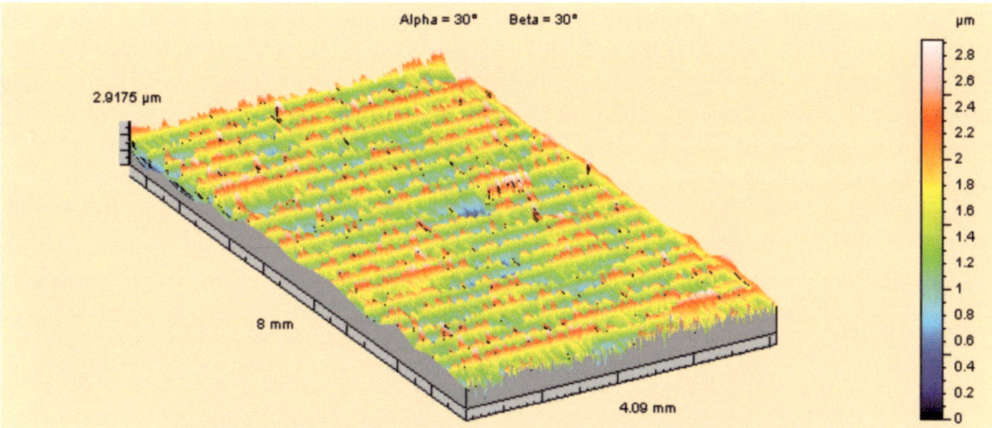

(a) Morphology of the polished surface

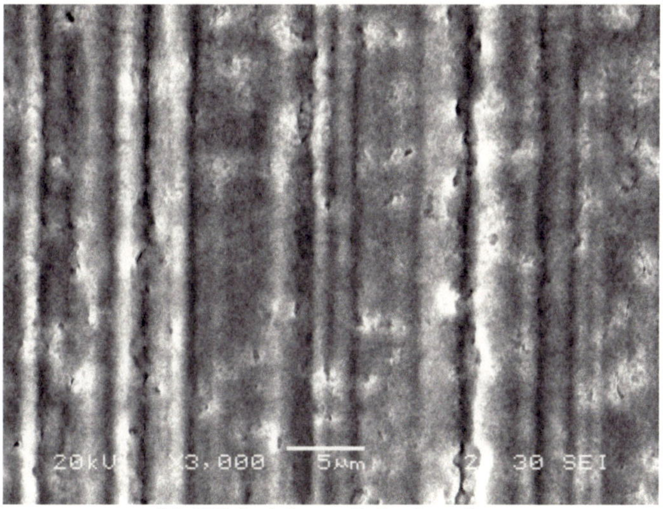

(b) SEM of the polished surface

Fig. 13. Morphology and its corresponding SEM of the polished surface

3.6 Laser polishing texture

Figure 14 showed the initial cross-section micrograph of the specimen before polishing. It could be seen that the surface of specimen was uneven as shown distinctly in Fig. 14(b) (dash line), which had the poor finish of mould surface in the tool machining.

Figure 15 illustrated the schematic diagram of cross-section of polished specimen, where A to D stood for the different zones affected by laser polishing. Its corresponding cross-section micrograph of polished specimen was shown in Fig. 16. The alteration could be observed obviously. The polished surface was significantly improved by laser polishing resulted in smoother surface with microstructure changed as shown in the left of Fig. 16(a). The zone A had been known to be a hardened layer with some of the spheroidized carbide particles enlarged to some extent as shown in the Fig. 16(a). As zone B, finer spheroidized carbide particles dispersed in the matrix of ferrite resulted in improving the higher properties of mold applications (cf. Figs. 16(a) and (b). It was well known the zone C was the heat affect zone (HAZ) as shown in Fig. 16(b) and (c). The zone D was the microstructure of DF2 steel. Contrasted zone C with zone D, it illuminated that the microstructure of zone C was nearly same as zone D, it indicated that the heat input into the base metal was too low to affect the mechanical properties of DF2 to large scale during the laser polishing. According to Fig. 16(a), it depicted the average thickness of the zone A and B was approximately 40 μm, where the average thickness of the zone A was about 10 μm. As a result, the polished surface was suitable for industry application.

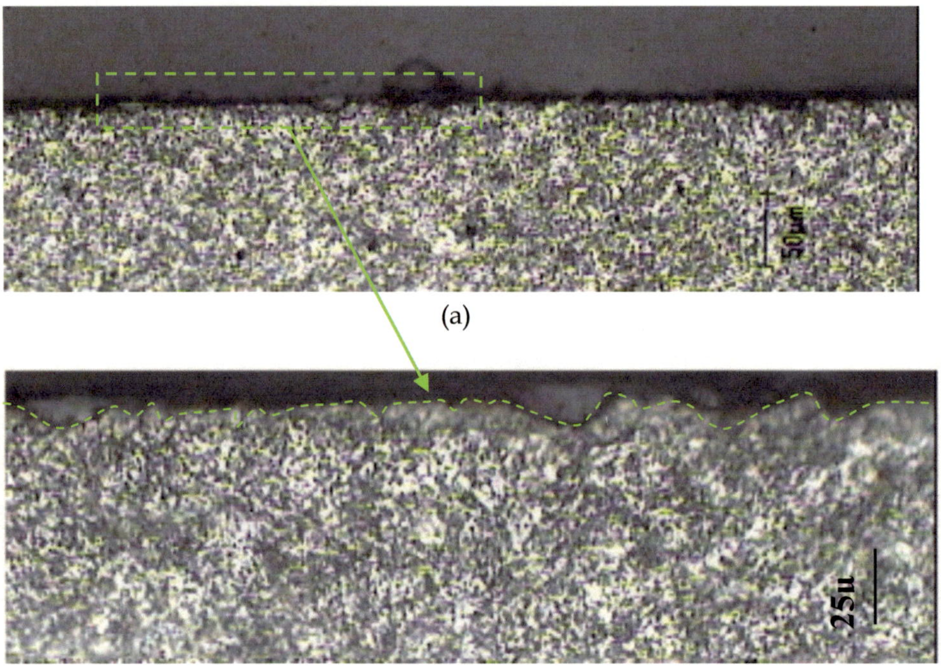

(a)

(b)

Fig. 14. Initial cross-section microstructure of the specimen

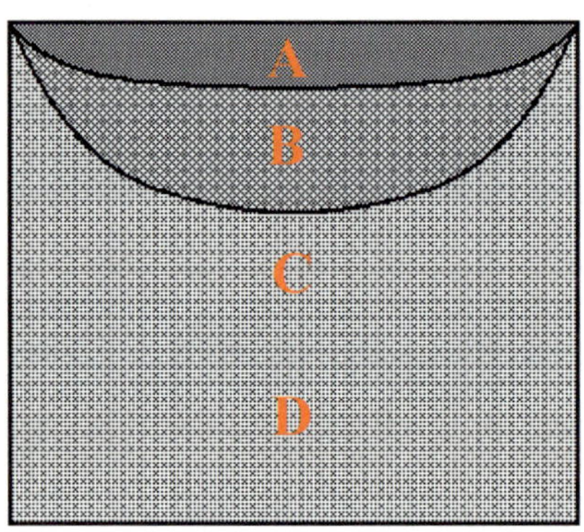

Fig. 15. Schematic illustration of cross-section of polished specimen

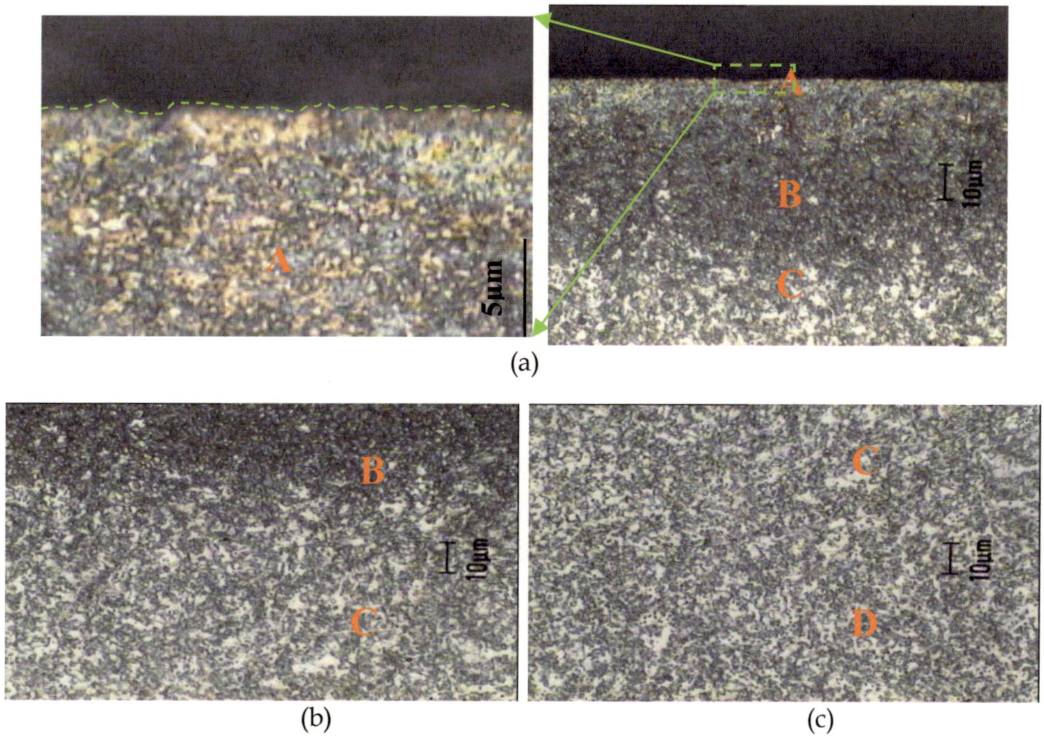

Fig. 16. Cross-section microstructures of polished specimen

Figure 17 showed that there were a plenty of micro-pits and melted substrate micro-balls scattered both on the polished surface (cf. Figs. 13(b) and 17) which could efficiently improve their load carry capacity and permit ease of release (Guo, 2009; Guo, 2007; Jastrzebski; 1987).

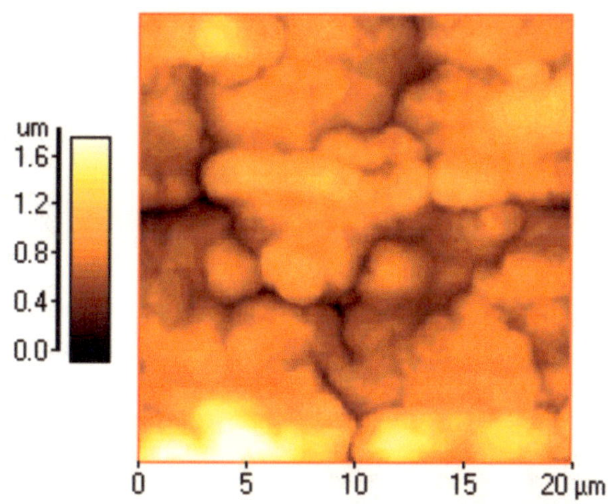

Fig. 17. Surface Textures of laser polished surface

4. Mechanism of laser polishing

4.1 Temperature field of laser polishing

The heat is assumed to be released instantaneously at time t=0 on the surface of the substrate. This causes a temperature rise in the material as follows (Duley, 1983; Janna,1986; Mills, 1995; Steen, 2003):

$$T - T_0 = \frac{Q}{\rho c (4\pi\alpha t)^{\frac{3}{2}}} \exp\left(\frac{-R^2}{4\alpha t}\right) \tag{1}$$

where ρ is the material density, C is specific heat, a is thermal diffusivity, λ is thermal conductivity, Q is the input energy.

When temperature distribution is quasi-steady state:

$$T - T_o = \frac{q_o}{2\pi\rho c\alpha R} \exp\left(-\frac{v}{2\alpha}(R + x)\right) \tag{2}$$

During the Nd:YAG laser polishing, q_0 can be expressed as

$$q_0 = \eta\left(N(f,v)\right) \times \frac{P}{\frac{\pi D^2}{4} \times PD} = \eta\left(N(f,v)\right) \times \frac{4P}{\pi D^2 \times PD} \tag{3}$$

Where

$\eta\left(N(f,v)\right)$ is the coefficient of laser polishing input energy, which is direct proportional to number of overlaps or pulse frequency. With number of overlaps increasing, or pulse frequency increasing with constant the feedrate, $\eta\left(N(f,v)\right)$ will be increased synchronously. With the velocity (feedrate) increasing, number of overlaps with the constant pulse frequency will be decreased correspondingly led to lower heat input, $\eta\left(N(f,v)\right)$ will be decreased accordingly.

Eq. (2) can be rewritten as:

$$T - T_o = \frac{2 \times \eta\left(N(f,v)\right) \times P}{\pi^2 D^2 \rho c \alpha R \times PD} \exp\left(-\frac{v}{2\alpha}(R + x)\right) \tag{4}$$

Define $\xi = \frac{1}{2\pi\rho c\alpha}$, then Eq. (4) can be written as:

$$T - T_o = \frac{\xi \times \eta\left(N(f,v)\right) \times P}{\frac{\pi D^2}{4} \times PD \times R} \exp\left(-\frac{v}{2\alpha}(R + x)\right) \tag{5}$$

Define

$$q = \frac{P}{\frac{\pi D^2}{4} \times PD} \tag{6}$$

then Eq. (6) can be written as

$$T - T_o = \frac{\xi \times \eta\left(N\left(f, v\right)\right) \times q}{R} \exp\left(-\frac{v}{2\alpha}\left(R + x\right)\right) \tag{7}$$

Define:

Dimensionless temperature:

$$\omega_T = \frac{T - T_o}{T_r - T_o} \tag{8}$$

where T_r is the reference temperature.

Define:

Dimensionless operating parameter:

$$\kappa = \frac{\eta\left(N\left(f, v\right)\right) q v}{4\pi\alpha^2 \rho c\left(T_r - T_o\right)} \tag{9}$$

Dimensionless x-coordinate:

$$\omega_x = \frac{vx}{2\alpha} \tag{10}$$

Dimensionless y-coordinate:

$$\omega_y = \frac{vy}{2\alpha} \tag{11}$$

Dimensionless z-coordinate:

$$\omega_z = \frac{vz}{2\alpha} \tag{12}$$

Dimensionless radius vector:

$$\omega_r = \frac{vR}{2\alpha} \tag{13}$$

Substituting these parameters into Eq. (7), then we have:

$$\omega_T = \frac{T - T_o}{T_r - T_o} = \kappa \frac{1}{\omega_r} \exp\left(-\omega_r - \omega_x\right) \tag{14}$$

or

$$\frac{\omega_T}{\kappa} = \frac{1}{\omega_r} \exp(-\omega_r - \omega_x) \tag{15}$$

Then the isothermal zone widths can be gotten just as fellows.

The maximum width of an isothermal enclosure is obtained by setting:

$$\frac{\partial \ln\left(\frac{\omega_T}{\kappa}\right)}{\partial \omega_r} = \frac{\partial \ln\left(\frac{\omega_T}{\kappa}\right)}{\partial \omega_x} \frac{\partial \omega_x}{\partial \omega_r} = 0 \tag{16}$$

where $\frac{\partial \omega_x}{\partial \omega_r} = \frac{\partial \omega_x}{\partial \sqrt{\omega_x + \omega_y + \omega_z}} = \frac{\omega_r}{\omega_x}$

Hence Eq. (16) will be transformed into:

$$\frac{\partial \ln\left(\frac{\omega_T}{\kappa}\right)}{\partial \omega_r} = \left(\left(-\frac{\omega_x}{\omega_r^{\ 2}} - \frac{\omega_x}{\omega_r} - 1\right)\frac{\omega_r}{\omega_x}\right)_m = \left(-\frac{1}{\omega_r} - 1 - \frac{\omega_r}{\omega_x}\right)_m = 0 \tag{17}$$

i.e.

$$\omega_{xm} = -\frac{\omega_{rm}^{\ 2}}{\omega_{rm} + 1} \tag{18}$$

Substituting ω_{xm} into Eq. (15), then we have:

$$\frac{\omega_{T_m}}{\kappa} = \frac{1}{\omega_{rm}} \exp(-\omega_{rm} - \omega_{xm}) = \frac{1}{\omega_{rm}} \exp\left(-\frac{\omega_{rm}}{\omega_{rm} + 1}\right) \tag{19}$$

Eq. (15) shows that the relationship between the laser polishing temperature and the polishing parameters, such as pulse energy, pulse duration, pulse frequency, velocity (feedrate) *etc*.. It can be seen that with the pulse energy and pulse frequency increasing, the polishing temperature will be higher. However, with the pulse duration and velocity (feedrate) increasing, the polishing temperature will be decreased. Furthermore, when the velocity (feedrate) is increased, $\eta(N(f,v))$ will be decreased synchronously, therefore, the polishing temperature will be decreased seriously. Furthermore, according to Eq. (13) and Eq. (19), it indicated that the velocity (feedrate) has more remarkably impact on the polishing temperature of the given points. Therefore, there is a certain heat-input threshold with a certain pulse feedrate, pulse energy, pulse duration and pulse frequency for laser polishing.

4.2 Mechanism of laser polishing

According to onset of laser polishing temperature, the laser feedrate (v) could be changed with the laser input energy (q), namely, when the laser input energy increases, the laser

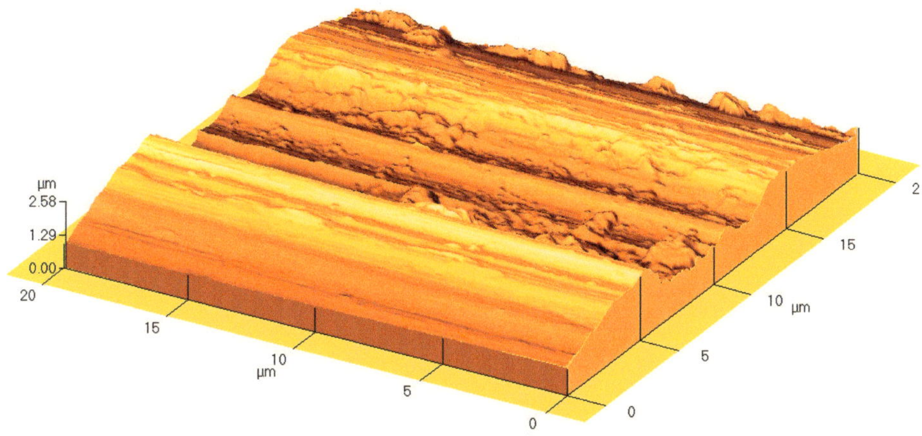

(a) Laser polished DF2 steel with Heat-input lower than threshold

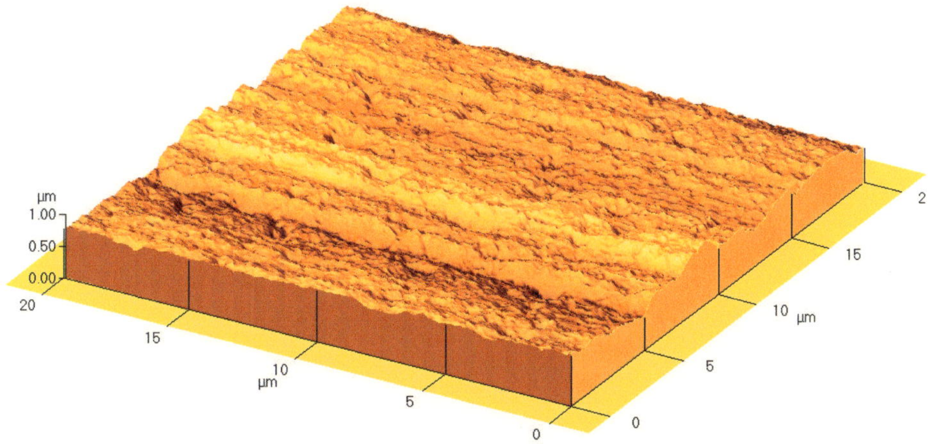

(b) Laser polished DF2 steel with Heat-input threshold

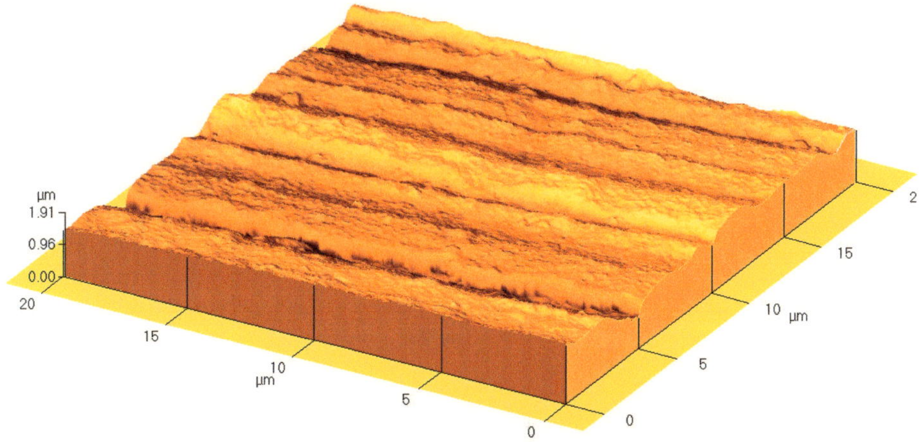

(c) Laser polished DF2 steel with Heat-input higher than threshold

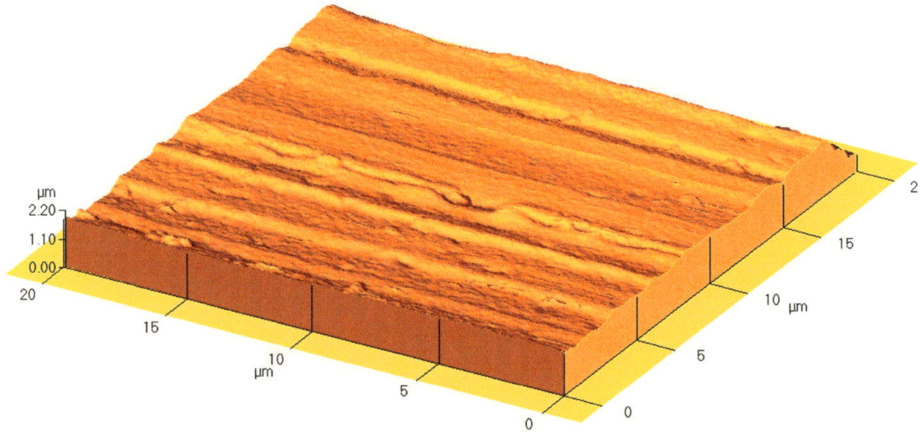

(d) Laser polished DF2 steel with Heat-input higher than (c)

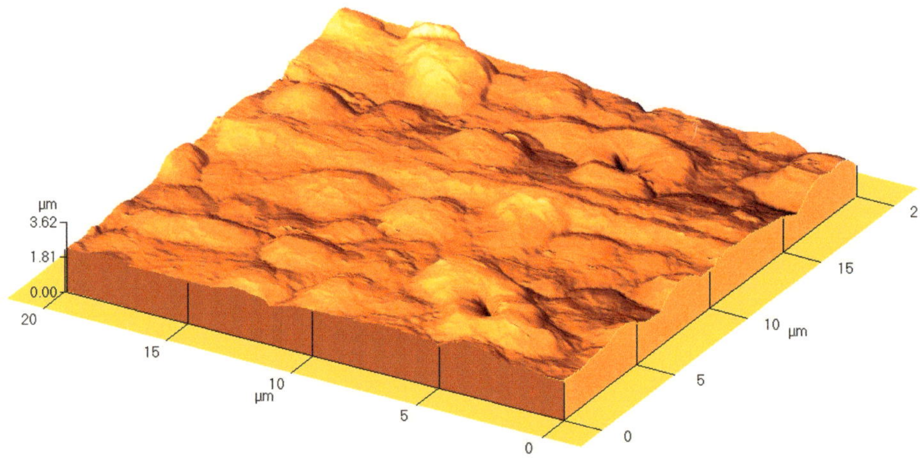

(e) Laser polished DF2 steel with Heat-input higher than (d)

Fig. 18. 3D morphologies of laser polished surface

feedrate should be increased correspondingly, vice versa. With a view to improving the efficiency of the laser polishing, saving the energy source, together with decreasing the effect on substrate (DF2), viz., lower laser input energy, the laser polishing feedrate should be higher and the laser input energy should be lower. The three dimensional morphologies of the laser polished surface with various heat-inputs measured by atomic force microscope (AFM) were shown as Fig. 18.

It illustrated that the surface of DF2 steel would not be polished basically (as shown in Fig. 18(a)), when the heat-input interaction with the DF2 steel surface was lower than the heat-input threshold, namely,

$$H_{input} < H_{threshold} \tag{20}$$

where

H_{input} stood for the heat-input interaction with the DF2 steel surface, $H_{threshold}$ stood for the heat-input threshold.

Figure 18(b) showed the morphology of laser polished surface with heat input threshold, namely, $H_{input}=H_{threshold}$. It indicated that the surface was polished successfully, and there were the large amount of micro-pits distributed on the polished surface which could efficiently improve their load carry capacity and permit ease of release (Guo, 2009; Guo, 2007; Jastrzebski, 1987). In addition, it was well known that the morphology of polished surface relied on the removal of surface material by evaporation and melting, and its subsequent re-solidification. Due to lower heat-input effecting on substrate, the mechanism of laser polishing during this procedure mainly relied on melting the base metal. When the DF2 steel surface was polished with higher heat-input than the threshold, namely, $H_{input}>H_{threshold}$, the morphology of the DF2 steel would be changed correspondingly as shown in Fig. 18(c). The mechanism of laser polishing at the moment was a combination of the base metal vaporizing and melting, the dominant factors of removing the surface material were evaporation and melting both, which was dominant factor relying on the heat-input. If the heat-input was higher, the metal evaporation would dominate the laser polishing procedure. Contrarily, the metal melting would dominate the laser polishing procedure. With the heat-input increasing, namely, $H_{input}>>H_{threshold}$, the mechanism of laser polishing would mainly lie on the metal evaporation shown as Fig. 18(d). With the further increasing of the heat-input, namely, $H_{input}>>>H_{threshold}$, the surface of DF2 steel would be destroyed thoroughly, shown as Fig. 18(e). There were lots of cracks, craters, protuberances on the polished surface resulted in coarser polished texture led to the poor properties of polished surface.

5. Conclusion

Successful polishing DF2 cold work steel was achievable by a YAG laser with the irradiating parameters set as P=1 J, feedrate=300 mm/min, PD=3 ms, *f*=20~25 Hz. Meanwhile, the temperature field of laser polishing was proposed as $T-T_o=\frac{\xi\times\eta\left(N(f,v)\right)\times q}{R}\exp\left(-\frac{v}{2\alpha}(R+x)\right)$ and the maximum width of an isothermal enclosure was $\frac{\omega_{T_m}}{\kappa}=\frac{1}{\omega_{rm}}\exp\left(-\frac{\omega_{rm}}{\omega_{rm}+1}\right)$. Moreover, the mechanism of laser polishing relied on the heat-input interaction with the base metal. When $H_{input}=H_{threshold}$, the mechanism of laser polishing was the metal melting; When $H_{input}>H_{threshold}$, the mechanism of laser polishing was a combination of metal evaporation and melting; When $H_{input}>>H_{threshold}$, the mechanism of laser polishing was the metal evaporation.

6. Acknowledgement

This work is supported by City University of Hong Kong Strategic Research Grant (SRG) No. 7002582.

7. References

Duley, W. W. (1983). *Laser Processing and Analysis of Materials*. New York and London: Plenum Press.

Guo, K. W. (2009). *Effect of Polishing Parameters on Morphology of DF2 (AISI-O1) Steel Surface Polished by Nd:YAG Laser*. Surface Engineering, Vol. 25, No. 3, pp. 187-195.

Guo, W. (2007). *Effect of Irradiation Parameters on Morphology of Polishing DF2 (AISI-O1) Surface by Nd:YAG Laser*. Advances in Materials Science and Engineering. Vol. 1, pp. 1-5.

Janna, W. S. (1986). *Engineering Heat Transfer*. Boston: PWS Engineering.

Jastrzebski, Z. D. (1987). *The Nature and Properties of Engineering Materials*. (3rd ed.). New York: John Wiley & Sons.

Kurt, S., Wolfgang, R. & Oskar, P. (1997). *Formation of paint surface on different surface structure of steel sheet*. Iron and Steel Engineer. Vol. 74, No. 3, pp.43-49.

Lugomer, S. (1990). *Laser Technology-Laser Driven Processes*. Englewood Cliffs, N.J. : Prentice Hall Inc.

Mills, A. F. (1995). *Heat and Mass Transfer*. Chicago: P. R. Donnelly & Sons Company.

Ryk, G., Kligerman, Y. & Etson, I. (2002). *Experimental investigation of laser surface texturing for reciprocating automotive components*. Tribology Transactions. Vol. 45, No. 4, pp.444-449.

Steen, W. M. (2003). *Laser Material Processing*. (3rd ed.). London: Springer-Verlag.

8

Laser Welding of Thin Sheet Magnesium Alloys

Mahadzir Ishak[1], Kazuhiko Yamasaki[2] and Katsuhiro Maekawa[2]
[1]Universiti Malaysia Pahang
[2]Ibaraki University
[1]Malaysia
[2]Japan

1. Introduction

Magnesium and its alloys are active materials, and the oxide can easily form when they react with air and moisture (Czerwinski 2002). In addition, magnesium and its alloys are flammable and require strict safeguards during the manufacturing process. These disadvantages make the processing of magnesium alloys into finished products more challenging. These drawbacks cause defects such as cracks, oxide inclusion, burn-through and voids both during and after processing. In order to minimize these defects, most processing methods must be performed at a certain temperature, unlike aluminum alloys and steels, which may be finished by cold working. Defect reduction can be achieved by minimizing the complexity of parts and the number of components produced. Furthermore, improved design techniques and production processing can eliminate defects in the finished products.

In order to produce finished products from several components, welding is unavoidable. However, welding magnesium alloys, especially thin sheets with thicknesses of less than 1.0 mm, is difficult. Furthermore, for electronics and communication devices, the weld width should be as small as possible. Possible methods of welding magnesium alloys include:

- Spot welding
- Arc welding
- Friction stir welding
- Laser welding

The first two methods are suitable for joining thick plates but not very thin sheets, because of the high heat input. Friction stir welding can join thin sheets, but it is difficult to weld complex parts and produces a small weld width. The most versatile and promising process for the fabrication of magnesium parts for electronics parts is laser welding.

The key advantage of laser welding is the ability to narrowly focus the laser beam on a small area, generating high intensity heat without any physical contact between welding hardware and the workpiece. The small area of the heat source can be rapidly scanned along the joint to be welded, and a narrow weld width can be obtained. Other advantages of laser welding include ease of automation and flexibility. Laser welding can be used with three-dimensional components and through transparent heat resistant glass in a vacuum or with inert gas in a closed box where the introduction of electrodes or tools is impossible. This flexibility can bring new possibilities for joint design and especially for components or parts that feature inaccessible surfaces.

Distortion and deformation of the thin sheet can be significantly reduced by laser welding because of its smaller spot size, lower heat input and improved penetration. These advantages lead to a narrow heat-affected zone (HAZ), which can limit thermal distortion and improve the metallurgical properties compared with arc welding methods. Besides, contamination from electrode materials can be eliminated, and the reduction of volatile alloy components such as zinc can be achieved (Duley 1999).

Currently two main laser systems are used in industry when laser welding sheet metals, namely the pulsed neodymium: yttrium garnet (Nd:YAG) laser and the carbon dioxide (CO_2) laser.

1.1 Types of lasers for magnesium welding

There are only two types of high power laser systems that are currently available for industrial use that are suitable for welding. These lasers are carbon dioxide (CO_2) and neodymium: yttrium garnet (Nd:YAG) lasers.

1.1.1 Neodymium: Yttrium Aluminum Garnet (Nd:YAG) laser

The active medium in the Nd:YAG laser is a solid crystal material of yttrium aluminum garnet (YAG) doped with neodymium ions, Nd^{3+}. The neodymium ions take the position of the yttrium ions in the garnet lattice in which they are roughly the same size. The ion concentration is around 1.0% by weight (Ion 2005). Neodymium: yttrium aluminum garnet ($Nd:Y_3A_{15}O_{12}$) is a solid structure that is used as a lasing medium for a solid-state laser.

This crystal produces infrared light with a wavelength of 1.06 μm. In addition, wavelengths of 940, 1120, 1320 and 1440 nm can be generated. One of the advantages of this laser is that its lower wavelength gives better absorption by metals. The other advantage is that it allows the use of an optical fiber to transfer the laser beam to a certain distance. The standard power outputs are 0.3- 3 kW, but advances in this laser can extend the maximum power available to 4.0 kW. This laser may be operated in continuous and pulsed modes.

1.1.2 Carbon dioxide (CO_2) laser

The CO_2 laser (carbon dioxide laser) is a laser based on a gas mixture as the gain medium, which contains carbon dioxide (CO_2), helium (He) and nitrogen (N_2). This laser is electrically pumped via a gas discharge, which can be operated with DC current, AC current (e.g. 20–50 kHz) or in the radio frequency (RF) domain. Nitrogen molecules are excited by the discharge into a vibration level and transfer their excitation energy to the CO_2 molecules when colliding with them. Helium serves to reduce the lower laser level and to remove the heat. Other constituents such as hydrogen or water vapor serve (particularly in sealed-tube lasers) to re-oxidize carbon monoxide (formed in the discharge) to carbon dioxide.

This laser produces infrared light with wavelengths from 9.6 μm to 10.6 μm. CO_2 lasers are usually used in industrial applications. Its output power is up to 50 kW, and the system is simple and reliable. The output efficiency defined as the ratio of output laser power to input electrical power can reach up to 10 %, which is efficient. However, it cannot be transferred via fiber optics.

1.1.3 Selection of laser source for welding magnesium alloys

Laser welding of metal in many industries uses both CO_2 and pulsed Nd:YAG lasers(Costaa, Quintinoa et al. 2003; Haferkamp, Ostendorf et al. 2003). Both lasers have unique advantages. However, one of the leading advantages of the Nd:YAG laser over the CO_2 laser in this specific application is the that lower wavelength gives better absorption by metals. However, the continuous mode of Nd:YAG possesses less power, which results in low power density. Thus, the laser usually is used in pulsed mode to raise the power and increase the power density.

The thicknesses of the metal welded in previous studies were mostly above 1.0 mm (Costaa, Quintinoa et al. 2003; Haferkamp, Ostendorf et al. 2003; Quan, Chen et al. 2008). The research on welding thin sheets of magnesium alloys lower than 1.0 mm is still deficient. Therefore, in this study, the pulsed Nd:YAG laser is chosen to weld thin sheets of magnesium alloy with thicknesses of 0.3 mm.

2. Nd:YAG laser system for welding magnesium alloys

2.1 Principles of the laser system

The laser that is used in this study is a pulsed Nd:YAG laser. The schematic of the laser optical system is shown in Fig. 2.1. It is a solid state laser in which the rod-shaped laser crystal is illuminated with visible light. The source of the energy is a plasma flash lamp (electrical discharge in a quartz tube filled with inert gas. The laser rod stores the pumping light energy for a short time at the pumped electron levels. The energy is emitted once again as infra red light with a wavelength of 1064 nm.

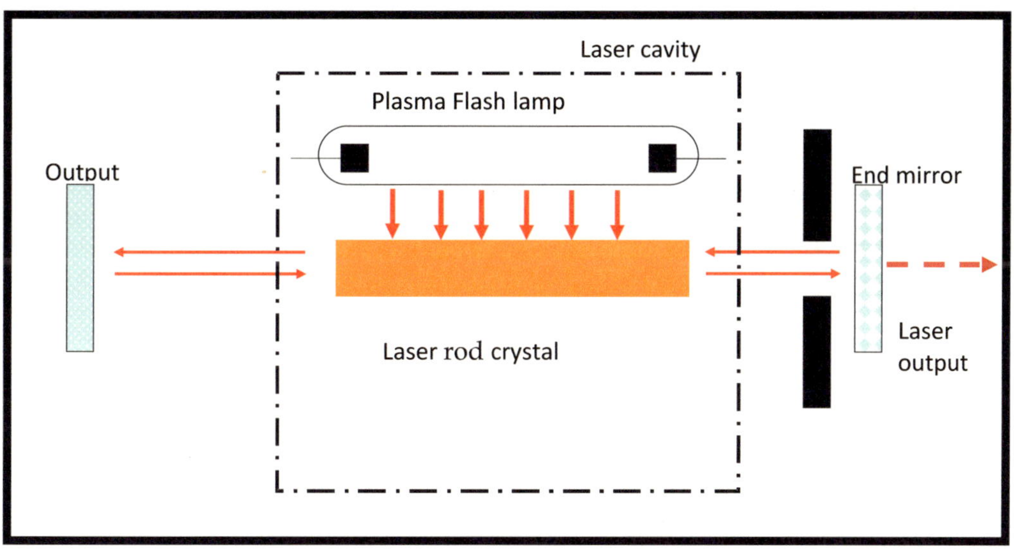

Fig. 2.1. Schematic illustration of Nd:YAG laser cavity

The optical resonator includes the laser crystal and two parallel mirrors, which enables multiple passages through the crystal. This can lead to induced light emission and therefore to coherent light. The laser radiation emerges through the partially reflecting out-coupling mirror.

2.2 Characteristic of the laser beam

The advantage of lasers compared with other powerful sources of light is their ability to concentrate beam energy onto a very small area, called the focus point. The distribution of the intensity in the application is another quality parameter. In this study the beam energy intensity is single mode, which results in a near-Gaussian distribution.

3. Laser parameters

The main parameter for a pulsed laser is the pulse energy, *E*, which can be adjusted by controlling the pulse duration, repetition rate and voltage. Other parameters such as the peak power and average power can be determined by the following equations.

$$P_p = \frac{E}{P_{avg}} \tag{3.1}$$

$$P_{avg} = E \times rep.rate \tag{3.2}$$

P_p, E, P_{avg} and *rep.rate* are the peak power, pulse energy, average power and repetition rate, respectively.

3.1 Laser material interactions

When a laser beam impacts a material, certain portions of the beam energy are reflected, absorbed and transmitted. The relation between the reflection *R*, absorption *A* and transmission *T* is as follow.

$$R + A + T = 1 \tag{3.3}$$

However, when a laser beam impacts a flat sheet of metal, a large portion of the energy is reflected, a small portion is absorbed, and no transmission occurs. Transmission usually only occurs for transparent materials such as glass, polymers, etc. When a flat metal is used, the reflected energy is lost to the environment. The remaining energy is absorbed in the metal as heat, which raises the temperature of the material under the beam and is conducted into the bulk of the material. Thus, for a metal surface the relation between R and absorption is:

$$R + A = 1 \tag{3.4}$$

The reflection percentage can be measured using a power detector. The reflection ratings of some metals, including the AZ31B alloy, are shown in Fig 3.1. It is apparent from the figure that the reflection of wrought AZ31B sheet metal with a thickness of 1.3 mm is 94 %. The reflection metric decreases to around 88% when the surface of the sheet metal is polished with SiC paper (#1000). Coating the sheet metal surface with silver nanopaste significantly reduces the reflection to around 36%. It is probable that the roughness and the coating material cause the decrease in the reflection and the increase in absorption. The absorptivity may be enhanced by many factors such as temperature, surface roughness, oxidation and changes in morphology (Duley 1998).

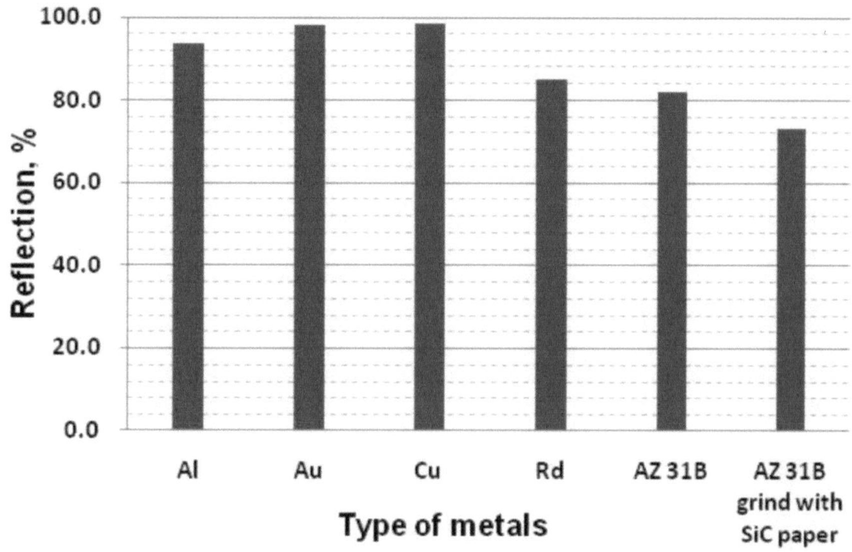

Fig. 3.1. Laser reflections of different materials (Ishak 2010)

3.2 Laser welding characteristics

Conduction mode welding occurs at a low energy density and involves the laser beam interacting with the surface of the sheet metal. The energy is absorbed by the atoms on the metal surface, which heat up and then transfer the heat to the atoms below the surface. Typically, conduction mode welding is useful when the energy density is relatively low. The process generates a wide weld bead. The cross-sections of conduction mode welds exhibit wide beads and relatively shallow depths. The size and shape of the weld bead depends upon the input energy and the material properties.

At high heat energy density, the metal surface rapidly heats up to the point where molten metal vaporizes at the center of the weld beam spot, which opens up a blind cavity (keyhole) in the molten metal. As the keyhole penetrates deeper into the metal, the laser light is scattered repeatedly within it, thus increasing the coupling of the laser energy to the workpiece. While the laser energy is applied, the keyhole is held open by vapor pressure, which prevents the keyhole wall from collapsing. A higher penetration depth and a narrower weld bead can be achieved with the keyhole mode.

3.3 Difficulties in laser welding of magnesium and its alloy thin sheet

The basic difference between the conduction mode and penetration mode is that the surface of the weld pool remains unbroken during conduction mode welding. The opening of the surface allows the laser beam to enter the melt pool in the keyhole mode and causes a break in the metal surface. In addition, openings in the molten metal result in gas entrapment during welding, which leads to the formation of pores in the weld.

The high intensity of the energy density in the keyhole mode can penetrate thicker metal compared with the conduction mode; however, to weld thin metal sheets the keyhole mode can cause cut or blow holes in the metal to be joined. Low heat input generates less distortion, and lower thermal distortion occurs with the conduction mode than with the keyhole mode. Based on these reasons the conduction mode is preferable to weld thin sheet magnesium alloy with thicknesses of 0.3 mm.

4. Bead on plate welding

The aim of bead on plate welding experiment is to study the laser welding properties of AZ31B magnesium-based sheet. Effects of welding parameters such as pulse energy, focal position, and scan speed on weld appearances were extensively investigated. This study could provide basic information regarding the laser parameters and weld appearances prior to lap fillet configurations.

4.1 Experimental method

The AZ31B magnesium-based sheet with a thickness of 0.6 mm was studied in this experiment. The sheets were cut into 20 x 30 mm specimens. Prior welding, the specimens were polished with SiC paper and cleaned using ethanol to remove oil film that existed on the material surface. In this study bead on plate welding was carried out.

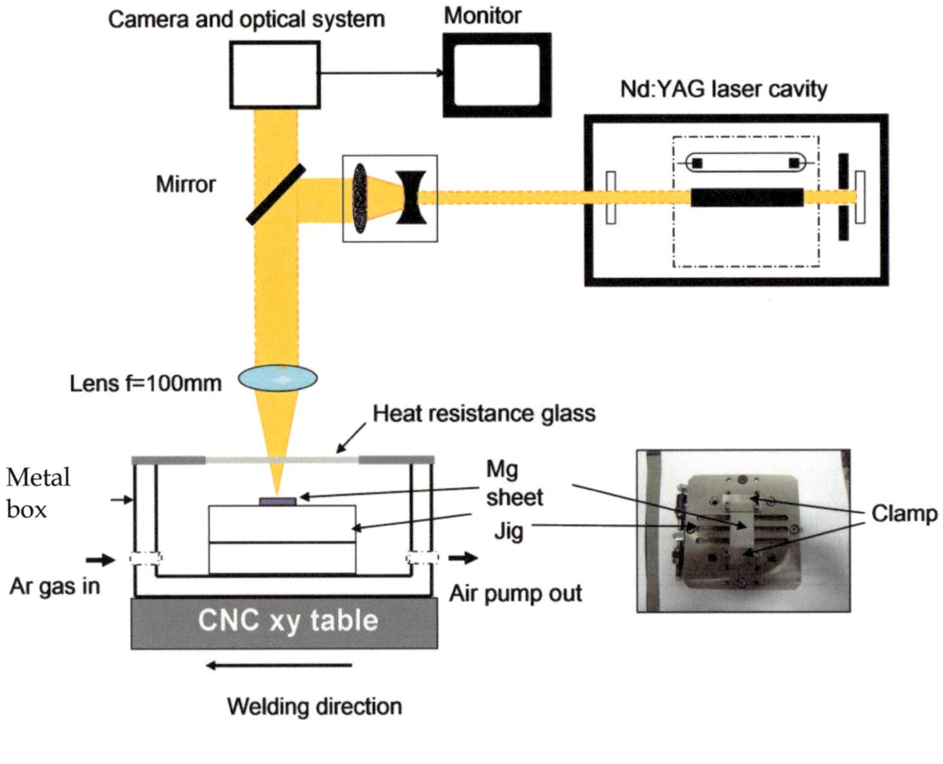

Fig. 4.1. Laser system and experimental set-up for bead on plate welding

A pulsed Nd:YAG laser with 1.06 μm wavelength and 100 mm focal length was used in this investigation. Spot size of a focused laser beam was measured about 0.4 mm. Fig. 4.1 illustrates the set-up for the bead on plate welding. The sheet was fixed on special jigs by clamping. The jig with a sheet was placed in a steel box having the top of the box was partially closed by a heat-resistant glass, where the laser beam is transmitted through the glass during welding. Argon gas was filled in for about 30 s as a shielding gas to prevent oxidation prior to welding, and gas flow was continued flow until welding finished. During the welding process, the laser beam was kept stationary while steel box was moved with a CNC x-y table. The laser beam was transverse the specimen surface, as shown in Fig.4.1.

The parameters used were laser pulse energy in the range of 1.0 J to 2.5 J, welding speed between 50 mm/min to 600 mm/min and beam defocusing in the range of -4.0 mm to +4.0 mm. Positive defocusing is defined as that the focal point above the top surface of specimen whereas negative one the focal point below the top surface. The repetition rate and pulse duration were constant at 80 Hz and 3.0 ms, respectively. After welding, the sample was cut into four parts, being mounted in a polymer resin. The cross-section was etched with an etchant (10 ml acetic acid diluted with 100 ml distilled water) for macroscopic observation. The penetration depth, bead width and undercut depth of the weld were measured by optical microscopy.

4.2 Results and discussions

4.2.1 Effect of focus point position

Fig. 4.2 shows the cross-sectioned appearance of bead on plate welding. By irradiating laser beam, many layers were formed from molten magnesium. The wider bead with shallow penetration depth shows that the type of welding is conduction mode welding rather than keyhole mode welding; conduction mode welding is resulted because of low energy density. Figure 4.3 shows the relationship between the focus position effect and penetration depth. The ranges can be varied from -4.0 mm to 3.0 mm in using this laser system. The focal point located just on the specimen surface produced the lowest depth. Increase or decrease in focus point location gradually increases the penetration depth. The highest depth penetration was achieved when the focus position is at a defocus point of -3.0 mm. (Zhu, Li

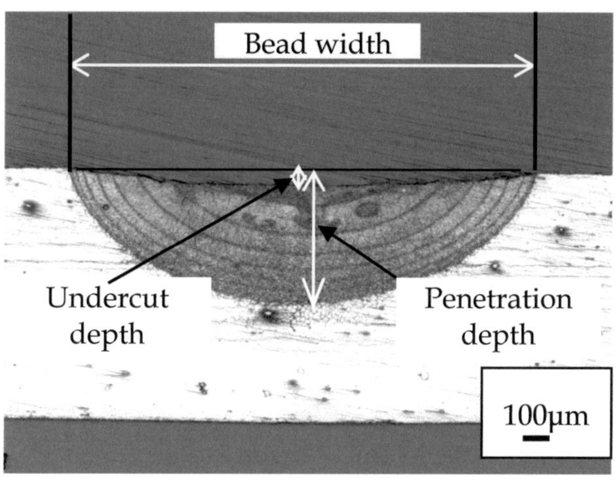

Fig. 4.2. Bead on plate weld appearance

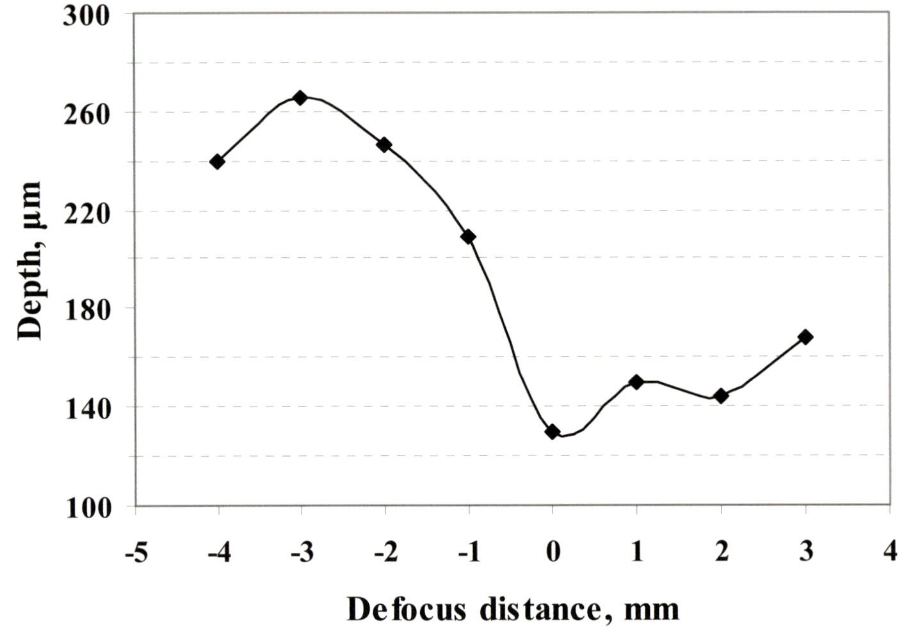

Fig. 4.3. Defocus effect on penetration depth

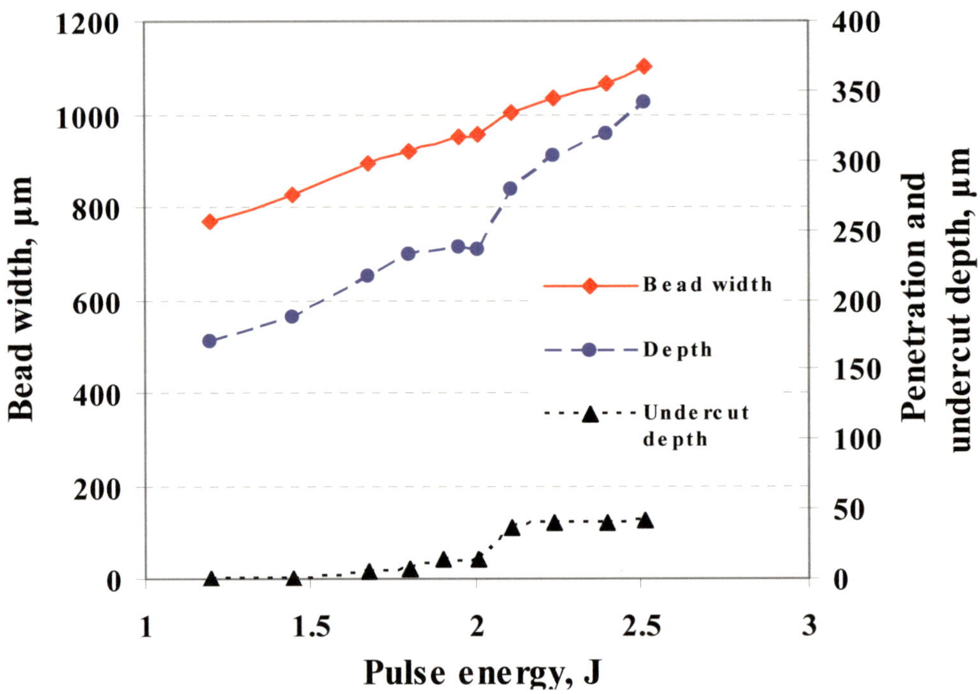

Fig. 4.4. Pulse energy effect on bead width, penetration and undercut depth

et al. 2005) reported that using CO_2 laser welding even at low power undercut occurs and can be avoided by defocusing. On the contrary, our results show that using Nd:YAG laser welding with different defocusing positions produced no effect on undercut. Thus, the defocus point at -3.0 mm was used throughout the investigation because it generated the maximum penetration depth.

4.2.2 Effect of pulse energy

Fig. 4.4 shows the effect of laser pulse energy on weld bead width, penetration depth and undercut depth when the scan speed was 300 mm/min. It is observed that the weld width becomes obviously larger when high pulse energy is used.

This indicates that the size of weld pool is increased with increasing pulse energy. There was no weld appearance observed when pulse energy was below than 1.2 J. Which mean that in this case, the threshold pulse energy of laser welding is about 1.2 J. Increase in pulse energy also resulted in deeper penetration depth; however, full penetration was not achieved for the 0.6 mm thickness specimen.

4.2.3 Effect of scanning speed

The weld bead width and penetration depth under different welding speeds with constant pulse energy of 1.8 J are depicted in Fig. 4.5. It is shown that the bead width and penetration depth increase when welding speed decreases. This result is consistent with the effect of laser pulse energy as was shown in Fig. 4.4. However, increase in pulse energy causes

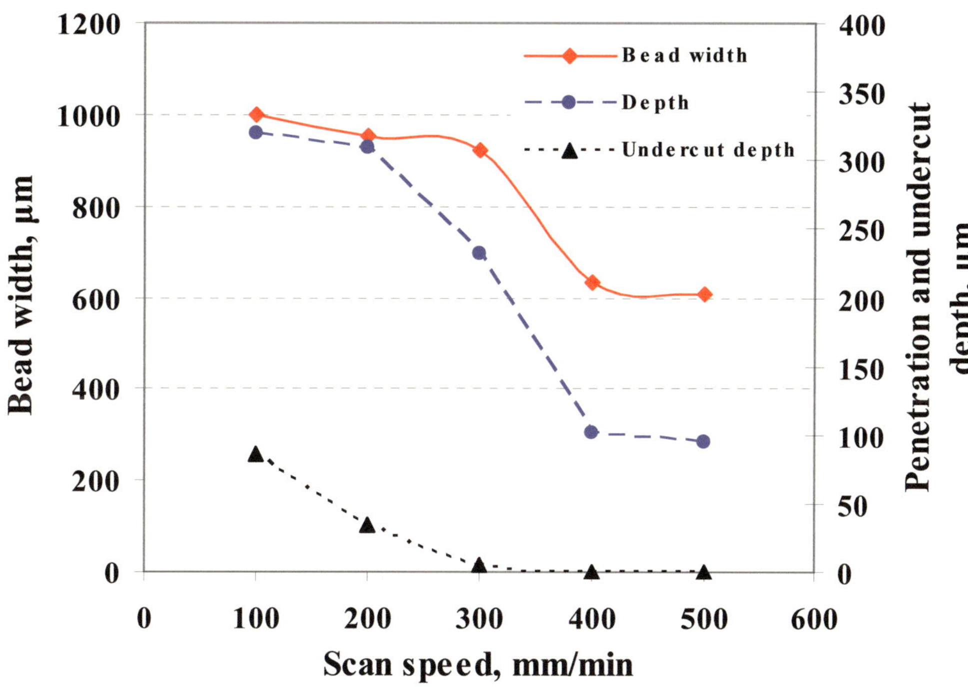

Fig. 4.5. Scan speed effect vs. width, penetration depth and undercut depth

deeper undercut depth. The same condition occurred when welding speed decreases, as shown in Fig. 4.5. When higher pulse energy or lower welding speed is used, the heat input per unit length will increase as equation 4.1.

$$E = \frac{E_p \times f}{v \times d_o} \tag{4.1}$$

Where, *Ep*=pulse energy (J), *f*=rep. rate (Hz), *v*=scan speed (m/min) and *do*=beam diameter. Higher heat input leads to more molten metal and large weld width. Higher heat input could evaporate more molten metal from the surface, so that undercut takes place.

4.2.4 Weld appearances and defects

Full penetration is more suitable for laser welding, but it was not achieved either by increase the pulse energy or low speed weld for this specimen thickness, 0.6 mm. When welding was carried out, scan speeds lower than 100 mm/min with pulse energy of 1.8 J, burn through took place especially at both edges of specimen and the weld appearance was very bad as shown in Fig. 4.6. The phenomenon was observed when higher pulse energy was over than 2.5 J at a welding speed 300 mm/min. The major concerns for welding magnesium alloys are the formation of cracks and pores. These defects could lead to low strength in welded products. As shown in table 4.1, a long crack was observed at scan speeds of 100 mm/min to 200 mm/min. At a scan speed of 50 mm/min, burn through occurred at both edges of the specimen together with a long crack presence at the middle of weld bead. Cracks appear at a scan speed 100 mm/min as shown in Fig. 4.7(a), where they it exist at the center of weld bead or in the fusion zone and relatively was long. The crack length reduces as scan speed increase to 200 mm/min, and then no macro cracks were observed when the scan speed was further increased from 300 mm/min to 500 mm/min. A lot of pores were observed in every scan speed when the weld was formed at constant pulse energy 1.8 J. Figure 4.8 shows the appearances of pores when weld at different scan speeds. The number of pores was reduced at 200 mm/min and 400 mm/min, where as many a lot of pores can be observed at a scan speed 300 mm/min. The remaining pores may be originated mainly from initial preexisting pores in the sheet. The growth of pores could result from the expansion and coalescence of the remaining pores, which produces larger pores in the weld bead. The pores formed during the first run of welding could be reduced by well controlled second run of welding (Zhao and Debroy 2001).

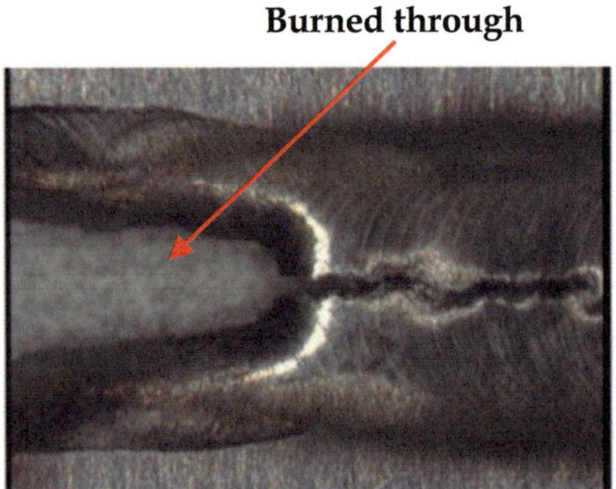

Fig. 4.6. Burned through and cracks at scan speed 50 mm/min

Scan speed (mm/min)	50	100	200	300	400	500	600
Weld condition	Burn through and long crack	Long crack	Cracks	Good	Good	Good	No bead

Table 4.1 Weld condition after welding with pulse energy 1.8 J at different scan speed

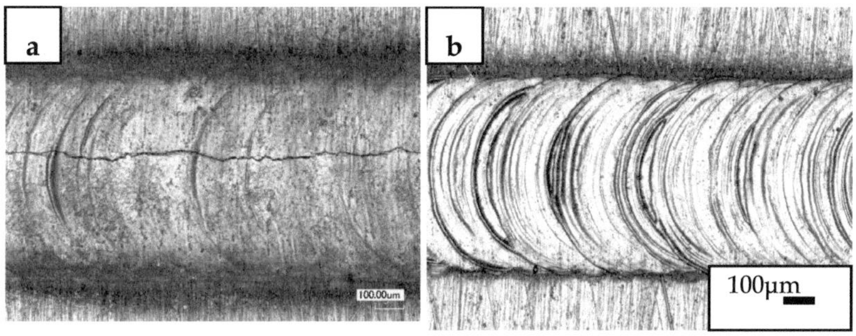

Fig. 4.7. Top view of bead scan speed a) 100 mm/min and b) 300 mm/min.

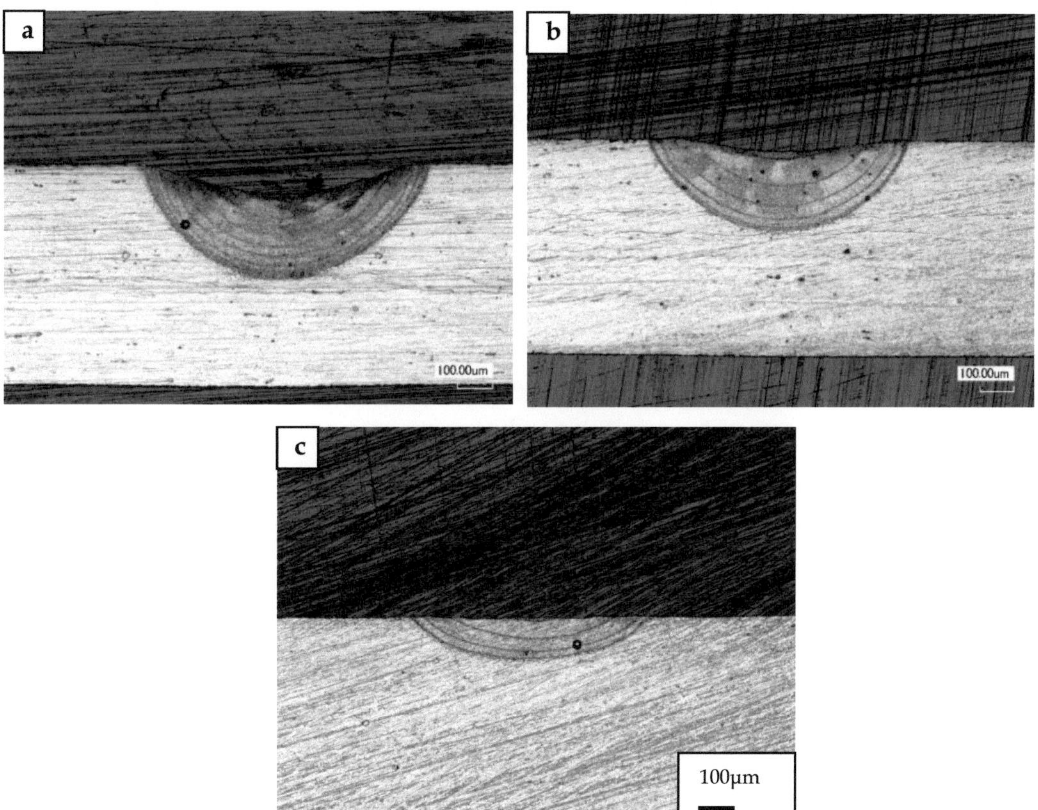

Fig. 4.8. Pores at scan speeds of a) 200 mm/min, b) 300 mm/min and c) 400 mm/min

5. Fillet welding of AZ31B magnesium alloy thin sheet

The welding of thin sheets is usually more problematic than welding thick sheet metal. Such problems are usually related to the high heat input of conventional arc welding processes. This high heat input leads to various problems such as cutting, burn through, distortion, porosity, cracking, etc. Thus, the selection of an appropriate welding process, procedure and technique are important in order to prevent these problems. Compared with arc welding, laser welding and electron beam welding are excellent methods that offer many advantages, such as narrow welds and impressive penetration depths. However, laser beam welding is the best choice because it can be used at ambient pressures and temperatures. However, the laser welding of thin sheet metals can still be problematic. Issues include the loss of material due to evaporation and inadequate control of heat, which leads to cutting and melt-through issues (Aghios, Bronfiu et al. 2001; Cao, Jahazi et al. 2006; G. Ben-Hamu 2007). There are plenty of reports on the welding of copper, stainless steel and aluminum alloys (Aghios, Bronfiu et al. 2001; Cao, Jahazi et al. 2006; G. Ben-Hamu 2007) in thin sheets of less than 1 mm thickness, but few studies have been published to date that focus on thin sheets of similar thickness in the case of magnesium alloys.

This chapter focuses on a Nd:YAG laser that was utilized to lap fillet weld AZ31B magnesium alloys thin sheet. The rationale behind the selection of AZ31B magnesium alloys focused on its unique properties. Magnesium alloys have a low density, high strength-to-weight ratio, high damping capacity and good recyclability; they have recently attracted more attention in many industries (Liming Liu and Changfu Dong 2006). However, processing magnesium alloys can be difficult due to defects such as cracks, pores and cuts (Ukita, Akamatsu et al. 1993; Leong, Sabo et al. 1999; Moon Jonghyun 2002; Toshikatsu Asahina 2005; Zhu, Li et al. 2005). This research focuses on the power density of the laser beam in conduction-mode welding applications, which results in a very stable weld pool. The absence of unstable fluid motion, which is usually present in keyhole mode welding, can lead to attractive weld qualities as well as good control of penetration depths. The advantages of conduction mode welding make it much more suitable for welding thin magnesium alloy sheets as thin as 0.3 mm.

5.1 Experimental

5.1.1 Material

Lap fillet welding by laser was performed on a 0.3 mm thickness of magnesium alloy sheet of AZ31B. The size of the sheet was 20 mm x 30 mm. Before welding, the sheets were polished with SiC paper and then cleaned with ethanol to remove oxide films, oil and dirt.

5.1.2 Nd:YAG laser

A pulsed Nd:YAG laser with a 1.06 μm wavelength, 0.4 mm beam spot diameter and 100 mm focal length lens were used for this experiment.

5.1.3 Experimental set-up

Figure 5.1 illustrates the set-up for the lap fillet welding experiment. Lap welding was performed by overlapping the two sheets, which were then joined by a laser at the edge side of the upper sheet. The distance between the clamps was 4 mm, and the edge of the upper sheet was equidistant from the two clamp arms. A specimen was fixed with jigs in a steel box and then located on a CNC X-Y table. The top of the box was partially closed by a heat resistant glass, through which the laser beam was transmitted during welding. Prior to welding, argon gas was pumped into the box at a flow rate of 20 L/min to prevent oxidation and continued flowing until the welding was finished.

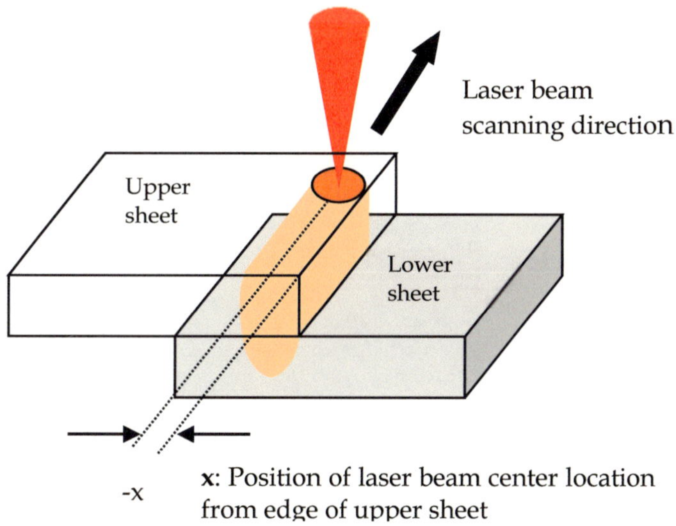

Fig. 5.1. Beam center location from edge of upper sheet (Ishak, Yamasaki et al. 2009)

5.1.4 Laser welding parameters

In this experiment, the two sheets were welded with different values of gap width from 0 to 100 μm, and the beam center located on the edge of the upper specimen (x=0). Next, the laser beam center location (x) was varied from the edge of the upper sheet towards the -x direction, as shown in Fig. 5.1: -0.1, 0, 0.1, 0.2 and 0.3 mm with varying scan speeds from 50 to 600 mm/min. The setting of the laser beam on a specific location on the upper sheet was determined by a camera using the precisely controllable CNC table to an accuracy of 1 μm.

As concluded from the bead on plate study described in section 4, defects such as undercuts, cracks and pores can be reduced by proper control of laser parameters such as the pulse energy and pulse duration. Therefore, the parameters for Nd:YAG laser welding such as pulse energy, pulse duration and repetition rate were fixed at 1.8 J, 3.0 ms and 80 Hz, respectively. The laser beam was scanned in the y direction during welding.

5.2 Results and discussions

5.2.1 Effect of gap size on weld geometry

The effect of the gap width between the two sheets on the weld appearance was investigated. Lap fillet welding was carried out at beam location x=0 with varying gap width. The laser parameters of pulse energy, pulse duration and scan speed remained constant at 1.8 J, 3.0 ms and 350 mm/min, respectively. The gap width and resulting welds are shown in Fig.5.2. An increase in the gap width to 35-55 μm generates a wide void, and further increase of the gap to 55-80 μm produces a wider and larger void. Gap widths of less than 35 μm significantly reduce the size and create a round shape void.

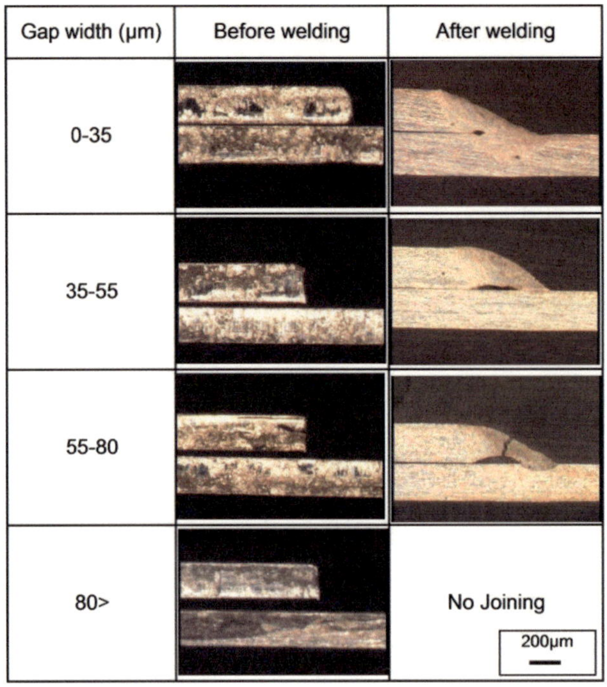

Fig. 5.2. Gap width effect on weld appearances at beam location x=0 (Ishak, Yamasaki et al. 2009)

Wider gaps over than 80 μm fail to create a joint. A gap wider than about 35-55 μm creates a long crack that starts at root of the two sheets. The crack then expands to the upper surface of the weld bead. Additionally, when the gap width exceeds 35 μm, the amount of molten metal is insufficient, and thus the bond width and penetration depth become small. These results suggest that the gap width is the most important factor in the occurrence of defects as well as in controlling the bond width and penetration depth of thin sheets. Therefore, to minimize the defects and optimize the bond width and penetration depth, the gap width must be restricted to less than 35 μm.

5.2.2 Effect of laser beam center position on weld geometry

To analyze the weld geometry of the sample, the average weld width and penetration depth were measured. The effects of the center beam location on the bond width and penetration depth are shown in Figs.5.3 and 5.4. The weld bond width and penetration depth decreased significantly with increasing distance of the beam center location from the edge of the upper specimen. The weld bond width and penetration depth decreased significantly with increasing distance of the beam center from the edge of the upper specimen. A larger distance results in a low heat concentration on the upper surface of the specimen. This occurs due to heat being dissipated to the surroundings because the magnesium alloy exhibits high thermal conductivity. Less heat is absorbed by the workpiece, resulting in a smaller melted weld area. An increased distance may also increase the size of voids at the root, as shown in Fig. 5.2. From this result, we conclude that the weld geometry in laser welding is very sensitive to beam center location distances on the order of 100 μm.

5.2.3 Microstructure

Figure 5.5 shows the weld microstructure obtained at x=0 with scan speeds of 250, 350 and 450 mm/min. Table 5.1 lists the mean grain size with different scan speeds and locations. Fine grains are apparent at the upper weld with an average grain size of 3.5 μm^2 ±2.3 for the 450 mm/min scenario. A mix of fine and medium-large grains with an average grain size of 17 μm^2 ±11 can be seen at the middle of weld bead, given a joint between the two sheets. These grain sizes are much smaller than those reported elsewhere, where grain sizes were as large as 20 μm in the fusion zone (Cao, M.Jahazi et al. 2006; Y.Quan 2008).

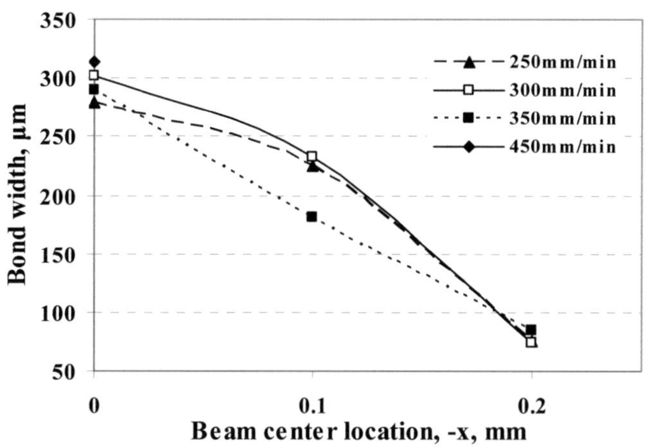

Fig. 5.3. Bond width at different beam location

Large elongated grains with an average grain size of 141 μm ±89 are apparent in region III, and non-uniform shapes with an average grain size of 13.5 μm ±9.0 can be seen in region IV. The microstructure at location IV is the same as that of the base metal. There are no significant changes in terms of grain size between scan speeds of 350 mm/min and 450 mm/min at each region.

With a scan speed of 250 mm/min, the average grain size in region I is about 27 μm ±18. An increase in the grain size is observed in region II with a grain size of about 53 μm ±31. However, we observed no changes in grain size in region III, while in region IV a larger grain size of about 26 μm ±12 was apparent at the bottom of the weld. The lengths of the long elongated grains decrease as scan speed increases, as listed in table 5.2.

We conclude that the grain size increases with increasing distance from the upper weld because of the changes in thermal cycling during welding. The grain size at the fused area

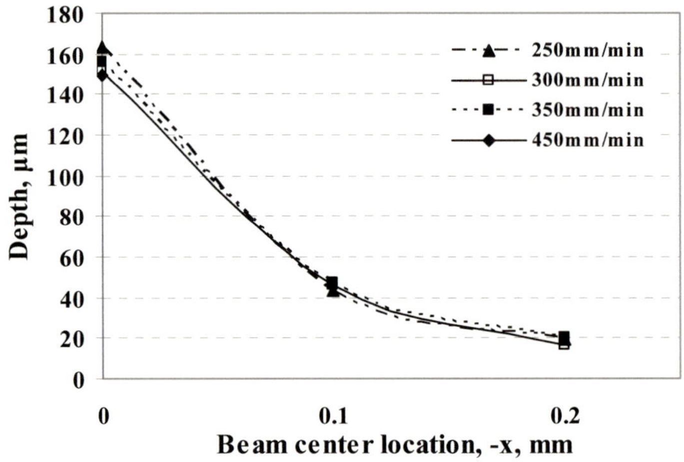

Fig. 5.4. Penetration depth at different beam location

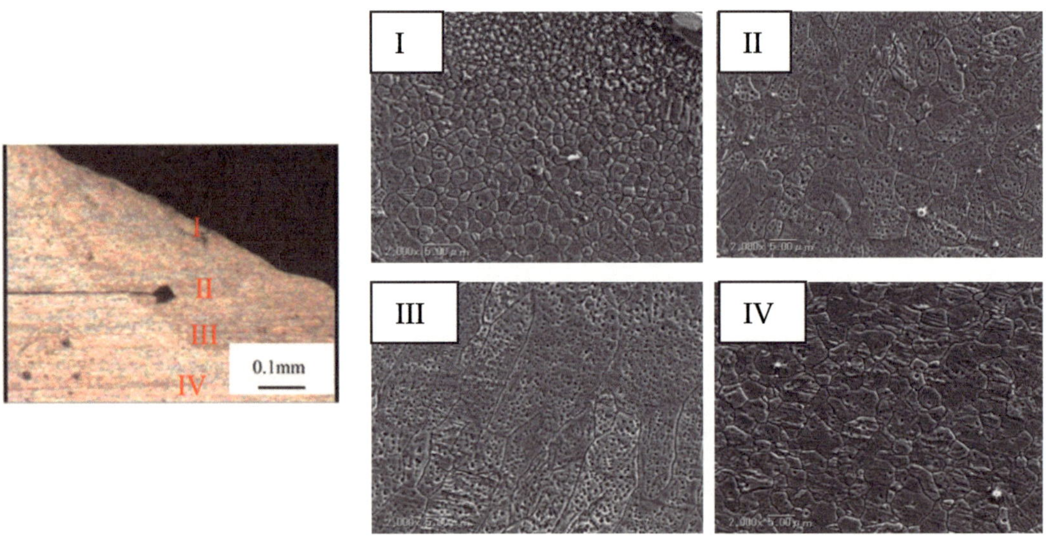

Fig. 5.5. Microstructure of welded joints at scan speed 450 mm/min at four different regions(Ishak, Yamasaki et al. 2009)

Mean grain size, µm				
Weld region	I	II	III	IV
250 mm/min	27± 18	53 ± 31	98 ± 57	26 ± 12
350 mm/min	4.0 ± 1.5	18 ± 15	160 ± 84	14 ± 10
450 mm/min	3.5 ± 2.0	17 ± 11	141 ± 89	13.5 ± 9.0

Table 5.1. Mean grain size at different scan speed and location

becomes larger at lower scan speeds. A low scan speed will result in impaired cooling. Moreover, since the cooling rate during solidification can be small, the time for grain coarsening will become longer. A large number of precipitated particles will also be apparent in the form of a very fine white deposit in the weld area. It has been reported elsewhere that the sole precipitation phase of AZ31 magnesium alloy is γ-$Mg_{17}Al_{12}$, and the white particles have been identified as $Mg_{17}(Al, Zn)_{12}$ (Quan, Chen et al. 2008).

Scan speed , mm/min	250	350	450
HAZ length	187 ± 14 (Area III and IV)	112 ± 8 (Area III)	83 ± 10 Area II

Table 5.2. HAZ length at different scan speed and location

5.2.4 Defects

5.2.4.1 Void at the root

Void at the root between the two sheets are apparent at almost all beam center locations and scan speeds. An increase in beam location distance makes the void size larger. Larger voids reduce the bond widths of the joints. By contrast, the size of voids can be reduced by selecting a high scan speed. The void size can be greatly reduced to around 0-40 μm at a beam location of x=0 and with a scan speed of 400-450 mm/min.

The void or discontinuity at the weld root probably occurred because of a lack of fusion (Yamauchi, Inaba et al. 1982; Nobuyuki, Masahiro et al. 2002; G.Rihar and M.Uran 2006; Z. Barsoum and A. Lundback 2009). This probably happens because of insufficient melting of the lower sheet, although the upper sheet can be heated at the welded root area as shown in Fig. 5.6. This insufficient melting was due to inadequate energy input and weld preparation throughout the welded root area (Yamauchi, Inaba et al. 1982; Nobuyuki, Masahiro et al. 2002; G.Rihar and M.Uran 2006; Z. Barsoum and A. Lundback 2009) . The narrow gap may also contribute to a lack of fusion. Furthermore, we believe that an oxide thin film probably formed on the surface of the lower sheet, although it had been polished using SiC paper of grit 1000 and cleaned using acetone. Magnesium is an active metal, and oxide can easily form on the surface when in contact with air after grinding (Z. Barsoum and A. Lundback 2009). This thin layer may have acted as a barrier, preventing heat transfer to the lower sheet.

5.2.4.2 Cracks

Cracks are apparent at lower speeds when x=0 and become smaller as the scan speed is increased. No visible cracks were observed at the beam center location x=0 with scan speeds of 400 to 450 mm/min. Furthermore, no macro cracks were observed on the surface of the upper sheet weld bead. The appearance of the cracks proved to be highly consistent across all specimens at low scan speeds. The crack profile indicates that cracking initiated from the root and then propagated into the weld metal but not to the center or upper surface of weld zone. From microstructure images, we conclude that the cracking occurred and propagated in the large grain area in a transgranular manner.

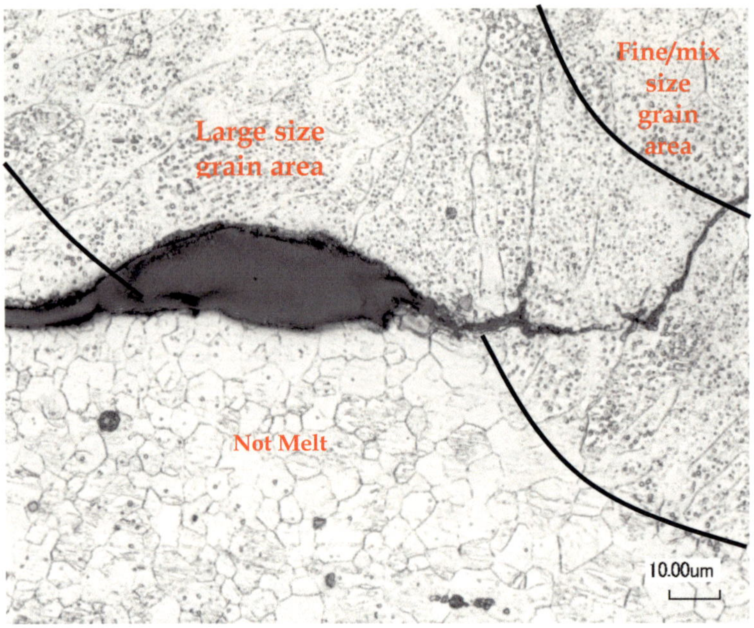

Fig. 5.6. Microstructure around void at root

(W.Zhou, Long et al. 2007)reported that a crack was observed during tungsten inert gas welding of AZ91D related to the liquation of the second phase or low melting point precipitates in the heat affected zone (HAZ) or partial melting zone (PMZ). This type of crack is known as "liquation cracking" due to the existence of a low melting point for intermetallic compounds at grain boundaries of the HAZ and PMZ. These effects can greatly decrease the strength of the weld (W.Zhou, Long et al. 2007). However based on EDS analysis there was no segregation of Zn, Al and Mn in that area could be detected. Secondary phases of low melting intermetallic compounds consisting of Mg and Al or Zn did not precipitate near the crack areas. No finer grains were apparent at the open crack, which was likely observed in reference (W.Zhou, Long et al. 2007). Furthermore, no significant hardness was detected in this area. This further shows that precipitation of second phase low melting intermetallic compounds did not take place in the crack region.

However, a peak of O was apparent near the crack's open surface. Although the surface of the thin sheet had been polished with SiC paper prior to welding, we believe that a thin oxide film probably still formed because magnesium is a very reactive metal. The crack probably initiated from the void at the root due to high stress concentrations, then propagated to the large elongated grain area along the thin oxide film.

During low scan speed welding, a wider area of large grains probably formed at the boundary of the fusion zone and base metal. This area was exposed to much lower temperature than the middle of the fusion zone; thus, oxide films were not properly broken or melted during welding. Furthermore, a low scan speed may result in high heat input, which consequently leads to higher stresses in the large grain area as the metal solidifies and contracts. Therefore, cracks could easily propagate throughout the oxide film area.

Higher scan speeds can reduce stress as well as leading to the formation of a wider fine equiaxed grain area. Subsequently, the large grain areas become much smaller, as shown in table 5.2. These fine equiaxed grains are less susceptible to cracking than larger ones because the stress is more evenly distributed among numerous grain boundaries.

5.3 Numerical simulation of lap fillet welding

This section present the simulation model of laser lap welding process utilize the efficient solution procedure of the finite element code ANSYS. The physical dimensions of the bond width, penetration depth and HAZ width were measured and compared with the prediction model.

5.3.1 Geometrical model description

A moving, pulsed laser beam vertically radiated on the edge of upper sheet. The laser beam center located on edge of upper specimen and laser were scanned towards y direction. The pulse energy is sufficient to form a joint on the specimens. The dimensions of the thin sheets and heat source are shown in Fig. 5.7. The thin sheets with thickness of 0.3 mm is overlapped each other.

The following assumptions were made in the formulation of the finite element model:

1. The laser beam is incident vertically towards the surface of the workpiece. The heat or laser beam center is located at edge of upper specimen.
2. The laser beam spot diameter on surface of specimen is 0.4 mm. Half of the spot beam hit on upper sheet and other half on lower sheet.
3. The workpiece is initially at 20 °C. Both the laser beam and the coordinate mesh are fixed and the work pieces moves in the positive y-direction with a constant velocity, v.
4. All thermophysical are considered to be constant.
5. The absorption of laser is fixed at 18 %. This based on the experimental results which are around 18-20 %.

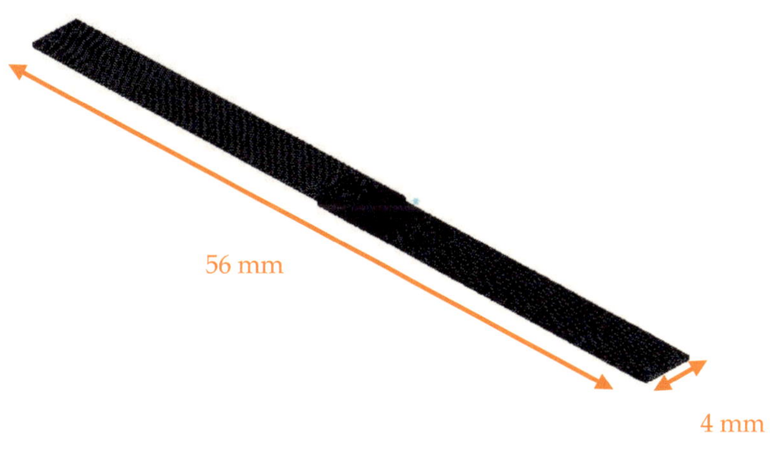

Fig. 5.7. Schematic drawing 3D geometrical model

6. The gap between the thin sheets is 30 μm and is divided into two layers which thickness of 15 μm each.
7. As pulsed laser welding involves very rapid melting and solidification, convective redistribution of heat within the weld pool is not a significant as it is in other processes where liquid pool is permanent. Convective flow of heat, therefore, is neglected.

The mesh structure of model is shown in Fig. 5.8. The mesh size is divided into 5 different zones. As the distance of specimens far away from weld area, the mesh becomes rougher. The finest mesh is defined at weld area where 84 mesh along the 4 mm width of specimen. Free mesh is defined at zone 1, while mapped mesh at zone 2, 3, 4 and 5. The thickness of mesh at zone 2 to zone 4 consists of 7 layers of one thin sheet, whereas, zone 1 only 1 layer. In this simulation, the pulse energy is fixed at 1.8 J, pulse duration 3.0 ms, repetition rate at 80 Hz. The simulation is done at three different scan speeds 250 mm/min, 350 mm/min and 450 mm/min. The simulation time was 0.53 s, 0.375 s and 0.2975 s for scan speed 250 mm/min, 350 mm/min and 450 mm/min, respectively.

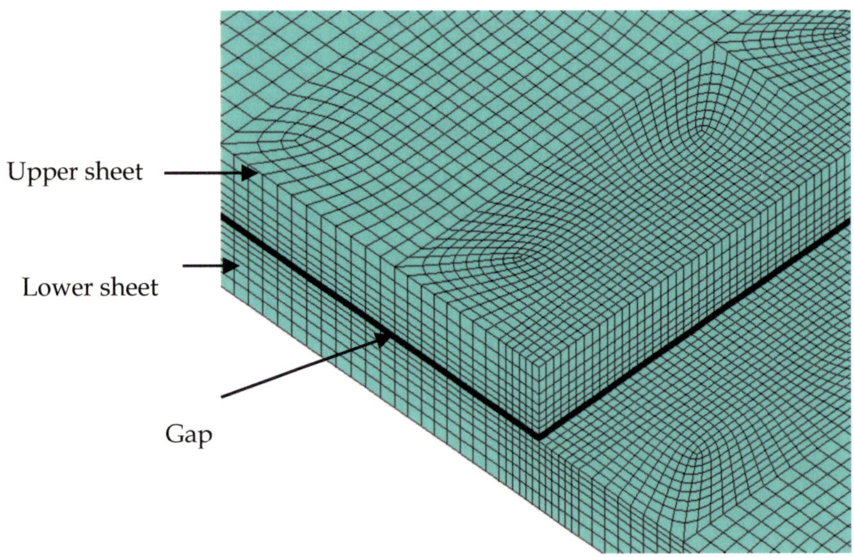

Fig. 5.8. Mesh structure at weld area

5.3.2 Numerical description

The temperature field caused by moving laser beam is transient in nature. Here, in order to simulate the process, a three-dimensional transient heat transfer model was constructed. The general three dimensional heat diffusion equations is of the form

$$\frac{\partial}{\partial x}\left(k_x\frac{\partial T}{\partial x}\right)+\frac{\partial}{\partial y}\left(k_y\frac{\partial T}{\partial y}\right)+\frac{\partial}{\partial z}\left(k_z\frac{\partial T}{\partial z}\right)+Q=\rho c\left(\partial\frac{\partial T}{\partial t}-v\frac{\partial T}{\partial x}\right) \tag{5.1}$$

where,

(x,y,z) = coordinate system attached to the heat source

Q = power generation per unit volume (W/m^3)

k_x, k_y, k_z = thermal conductivity in the x, y, and z directions (W/mK)
c = specific heat capacity (J/kgK)
ρ = density (kg/m³)
t = time (s)
v = velocity of workpiece (m/s)

Laser heat input was on the surface as heat flux. Heat losses due to convection and radiation are very small (Yeung and P.H 1999) and hence also there were not considered in this analysis, as they make calculation non-linear, thus causing convergence problems and increasing computing time by a great amount.

The finite element code ANSYS provides the convenient means of numerically modeling pulsed laser welding. The solution technique is chosen depending on the type of problem. In the present case, the thermal history of lap fillet weld is required, so a transient thermal analysis must be performed. This requires an integration of the heat conduction equation with respect to time. In the finite element formulation, this equation can be written for each element as follow (Frewin and Scott 1999);

$$[C(T)\{\dot{T}\} + [K(T)\{T\} + \{V\} = \{Q(T)\} \quad (5.2)$$

where

$[K]$ = conductivity matrix
$[C]$ = specific heat matrix
{T} = vector or nodal temperature
$\{\dot{T}\}$ = vector of time derivative of {T}
$\{V\}$ = velocity vector for the moving workpiece
{Q} = nodal heat flow vector

This equation is simply the vector and matrix equivalent of Equation 4.4.

5.3.3 Numerical result and discussion

The predicted temperature distributions from numerical analysis at different scan speeds are shown in through Fig. 5.9. It is noted that the temperature distribution has a parabolic form, which is the consequence of the Gaussian energy profile. The maximum temperature is recorded at the central axis of laser beam. It is explained by the fact that the heat source is applied on the thin sheet surface at the edge of upper sheets.

The maximum temperature decreases as scan speed increases as a result of lower heat input. At scan speed of 250 mm/min the melt area are larger than scan speeds of 350 mm/min and 450 mm/min. It is also noted that the area of temperature between 500 °C until 575 °C becomes smaller as scan speed increases. This result because of heat input decreases as scan speed increases. From the results, it shows that the weld area slightly larger at starting (starting calculation time) and end point (end calculation time) of specimen compare with at the middle of the specimen. It shows similar tendency with the experimental results where wider weld zone and deeper penetration obtained at edge compare with at the middle of the specimen. This occurs because of lower heat concentration at middle than at the edge of specimen. It is due to heat easily dissipated away as a result of high conductivity at the middle of specimen.

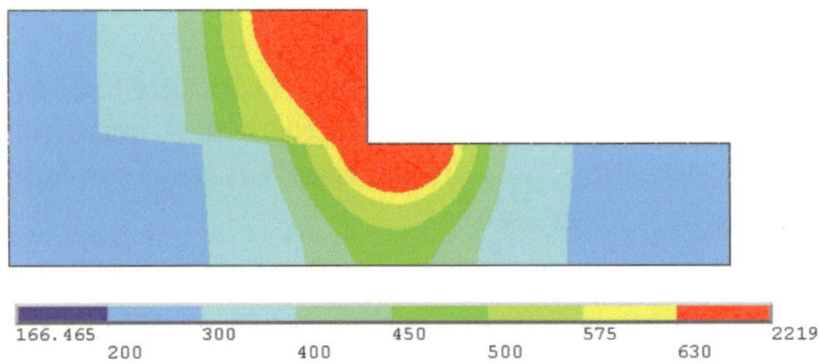

Fig. 5.9. Result of numerical simulation at scan speed 450 mm/min at t=0.2975s

Fig. 5.10 shows the comparison between the numerical simulation result and cross section of experimental results. It shows that at higher scan speeds, the weld geometry between simulation and the experimental results are almost similar. Fig. 5.11 shows quantitative comparison between numerical and experimental results of fusion zone width, penetration depth and HAZ width at weld root. The fusion zone area is the area recorded temperature from melting point. Meanwhile, the HAZ width is measured at weld root with temperature between 500° C until 575 °C. The range of HAZ temperature is based on grain growth studies of AZ31B (Mehtedi, Balloni et al. ; Ben-Artzy, Shtechman et al. 2000; Maryaa, Hector et al. 2006; Sun, Wu et al. 2009). The results of comparison between numerical calculation and experimental results show a reasonable agreement. At lowest scan speed of 250 mm/min, numerical result shows larger fusion width compare with experimental result. The different occurs because of larger defects (void and crack) at the weld root, and occurrence of severe undercut which result in smaller weld width of experimental result. The defects improve as scan speed increases. As a result, good agreement can be obtained at scan speed of 350 mm/min and 450 mm/min.

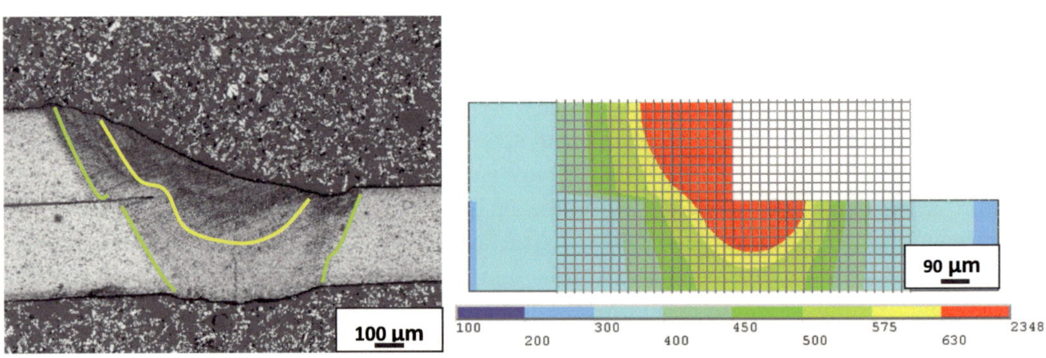

Fig. 5.10. Comparison between experimental and numerical result at scan speed 250 mm/min

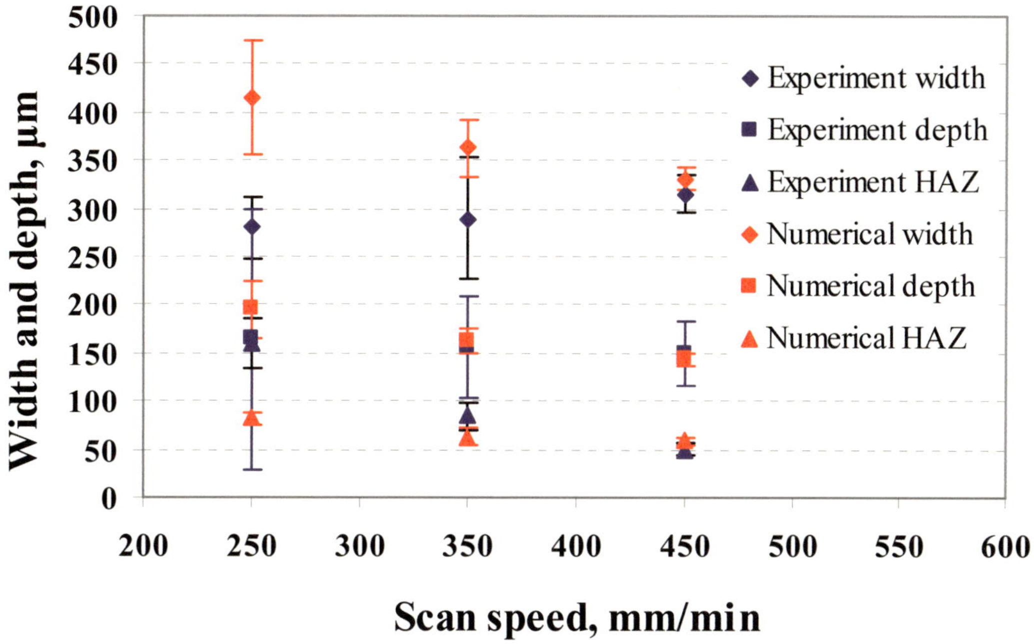

Fig. 5.11. Comparison of weld geometry and HAZ width between experimental and numerical results at different scan speeds

6. Conclusion

We investigated the influences of welding parameters, such as beam center location, laser scan speed and gap size, on weldability in the context of lap fillet welding of AZ31B magnesium alloys with a thickness of 0.3 mm with a pulsed Nd:YAG laser. The following conclusions can be drawn from this work:

For bead on plate (BOP) welding undercut, cracks and pores can be greatly reduced when the process parameters were properly chosen to weld bead on the AZ31B magnesium alloy sheet in the Nd:YAG laser welding system. The depth of penetration decreases as focus point located on surface of thin sheet magnesium alloys. The defects such as undercut, cracks and pores are significantly reduced as pulse energy at 1.8 J, repetition rate at 80 Hz, pulse duration 3.0 ms and scan speed range from 300 to 500 mm/min.

The gap width exerts a major influence on the occurrence of joints and defects, such as cracks and voids, between the two sheets. A gap width of less than 35 μm significantly reduces the number of cracks and voids and also increases the penetration depth and bond width.

Voids and discontinuities at the root occurred because of lack of fusion. A narrow gap oxide layer on the lower surface acted as a heat barrier to prevent melting of the lower sheet. High scan speeds significantly improved the number of voids and discontinuities at the root. Cracks at the weld originated from voids at the root, which propagated in a transgranular manner within the large grain area. No segregation of Al, Mn and Zn occurred at the crack opening, whereas a high content of oxide may have caused the crack/discontinuity. Higher scan speeds significantly reduced the number of defects because of the narrower large grain area and the wider fine grain area.

Lap fillet welding of the thin sheet produced a weld bead width of less than 1 mm, with no melt-through and good weld bead appearance, when the beam center location was on the edge of the upper specimen and when we used scan speeds of 350 to 450 mm/min. Under the optimized conditions, a large fine grain size area can be obtained from the top to the middle of the weld bead, and the HAZ becomes very narrow.

The weld geometry (weld width, depth and HAZ width) of numerical simulation results show reasonable agreement with experimental results at high scan speed of 350 mm/min and 450 mm/min.

7. References

Aghios, E., B. Bronfiu, et al. (2001). "The role of the magnesium industry in protecting the environment." Journal of Materials Processing Technology 117: 381-385.

Ben-Artzy, A., A. Shtechman, et al. (2000). "Deformation characteristics of wrought magnesium alloys AZ31, ZK60." Magnesium Technology: 363-374.

Cao, X., M. Jahazi, et al. (2006). " A review of laser welding techniques for magnesium alloys." Journal of Materials Processing Technology 171: 188-204.

Cao, X., M.Jahazi, et al. (2006). "A review of laser welding techniques for magnesium alloys." Journals of Materials Processing Technology 171: 188-204.

Costaa, A. P., L. s. Quintinoa, et al. (2003). " Laser beam welding hard metals to steel." Journal of Materials Processing Technology 141: 163–173.

Czerwinski, F. (2002). "The oxidation behavior of an AZ91D magnesium alloy at high temperatures." Acta Materialia 50: 2639–2654.

Duley, W. W. (1998). Laser welding, A Wiley-Interscience publication.

Duley, W. W. (1999). Laser Welding, John Wiley & Sons, Inc.

Frewin, M. R. and D. A. Scott (1999). "Finite element model of pulsed laser welding." Welding research supplement: 15s-21s.

G. Ben-Hamu, D. E., and C. E Cross, Th. Bollinghaus, , . (2007). "The relation between microstructure and corrosion behavior of GTA welded AZ31B magnesium sheet." Materials Science and Engineering A 452-453: 210-218.

G.Rihar and M.Uran (2006). "Lack of fusion-Characterisation of indications." Welding in the world issue ½ 50(1/2): 35-39.

Haferkamp, H., A. Ostendorf, et al. (2003). Laser beam welding of magnesium alloys a process at the threshold to the industrial manufacturing. International Conference on Leading Edge Manufacturing in 21st Century: 891-896.

Ion, J. C. (2005). Laser Processing of Engineering Materials, Elsevier.

Ishak, M. (2010). Study on Lap Fillet Welding of AZ31B Magnesium Thin Sheets using Pulsed Nd:YAG Laser. Graduate School of Science and Engineering, Ibaraki University, Japan. Phd thesis.

Ishak, M., K. Yamasaki, et al. (2009). "Lap fillet welding of thin sheet AZ31magnesium alloy with pulsed Nd:YAG laser." Journal of Solid Mechanics and Materials Engineering 3(9): 1045-1056.

Leong, K. H., K. R. Sabo, et al. (1999). "Laser beam welding of 5182 Aluminum alloy sheet." J. Laser Applications 11(3): 109-118.

Liming Liu and Changfu Dong (2006). "Gas tungsten-arc filler welding of AZ31 magnesium alloy." Materials letters 60: 2194-2197.

Maryaa, M., L. G. Hector, et al. (2006). "Microstructural effects of AZ31 magnesium alloy on its tensile deformation and failure behaviors." Materials Science and Engineering, A418: 341-356.

Mehtedi, M. E., L. Balloni, et al. "Comparative study of high temperature workability of ZM21 and AZ31 magnesium alloys." Metallurgical science and technology: 23-30.

Moon Jonghyun, K. S., Mizutani Masami, and Matsunawa Akira (2002). "Lap welding characteristics of thin sheet metals with combined laser beams of different length (In Japanese)." Japan welding society 20: 468-476.

Nobuyuki, A., T. Masahiro, et al. (2002). "Welding of Aluminum Alloy with High Power Direct Diode Laser, ." Trans. JWRI 31(2): 157-163.

Quan, Y. J., Z. H. Chen, et al. (2008). "Effects of heat input on microstructure and tensile properties of laser welded magnesium alloy AZ31." Mater. Charact. 59(10): 1491-1497.

Sun, P.-h., H.-y. Wu, et al. (2009). Deformation Characteristics of Fine-grained Magnesium Alloy AZ31B Thin Sheet during Fast Gas Blow Forming,. Proceedings of the International MultiConference of Engineers and Computer Scientists 2009 (IMECS 2009).

Toshikatsu Asahina, H. T., Hitaka Itoh and Shigekazu Taguchi (2005). Some characteristics of pulsed YAG laser welds of Magnesium Alloys. Nihon University research report 38, Nihon University.

Ukita, S., T. Akamatsu, et al. (1993). "The welding conditions of very thin aluminum sheet of high welding." Japan welding society 11: 361-364.

W.Zhou, T. Z. Long, et al. (2007). "Hot cracking in tungsten inert gas welding of magnesium alloy AZ91D." Mat. Sci. and Tech. 23 (11): 1294-1299.

Y.Quan, S. C., Z. Yu, X.S. Gong, M.Li, (2008). "Characteristics of laser welded wrought Mg-Al-Mn alloy." Materials Characterisation 59: 1799-1804.

Yamauchi, N., Y. Inaba, et al. (1982). "Formation mechanism of lack of fusion in MAG welding." Japan welding society (in Japanese) 51(10): 843-849.

Yeung, K. S. and T. P.H (1999). "Transient thermal analysis of spot welding electrodes" Welding Journal 78(1): 1s-6s.

Z. Barsoum and A. Lundback (2009). "Simplified FE welding simulation of fillet welds-3D effects on the formation residual stresses." Eng. Fail. Anal. 16(7): 2281-2289.

Zhao, H. and T. Debroy (2001). "Pore Formation During Laser Beam Welding Of Die-Cast Magnesium Alloy AM60B —Mechanism And Remedy." Weld. Research Supp: 204s-210s.

Zhu, J., L. Li, et al. (2005). "CO2 and diode laser welding of AZ31 magnesium alloy." Applied Surface Science 247 pp. 300-306.

9

Micro-Welding of Super Thermal Conductive Composite by Pulsed Nd:YAG Laser

Mohd Idris Shah Ismail[1,2], Yasuhiro Okamoto[1] and Akira Okada[1]
[1]*Graduate School of Natural Science and Technology, Okayama University*
[2]*Department of Mechanical & Manufacturing Engineering, Universiti Putra Malaysia*
[1]*Japan*
[2]*Malaysia*

1. Introduction

The diffusion of generated heat in the electronic devices is an important issue. The heat would be diffused from electronic devices by passive strategies, which would be carried out by the use of high thermal conductivity materials as a heat sink. The development of advanced materials with the superior high-thermal properties and the high strength-to-weight ratio has led to new metal matrix composites (MMCs) as a great attractive material in the electrical and electronic industries. Aluminum and its alloys are widely used for the manufacturing of MMCs, which have reached the industrial stage in some areas (Barekar et al., 2009). In order to manufacture practical components from MMCs, a technique for joining MMCs to other similar composites or to monolithic materials is strongly required. Therefore, the development of reliable and economic joining technique is important for extending the applications of MMCs. It is well-known that laser welding is the most flexible and versatile welding technology, and it has succeeded in the welding of MMCs (Niu et al., 2006; Bassani et al., 2007).

Recently, a super thermal conductive (STC) aluminum-graphite (Al-Gr) composite with the high thermal conductivity and the low coefficient of thermal expansion (CTE) was developed (Ueno et al., 2009). The properties of thermal conductivity versus thermal expansion coefficient are summarized in Figure 1, where the upper ellipse zone shows the STC composite materials, and the other of conventional thermal conductive materials such as Cu, Al, Al-SiC, Cu-W, Cu-Mo, AlN and Si are also indicated. It is difficult to weld STC Al-Gr composite material compared with existing MMCs, since the graphite material only can be melted under the high pressure with the high temperature (Kirillin & Kostanovskii, 2003). In the laser welding, the use of standard pulse profile is limited to joint this material, since the uncontrolled heat input generates an overshoot, which leads to undesirable welded joints. The control of heat input by a pulse waveform is very important to achieve a suitable welding penetration, preventions of overheating and unacceptable welding defects.

In this study, the welding of STC Al-Gr composite was experimentally and numerically investigated by using a pulsed Nd:YAG laser with the control of pulse waveforms, which can provide a well-controlled heat input with high energy density. These investigations have led to an optimum welding condition proposed for a pulsed laser welding with minimum weld defects. The experimental work was carried out in two sections, namely the bead-on-plate welding of STC Al-Gr composite, and the overlap welding of pure aluminum and STC Al-Gr composite. The welding characteristics of STC Al-Gr composite were discussed by the observation of joint part with optical microscope, scanning electron microscope (SEM) and energy dispersive spectroscopy (EDS). Moreover, the temperature distributions in the laser micro-welding process were numerically analyzed to discuss the proper heat input. The weld strength was also evaluated by a shearing test for the overlap welding with and without the control of pulse waveform.

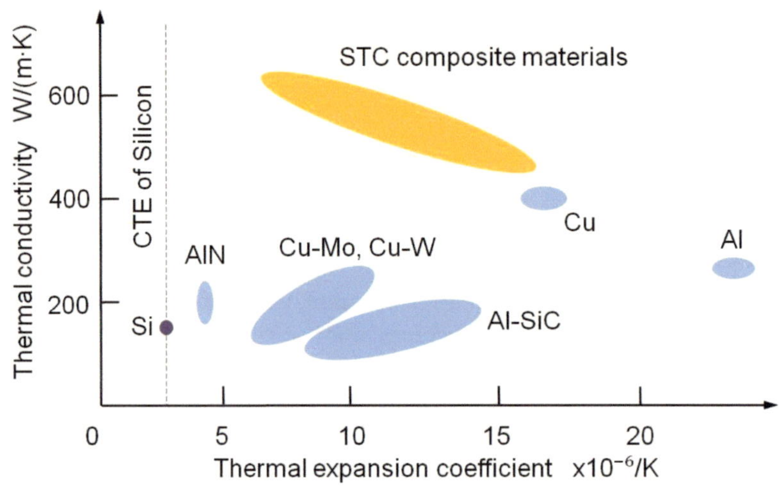

Fig. 1. Thermal conductivity and thermal expansion coefficient of STC composites and other materials

2. Background and overview

Each material has its own unique properties. In many applications, it has been found that there is a need for the combination of several properties together. These desired material property combinations can be achieved by the development of new composite materials. Composite materials are usually classified by the type of material used for the matrix. The four primary categories of composites are polymer matrix composites (PMCs), metal matrix composites (MMCs), ceramic matrix composites (CMCs), and carbon-carbon composites (CCCs) (Kutz, 2002). In this study, the viewpoint was MMCs, which having the properties of lightweight, high specific strength, wear resistance, and a low coefficient of thermal expansion.

Aluminum has many properties such as high thermal and electrical conductivity, high corrosion resistance, low cost and lightweight that make it ideal for use in MMCs. In addition, aluminum can be recycled and offers intriguing environmental and economical opportunities (Cannillo et al., 2010). From the chemical point of view, the aluminum reacts with carbon to form only aluminum carbide (Al_4C_3) intermetallic compounds as shown in the Figure 2. The Al_4C_3 is known as the only stable intermediate in the Al-C binary system. However, it is very brittle at the ambient temperature (Lu et al., 2001). Based on the Al-C phase diagram (Massalski et al., 1990), it can be seen that the coexistence of liquid aluminum and solid carbon requires the existence of an intermediate phase Al_4C_3. At equilibrium phase, it is impossible to directly dissolve solid carbon into liquid aluminum unless the temperature is above the melting point of Al_4C_3 (Lu et al., 2001). Graphite, in the form of fibers or particulates has been recognized as a high-strength and low density material (Barekar et al., 2009). The material combination of aluminum and graphite could produce an advanced aluminum-graphite (Al-Gr) composite, which has an unique thermal properties, due to to the opposite thermal expansion coefficients of aluminum and graphite.

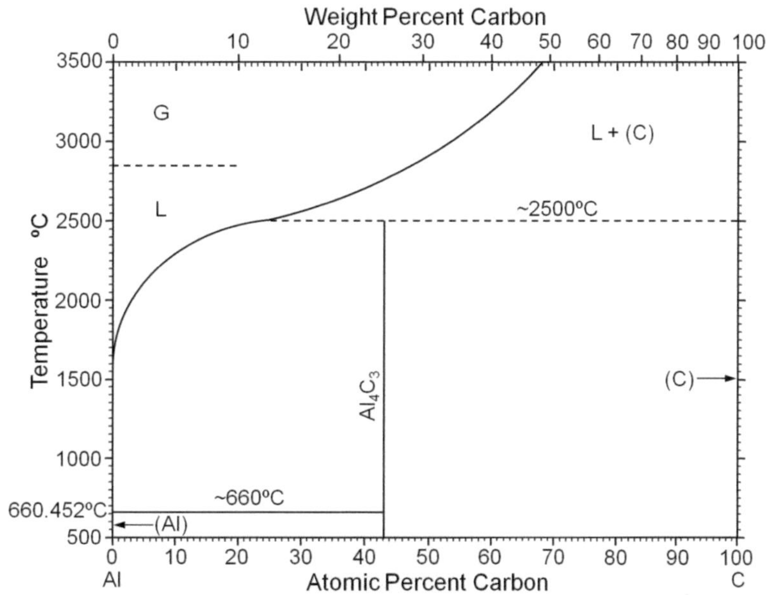

Fig. 2. Al-C phase diagram

Recently, a super thermal conductive(STC) Al-Gr composite was developed, which have been prepared by utilizing a pulsed electric current sintering method (Ueno et al., 2009) to realize the properties of higher thermal conductivity and lower thermal expansion coefficient. The material properties of STC Al-Gr composite are shown in Table 1. It shows that the thermal conductivity, thermal expansion coefficient and bending strength are anisotropic due to the directional properties of graphite and their designed orientation in xy plane-direction (top view) and xz thickness-direction (section view) as shown in Figure 3.

The demand for lighter and thinner products has led MMCs to be manufactured, especially in welding process. The welding process of MMCs has attracted various industries by the reduction of material weight and costs, and improvement of design flexibility. Therefore, the development of efficient, flexible and reliable joining technique is significantly important to increase the potential of MMCs. Compared with the conventional fusion welding techniques, laser welding has numerous advantages such as narrow heat-affected zone (HAZ), good quality of weld bead, precise positioning and control of irradiation beam and its movement, the ease of automation and the high production speed (Behler et al., 1997; Pan et al., 2004). Laser welding is characterized by intermittent laser beam powers that would allow melting and solidification consecutively. The great advantage of laser welding is keyhole effect by very high power densities involved in the laser welding, which makes laser welding technique widely used in many industrial productions.

Material properties		Value
Thermal conductivity	k_{xx} , k_{yy}	450 W/(m·K)
	k_{zz}	40 W/(m·K)
Specific heat	c	810 J/(kg·K)
Density	ρ	2450 kg/m³
Thermal expansion coefficient	α_{xx} , α_{yy}	17×10^{-6}/K
	α_{zz}	8×10^{-6}/K
Bending strength	M_{xx}	65 MPa
	M_{zz}	7 MPa
Tensile strength	σ	49 MPa
Young modulus	G	31 GPa
Electrical resistivity	R	17×10^{-8} Ωm

Table 1. Material properties of STC Al-Gr composite

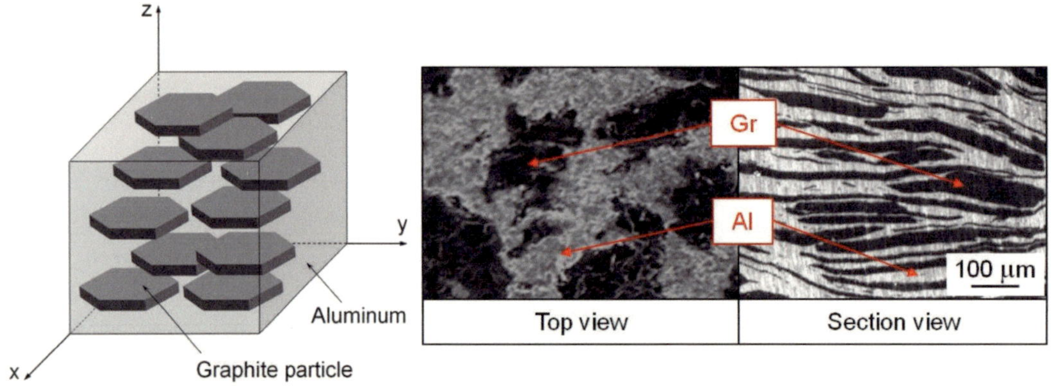

Fig. 3. Schematic illustration and optical micrographs of STC Al-Gr composite

Since the STC Al-Gr composite is a new composite in MMCs, no work has been reported concerning the welding behaviour on this composite by laser welding technique. There are several published papers related to the laser welding of MMCs, especially in the composites with combination of Al and SiC. Niu et al. (2006) reported that the welding of Al-SiC can be successfully realized by using laser welding. It is possible to control the geometry of weld bead by precisely controlling the laser output parameters (Yue et al., 1996). However, the reaction between SiC and Al in the weld zone generates a formation of needle-like brittle Al_4C_3, which can degrade the mechanical properties of Al-SiC composite. It dissolves in aqueous environments resulting in a loss of integrity to the weld zone (Lienert et al., 1993).

In addition, the Al_4C_3 formation reduces the toughness of the weld bead, even it results in the increase of hardness (Bassani et al., 2007). Therefore, preventing the formation of the Al_4C_3 during welding process is important for successful welding of Al-SiC composites (Wang et al., 2000).

There are severe heterogeneity of the material structure and the great difference in both physical and chemical properties between the aluminum matrix and graphite particles. This point would make the welding of STC Al-Gr composites to a challenging technological problem for applying the laser welding technique in the industrial application. In this study, the pulsed Nd:YAG laser was used as a laser source. Square shape pulses are the standard output of this laser with the constant power distribution during its duration time. By utilizing the ability of pulsed Nd:YAG laser to shape the temporal power profile of each pulse offers the high flexibility in optimising the weld parameters. This pulse waveform allows the control of penetration, melt pool geometry and keyhole formation. It also has been reported that pulse waveform to be an effective method to reduce or eliminate the weld defects (Zhang et al., 2008).

3. Experimental procedures

A schematic diagram of experimental setup is shown in Figure 4. In this study, a pulsed Nd:YAG laser (LASAG SLS200 CL8) of 1064 nm in wavelength was used as a laser source. The laser beam was collimated to 15 mm in diameter and delivered by an optical fiber of 50 μm core diameter. The collimator was installed between the optical fiber and the bending mirror, and the collimated laser beam was focused on the specimen surface by a lens of 50 mm in focal length. In order to avoid the back-reflection of incident laser beam, the processing head was aligned by 10 degrees to the perpendicular axis of the specimen surface. The irradiation experiments were carried out in two sections, namely the bead-on-plate welding of STC Al-Gr composite, and the overlap welding of pure aluminum and STC Al-Gr composite. The welding experiments were done in a shielding gas of nitrogen with 13L/min flow rate. The stage controller could determine the movement of X-Y-Z stage, and it also synchronized the laser pulse.

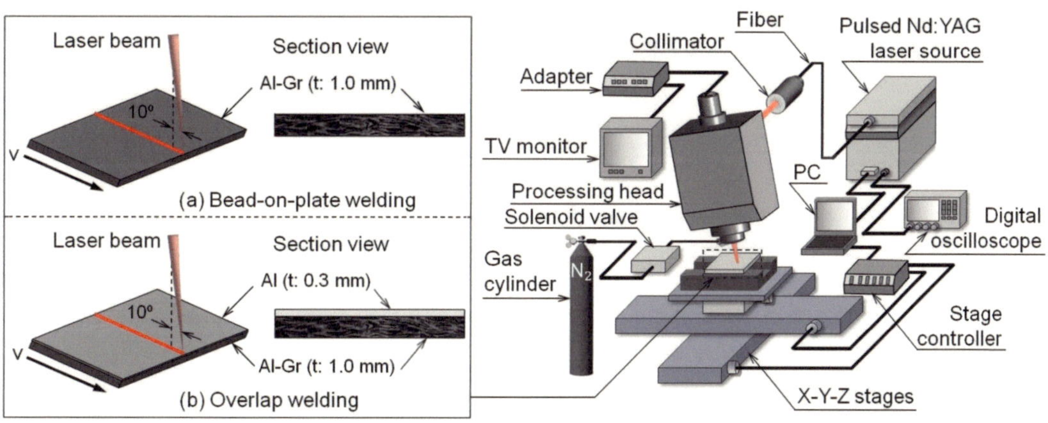

Fig. 4. Schematic diagram of experimental setup

After the laser welding, the sectioned surface of welded specimen was ground, polished and etched for the observation of weld bead by an optical microscope, scanning electron microscope (SEM) and energy dispersive spectroscopy (EDS). In addition, the shearing test was carried out to measure the shear strength of the overlap welded joints with a Shimadzu EZ-L test machine. The specimen for shear strength of overlap welding was designed with five seam lines and 1.5 mm distance between each line as shown in Figure 5. The cross-head speed was set to 0.5 mm/min. The specimen was gripped by the clampers, which are placed in the fixture blocks. Then, a shear load was slowly increased at the suitable increments by the mechanical lever system until the welded joint of specimen was fractured. The shear strength was calculated by using fracture load and welding area. The value of shear strength was the average of three specimens. Then, the fracture surfaces were examined with SEM and EDS analysis.

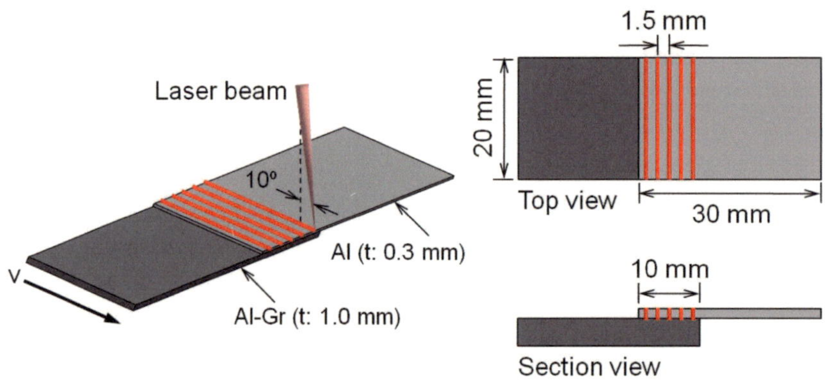

Fig. 5. Configuration of shearing test specimen

4. Thermal analysis by finite element method

In this study, the further analysis of the welding phenomenon was discussed by the heat conduction analysis with finite element method (FEM) to simulate the thermal process of laser welding. The analysis model is based on the fundamental heat transfer for the laser welding process. Based on the first law of thermodynamics, the heat flow in a three-dimensional solid can be expressed as equation (1) (Kannatey-Asibu, 2009)

$$\rho \cdot c(\theta)\frac{\delta\theta}{\delta t} = \frac{\delta}{\delta x}\left[k(\theta)\frac{\delta\theta}{\delta x}\right] + \frac{\delta}{\delta y}\left[k(\theta)\frac{\delta\theta}{\delta y}\right] + \frac{\delta}{\delta z}\left[k(\theta)\frac{\delta\theta}{\delta z}\right] + Q_v \quad (1)$$

where ρ, $c(\theta)$, $k(\theta)$ and Q_v are the material density, the temperature dependent specific heat, the temperature dependent thermal conductivity and the volumetric heat source term which varies with a laser power. $\theta=\theta(x,y,z,t)$ is the resulting three-dimensional and temporal temperature distribution in the material. t is time, and x, y, z are the spatial Cartesian coordinates.

Figure 6 shows the developed finite element models for bead-on-plate welding and overlap welding. Fine mesh resolution is given at and near the heat source, while a fairly coarse mesh density is considered at the region far from the heat source. A portion of the specimens was designed in the analysis model in order to reduce the calculation time. In addition, the half model with symmetric conditions is used in the analysis. As the heat source of laser beam is symmetric in the x-z plane, only half the heat source is considered. The heat source comprises a Gaussian plane heat source on the top surface and a conical shape heat source along the thickness of the specimen. To simplify the analysis, it was assumed that the alignment of laser spot was perpendicular to the specimen surface and a single spot of laser irradiation was utilized in the heat conduction analysis. The convective heat transfer condition of air was considered after the set time of laser irradiation. Except for a laser beam irradiated area, the convective heat transfer condition of air was also assumed. The main FEM analytical conditions are shown in Table 2.

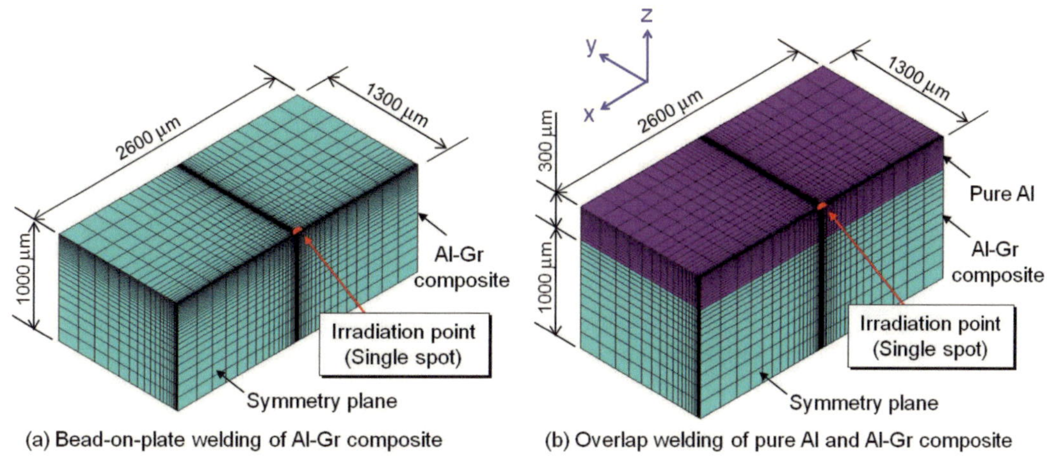

Fig. 6. Finite element models for investigation of temperature distributions

Parameter/condition		Bead-on-plate welding	Overlap welding
Laser power	P	20, 30, 40 W	400 W
Pulse width	τ	1.5 ms	2.7, 3.0, 5.0 ms
Beam diameter	d	50 µm	
Heat transfer coefficient	h	10 W/(m^2·K)	
Room temperature	θ_{room}	293 K	

Table 2. FEM analysis conditions

5. Bead-on-plate welding

In the bead-on-plate welding experiment, the STC Al-Gr composite of 1 mm thickness was used as the specimen. The specimen was mounted by clamping fixture with the alumina-ceramic plates, which are located on the bottom surface of specimen to minimize the heat loss during welding experiments. The rectangular shape of laser pulse was fixed as the standard output of laser source. The experiments were conducted with laser power up to 50 W, pulse width from 1 to 5 ms and scanning velocities varied between 12 to 72 mm/min.

5.1 Influence of welding parameters

In the first experiment, the welding condition on the top surface of Al-Gr composite was investigated by changing the power of laser pulse. The pulse width, pulse repetition rate and scanning velocity were kept constant at 1 ms, 5 Hz and 12 mm/min, respectively. The results of this experiment are shown in Figure 7. It can be noticed that the laser power significantly affected the welding condition. As shown in figure, when the laser power is insufficient high, the molten zone was discontinuous. At the lower 20 W of laser power, the molten zone can be observed only on the aluminum material, while the graphite material is non-molten phase. It is consider that the insufficient laser energy would not react the graphite material.

It is assumed that too high laser power leads to the generation of weld defects with increasing the power density. In practice, the power density below 10^7 W/cm^2 is generally advisable to avoid severe ejection of molten material (Cao et al., 2003). At the 50 W of laser power it seems too high power density resulting blowholes and a poor quality of weld surface. The power density on the specimen surface can be reduced to prevent the porosities at the weld pool surface by reducing the laser power. Therefore, the acceptable condition of weld surface is obtained between 30 W to 40 W of laser power. However, since the graphite element only can be melted under the high pressure with high temperature (Kirillin & Kostanovskii, 2003), the existence of this element in the molten zone of Al-Gr composite is important to identify. It might be considered that the required laser power to generate molten zone in this composite is relatively smaller than the welding of pure aluminum, even it has a higher thermal conductivity.

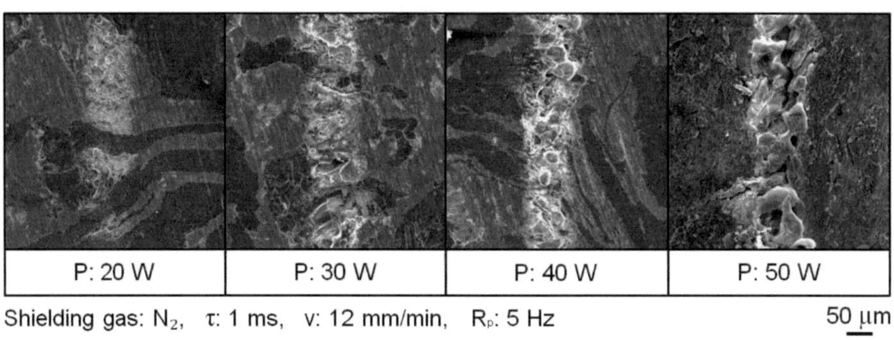

Fig. 7. Influence of laser power on the top surface of weld bead

In order to identify the presence of graphite element in the molten zone, the energy-dispersive X-ray spectroscopy (EDS) technique was used for the elemental analysis. Figure 8 shows the distribution map of elements Al and C on the top surface of molten zone. As shown in the figure, the element Al is the main composition detected in the molten zone. However, the EDS mapping shows that only small particles of graphite were observed on the top of the solidified aluminum. Judging from the distribution maps of elements, it can be presumed that the graphite was not mixture with aluminum, and it was ruptured into small particles. Even the aluminum material of Al-Gr composite was melted with the low laser power, the graphite material could not be melted. Because, the graphite only can be melted at the high temperature (θ_{melt}: 4765 K) under the high pressure (F_{melt}: 100 bar) (Joseph et al.,

2002) compared with the melting point of aluminum (θ_{melt}: 933 K). In addition, since the graphite is soft and brittle material, the graphite would be smashed only by direct irradiated laser beam to form the small particles during keyhole welding process.

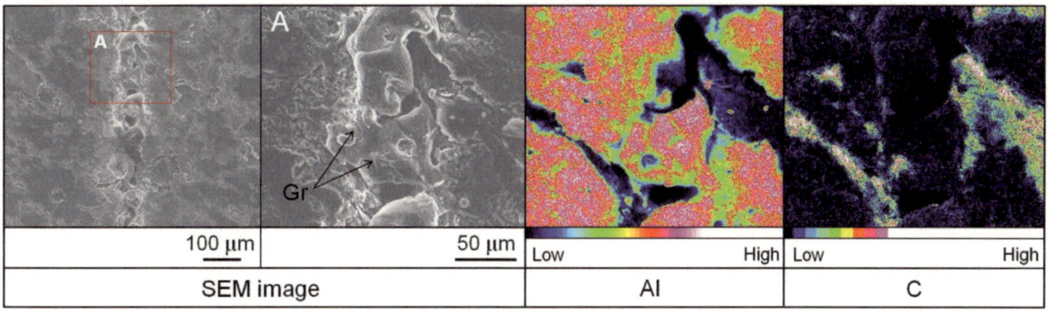

Shielding gas: N_2, P: 40 W, τ: 2 ms, R_p: 5 Hz, v: 12 mm/min

Fig. 8. SEM photographs and element mapping of Al and C

In the second series of welding experiments, the influence of pulse width was investigated. Figure 9 shows the appearance of weld beads under the various pulse widths at a constant laser power. It shows that the longer pulse width would generate more molten volume because the weld pool remains for an extended period of time. The widest welds are generated by the longer pulse width, which is considered due to the longer heating time. As a result, the irregular shapes and large size of cavities or blowholes are generated periodically when the pulse width was more than 1 ms. The tendency of blowhole occurrence and weld bead imperfections were drastically increased in the case of longer pulse width due to instabilities of the keyhole. In addition, the adverse effect of long pulse width on the weld penetration was believed due to a higher proportion of heat conducted laterally into the specimen. The results reflect that the higher energy input would enlarge processed zone. As mentioned later in section 5.2, based on the thermal analysis by FEM, a very short pulse width is enough to initiate the melting of aluminum and the evaporation of graphite on the welding of Al-Gr composite.

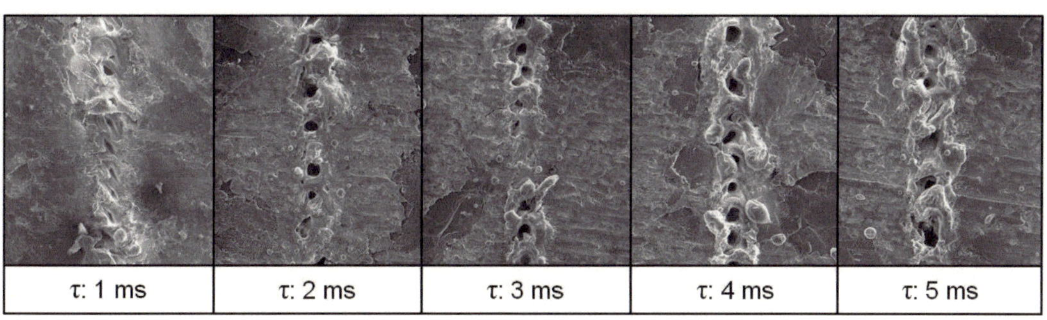

Shielding gas: N_2, P: 30 W, v: 12 mm/min, R_p: 5 Hz 50 μm

Fig. 9. Influence of pulse width on the top surface of weld bead

In order to keep the continuity of the penetration, the overlapping ratio was kept constant at 20 %. An increase in scanning velocity would increase the pulse repetition rate in the pulsed laser welding. The influence of scanning velocity on the top surface of weld bead was shown in Figure 10. It is noted that with the decrease of scanning velocity led to the enhancement of the molten zone surface, in which the weld seam surface showed an uniform surface ripple formation without blowholes. It can be seen that the effect of the remaining energy of the previous pulse is significant to interact the Al-Gr composite by the subsequent laser pulse. However, the size of molten zone was wider at the higher scanning velocity with the blowholes and poor quality of weld surface. Because the scanning velocity matches an opposite with the heat input, and the increasing of the scanning velocity means the decreasing of the average power per unit weld length exerted on the welding line, thereby producing a small amount of intermixed melt. Therefore, the lower scanning velocity is required to produce acceptable welded joints and maintain welding quality.

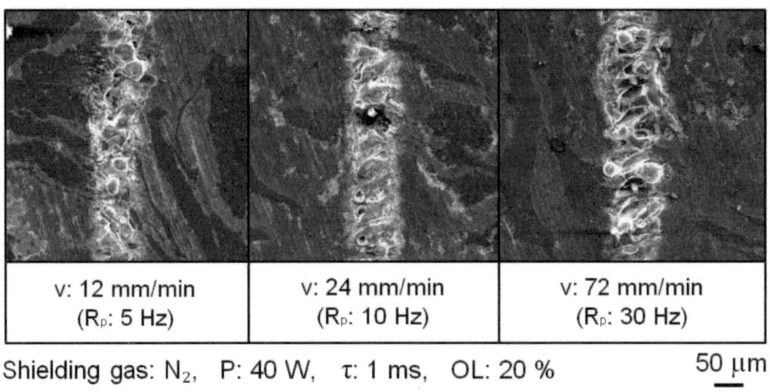

Fig. 10. Influence of scanning velocity on the top surface of weld bead

5.2 Temperature field induced in STC Al-Gr composite

A thermal analysis was carried out, in order to study the temperature field induced in the welding process of Al-Gr composite. Figure 11 and 12 show the influence of laser power and pulse width on temperature distribution of Al-Gr composite spatially and temporally. It can be clearly clarified that the absorbed energy on the Al-Gr composite was quickly removed in x-direction compared to the z-direction due to the higher thermal conductivity in x-direction (k_{xx}: 450 W/(m·K)), which is more than ten times of the thermal conductivity in z-direction (k_{zz}: 40 W/(m·K)). It can be seen that the increase of laser power and pulse width could enable an elevated temperature. This can be described that an enhanced laser energy input is absorbed by Al-Gr composite. In particular, when the low laser power of 20 W and 30 W are applied, the maximum temperatures are approximately 2400 K and 3500 K, respectively, which reaches to the melting and evaporation temperature of aluminum material below evaporation temeparture of graphite material. At the higher laser power of 40 W, the evaporation of graphite material occurs because the peak temperature of 4607 K exceeds the evaporation temperature of graphite. However, the influence of pulse width shows the slight increasing on the temperature after the melting of aluminum or evaporation of graphite was achieved. According to the thermal analysis results, it can be concluded that the laser power is significantly influenced on the evaporation of graphite. Judging from these results, a very short pulse width is required to initiate the melting of aluminum and the evaporation of graphite on the welding of Al-Gr composite.

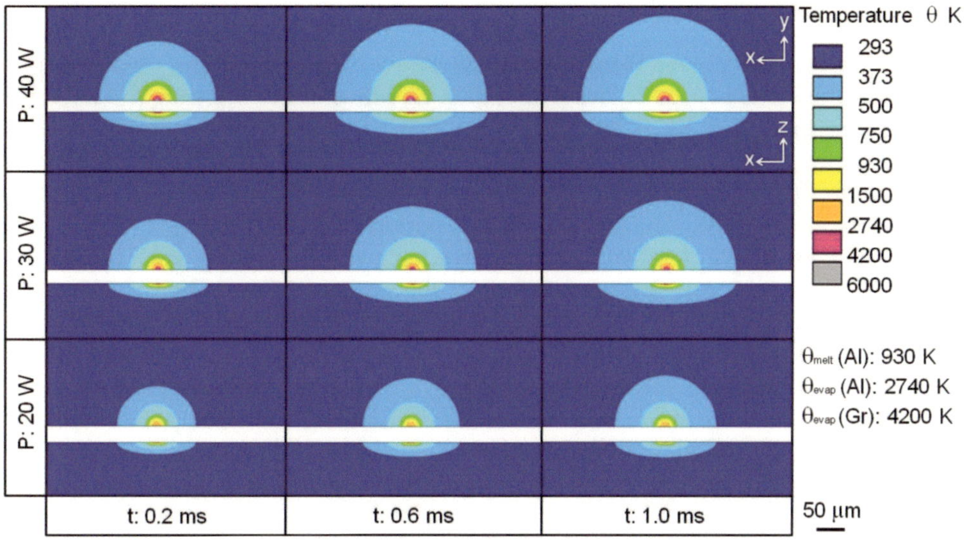

Fig. 11. Spatial temperature distributions in bead-on-plate welding

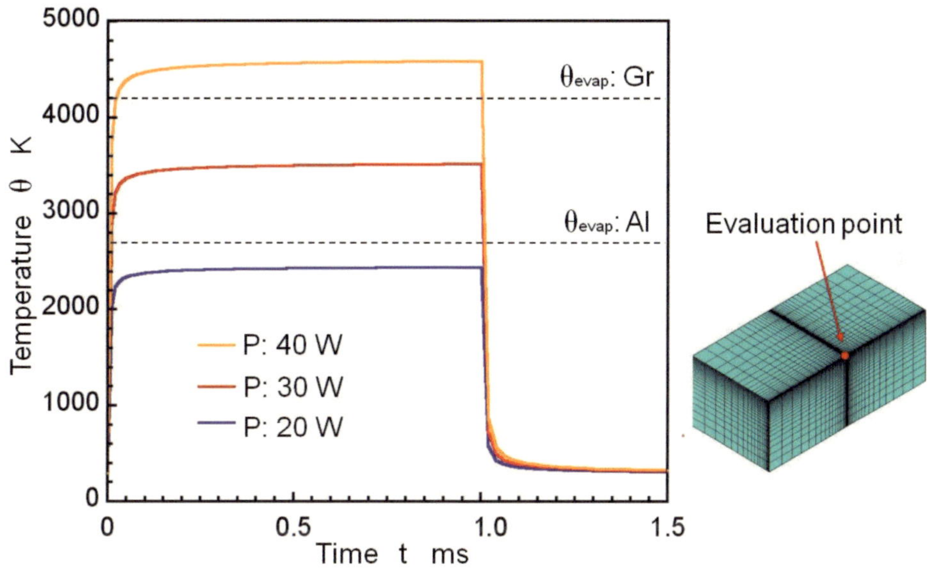

Fig. 12. Temperature histories in bead-on-plate welding

6. Overlap welding

In the overlap welding experiment, the pure aluminum sheet of 0.3 mm thickness and the STC Al-Gr composite of 1 mm thickness were used as the specimens. In order to achieve an optimum welding condition, the rectangular and controlled pulse waveforms of laser pulses were utilized as the output of laser source.

6.1 Rectangular pulse waveform

The variations of laser power from 300 to 550 W with constant 1 ms pulse width were irradiated on the specimen of 0.3 mm thickness aluminum and 1.0 mm Al-Gr composite plate as shown in Figure 13. It could be seen that the laser pulse energy less than 400 mJ/pulse was insufficient to form a molten geometry for joining both materials, in which the penetration depth was shallow. The higher pulse energy was crucial to achieve the sufficient penetration depth and control the formation of molten geometry. However, the cross-section view above 450 mJ/pulse showed signs of porosity and bump defects. Moreover, the undercut defect was observed at the higher pulse energy of 550 mJ/pulse. Judging from these observation results, it was confirmed that the higher laser power (more than 400 W) causes destructive effects, and lower laser power (less than 400 W) will restrict the penetration depth and joining. From these viewpoints mentioned above, the next welding experiments were carried out under the constant 400 W laser power with various pulse widths in the rectangular pulse waveform.

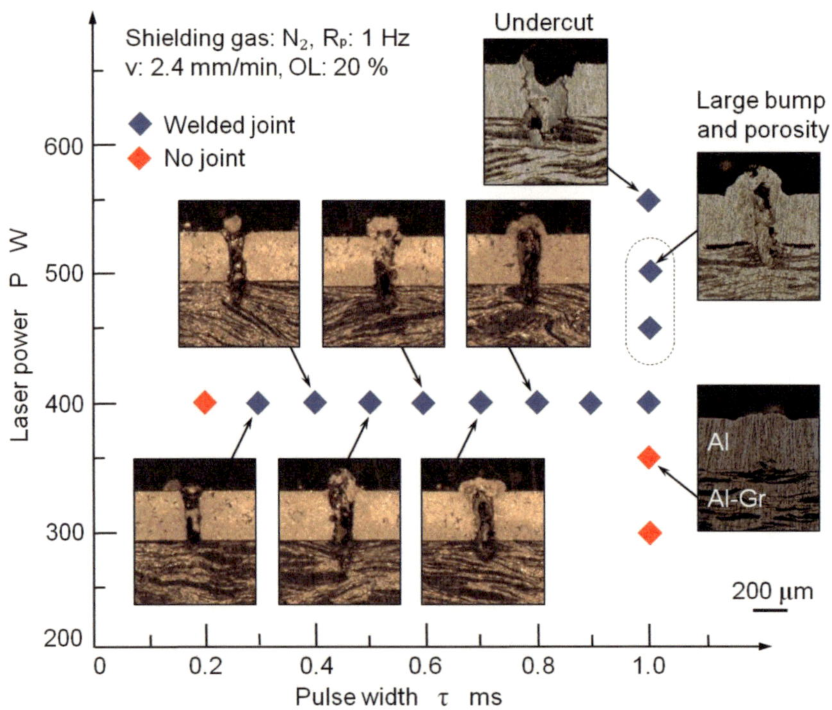

Fig. 13. Welding results with rectangular pulse waveform

Figure 13 shows welding results with the rectangular pulse waveform for various pulse widths and laser powers. As shown in the figure, the width of laser pulses had a significant role on the penetration depth. The weld joint between aluminum plate and Al-Gr composite can be seen at the pulse widths more than 0.3 ms. The penetration depth and bead width gradually increased with increasing the pulse width. However, these increments led to the increase of bump size in the weld joint. On the other hand, the penetration depth increased gradually with increasing the laser pulse energy, and a joining part between pure aluminum plate and Al-Gr composite was obtained with porosity and bump defects. According to the observation results of cross-section in the welding with rectangular pulse waveform, the pulse width of 0.6 ms was better condition with the deeper penetration, smaller porosities and bumps.

6.2 Controlling pulse waveform

As mentioned in the above observation results of normal rectangular pulse waveform, the bump and porosity defects were the major problem in the overlap welding between pure aluminum and Al-Gr composite. In order to overcome these problems, a controlled pulse waveform is discussed, since it is considered that an appropriate controlled laser pulse waveform could generate a better welded joint. Figure 14 shows the controlled pulse waveform named as a spike pulse waveform. The spike pulse waveform is divided into two phases. At the phase one, the laser power of 400 W and pulse width of 0.6 ms were selected according to the previous experimental results of rectangular pulse waveform. Phase two is a subsequent function of the phase one by adding the heat to melt the bump generated in the phase one. This re-melting process is intended to remove the porosity in order to obtain a better joining state. Therefore, the experiments were carried out to discuss an appropriate value of laser power P_2 and time period t_2 at the phase two of spike pulse waveform.

From the cross-section observation as shows in Figure 15, it was seen that the porosity and bump have remained at a range of laser power 50-125 W, while an undercut was generated at the laser power more than 200W. The observation results of laser power 150 W and 175 W showed acceptable joining conditions, and 175 W was selected as the laser power of phase two (P_2) for the deeper penetration and less porosity defect. The next experiment was carried out to define an optimum pulse width to set the time period t_2 at the phase two of spike pulse waveform as shown in Figure 14.

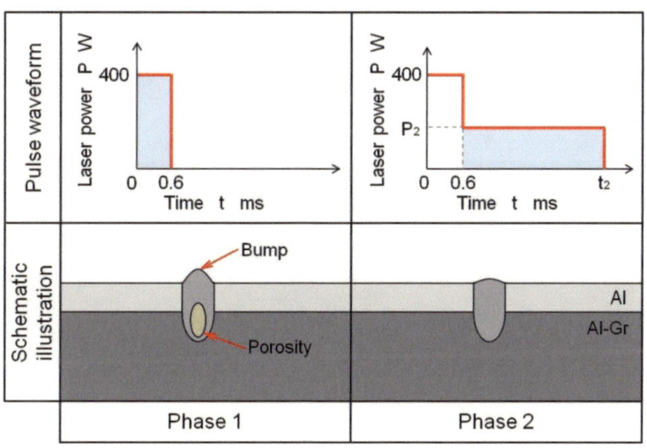

Fig. 14. Schematic illustration of welding process with spike pulse waveform

The pulse width was set less than 5 ms due to the limitation of maximum pulse width in the laser system used in this study. Figure 16 shows the cross-section views for various pulse widths under the same laser power of phase two to define the appropriate value of pulse width at the phase two for the spike pulse waveform. An appropriate irradiation time is necessary to melt and reduce a bump during a welding process. Within the range 1.5-2.5 ms of pulse width (t_2), the results show that the welding defect of bump was not removed. Moreover, it could be cleared that the longer irradiation time (t_2: 3.5-5.0 ms) generated the larger size of bump. On the other hand, the pulse width t_2 of 3 ms showed an acceptable penetration depth. Judging from these results, 3 ms is selected as an appropriate value of t_2 at the phase two in the case of spike pulse waveform due to the stable penetration and smaller bump without porosities.

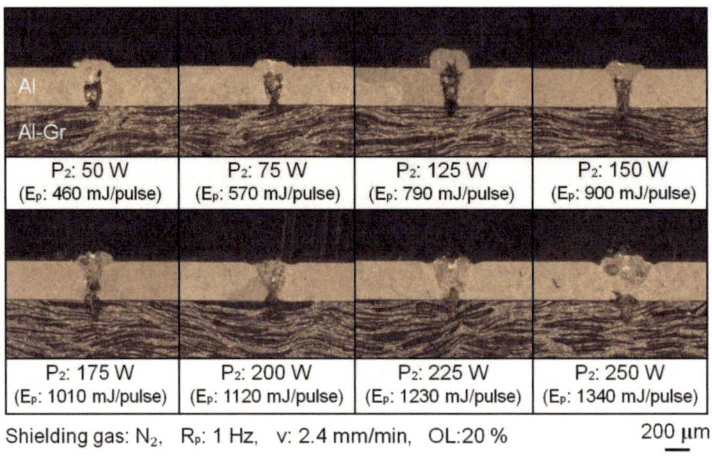

Fig. 15. Influence of laser power (P_2) in phase 2 of spike pulse waveform

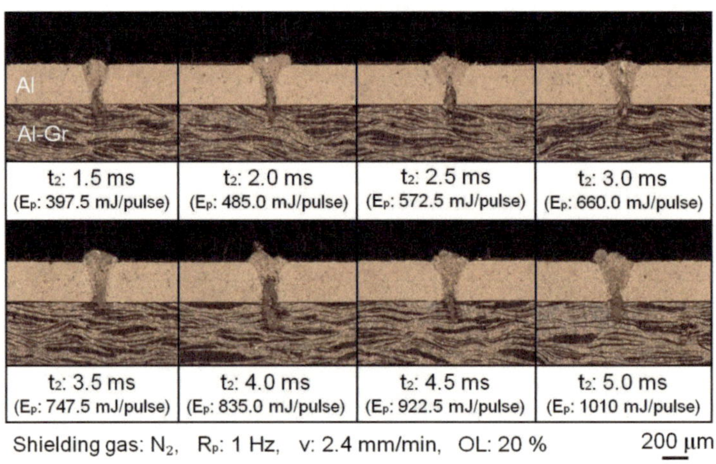

Fig. 16. Influence of pulse width (t_2) in phase 2 of spike pulse waveform

In addition, the elemental analysis was performed to observe the element distribution at the welded parts using a SEM equipped with the energy dispersive spectroscopy (EDS). Figure 17 showed the results of EDS mapping analysis in the case of spike pulse waveform. The shape of molten zone at the Al-Gr composite can not be seen clearly because of too much carbon particles on the molten zone, which can not be melted during the welding process. It could be seen that the coarse carbon located at the welded joint between pure aluminum and Al-Gr composite. It is considered that the severe convection of graphite particles with aluminum materials might be occurred in the molten zone. In other words, when Al-Gr composite is evaporated, an over-pressure is developed in the keyhole. This phenomenon would cause graphite particles pushed up from the prior zone in Al-Gr composite to the zone in the pure aluminum with melting phase, and finally the carbon particles would be redistributed during the re-solidification. However, it could be seen that a lack of fusion defect was generated by the spike pulse waveform. It could be detected by using an accurate microscopic observation in the high magnification condition. The lack of fusion defect in the molten zone can be identified by its string-like appearance, and it had randomly oriented curvature. It is considered that the lack of fusion defect is not pore, since the carbon element could be seen in this area.

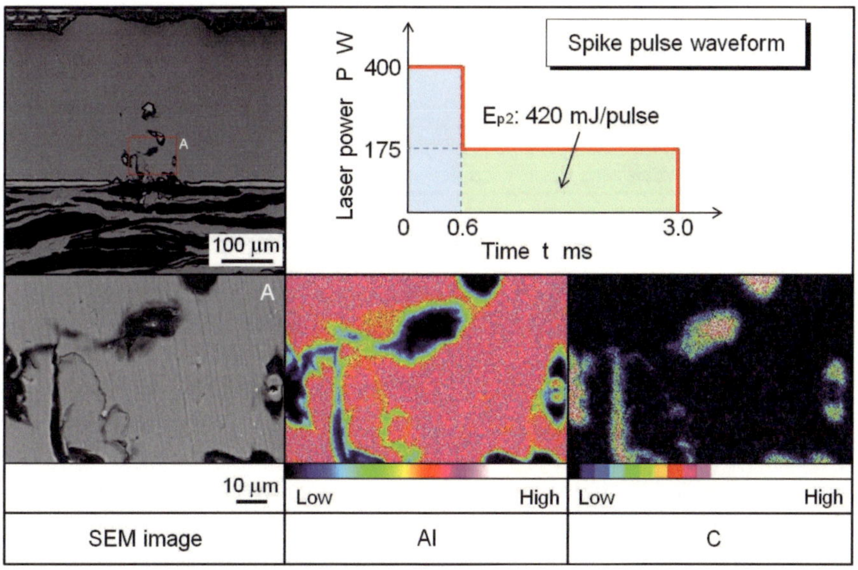

Fig. 17. SEM images and EDS mappings with spike pulse waveform

From the observation results by the spike pulse waveform, the use of controlled pulse waveform has a positive effect on molten zone to remove the porosity and minimize the bump. However, the rapid cooling after welding process would generate the coarse carbon particles and lack of fusion defect in the molten zone. Therefore, an improvement of heat input at the end of laser pulse is necessary by introducing an approach of ramp-down on the phase two to relieve the internal stress during re-solidification process, which would generate a better welded joint. The new controlled pulse waveforms are expressed as the annealing pulse waveform, and the trailing pulse waveform is also shown in Figure 18, in which the setting profile and the actual signal of three controlled pulse waveforms are also shown. Since the phase two of spike pulse waveform generated a stable weld bead compared with the rectangular pulse waveform, the amount of energy at the phase two for the annealing and the trailing pulse waveforms was conducted under the same energy of 420 mJ. The main difference between the annealing and the trailing pulse waveforms are the laser power and irradiation time during the phase two. The annealing pulse waveform has the higher laser power with the shorter interaction time, while the trailing pulse waveform has the lower laser power with the longer interaction time.

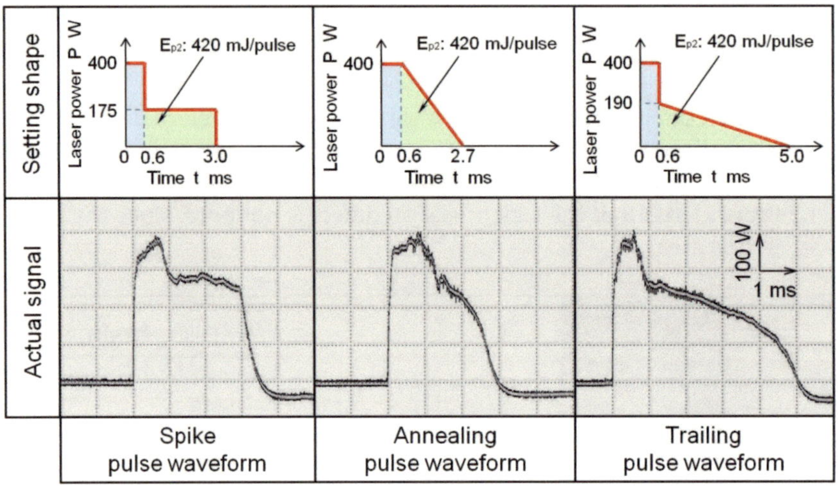

Fig. 18. Setting shape and actual signal of controlled pulse waveform

The difference of temperature change by spike, annealing and trailing pulse waveforms was investigated with the thermal calculation. Spike, annealing and trailing pulse waveforms are finished at the time of 3.0 ms, 2.7 ms and 5.0 ms, respectively. The temperature distributions were similar until 0.6 ms for three pulse waveforms, since they have similar pulse shape (laser power and pulse width) at the phase one. The main difference appeared at the pulse shape of phase two even under the same energy (E_{p2}: 420 mJ). Figure 19 and 20 show the temperature histories and distributions by these three controlled pulse waveforms. It can be seen that spike pulse waveform shows the constant temperature distribution during the phase two. Therefore, it is considered that the rapid cooling at 3 ms during the solidification generates internal stress, and it might cause the lack of fusion defect. From this disadvantage of spike pulse waveform, the slow cooling process is required during the solidification phase. As shown in the figures, the annealing pulse waveform shows the rapid decreasing of temperature distribution at the phase two. On the other hand, the trailing pulse waveform indicates the gradual decreasing of temperature distribution at the phase two, which means that slow cooling could be realized. In other words, the longer time of laser irradiation is useful at the phase two in order to overcome the lack of fusion problem.

SEM images and EDS mappings of cross-section with annealing and trailing pulse waveforms are shown in Figure 21. In the SEM image, the shape of molten zone can be seen clearly if there are existences of aluminum element on the Al-Gr composite material. Compared with the spike pulse waveform, the size of carbon particles was much smaller in the case of annealing pulse waveform. However, it shows that the lack of fusion was appeared around the carbon particles. In the case of trailing pulse waveform, it can be seen that the specimens have been molten well with an acceptable weld bead state without lack of fusion defect. It also shows that a better weld bead state was obtained by applying the trailing pulse waveform. Furthermore, the sufficient long time during welding process. In other words, the slow cooling process during re-solidification is necessary to avoid the appearance of carbon particles and minimize the lack of fusion defect in the molten zone.

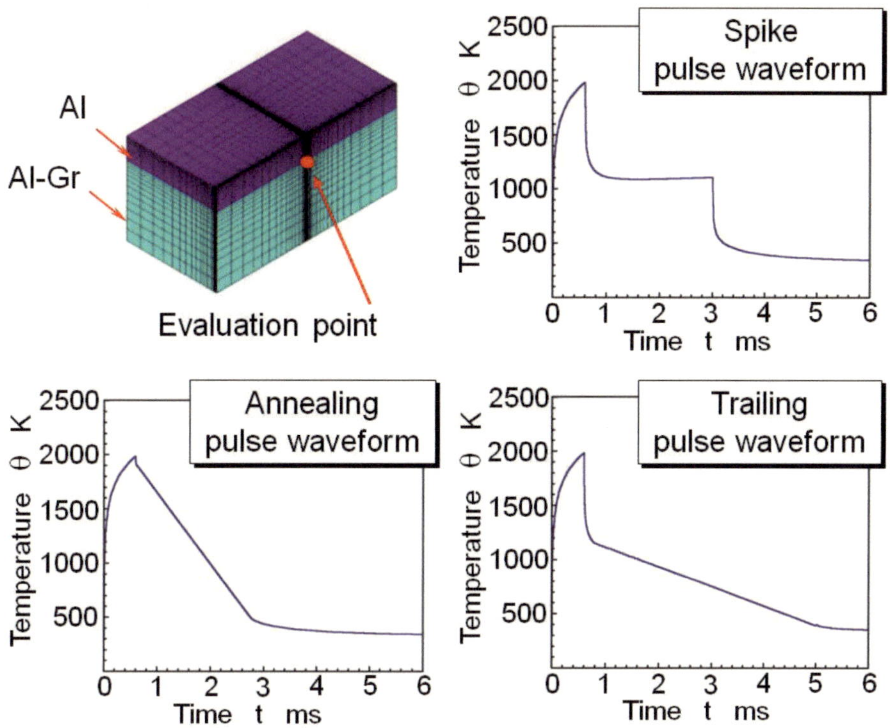

Fig. 19. Temperature histories by three controlled pulse waveforms

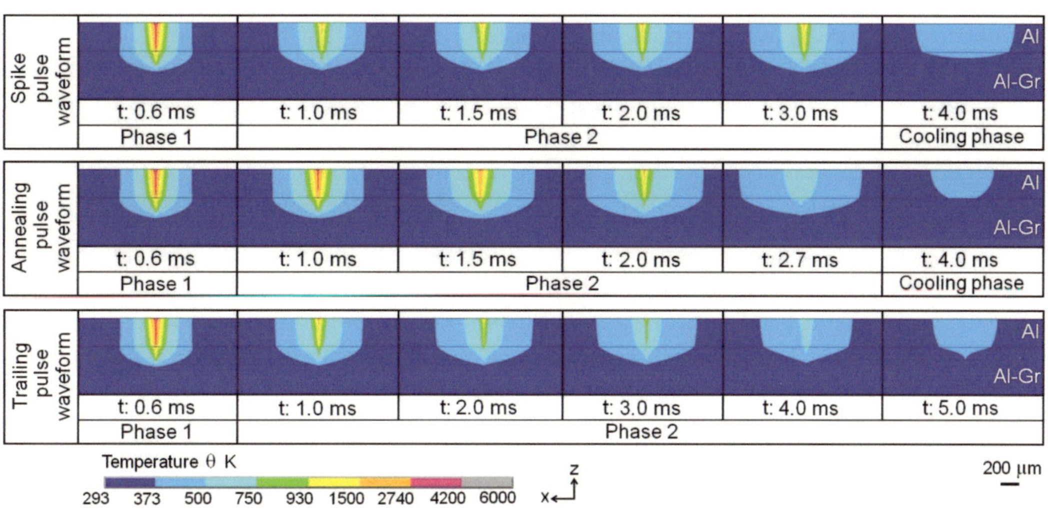

Fig. 20. Spatial temperature distribution by three controlled pulse waveforms

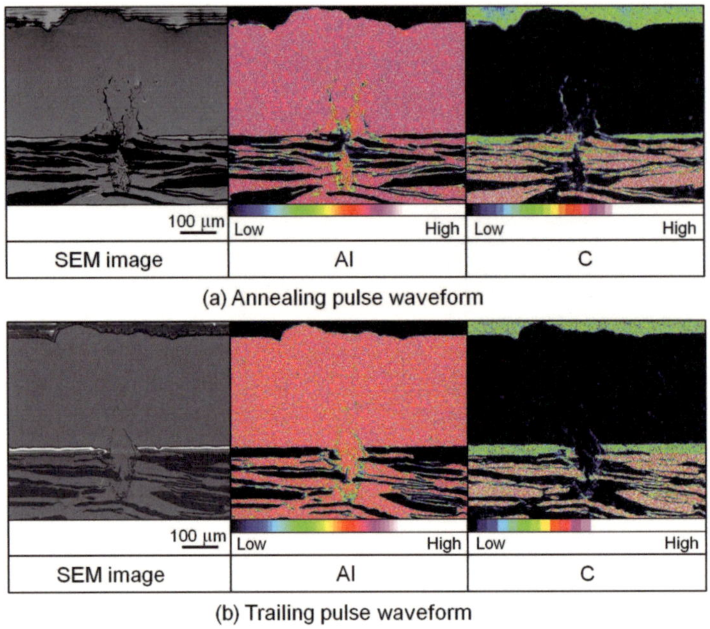

Fig. 21. SEM images and EDS mappings of cross-section with (a) annealing and (b) trailing pulse waveforms

6.3 Evaluation of mechanical strength

In order to evaluate the weld strength on the overlap welding with and without the control of pulse waveform under the constant laser pulse energy (E_P: 660 mJ/pulse), the shearing test was carried out. Figure 22 shows the shear strength for various laser pulse waveforms. As shown in the figure, the weld joint with controlled trailing pulse waveform indicated the greater weld strength compared to the uncontrolled pulse waveform. The lower strength of the uncontrolled pulse waveform is attributable to the fact that the existence of porosity reduced the strength of weld joint. Meanwhile, it shows that the spike pulse waveform was less significant to increase the weld strength compared to the uncontrolled pulse waveform. It can be noted that the lack of fusion defect was affected on the lower weld strength. Therefore, it is cleared that the the appropriate controlled laser pulse configurations are effective to improve the weld joint between aluminum and STC Al-Gr composite.

Figure 23 shows the fracture part on the top surface of Al-Gr composite after the shearing test. As can be seen from the figure, the fracture with uncontrolled pulse waveform occurred at the weld bead boundary in the Al-Gr composite and the presence of groove could be observed. It could be noticed that the existence of groove defect clearly influenced the weakness of weld joint strength. However, the fracture in the welding with the control of pulse waveform appeared inside the weld bead, which is located at the interface between pure aluminum and Al-Gr composite without the groove defect. In other words, the interfaces of aluminum and STC Al-Gr composite were expected to be the weak points of the weld joint. It is also considered that a weld joint without weld defects would increase the weld strength.

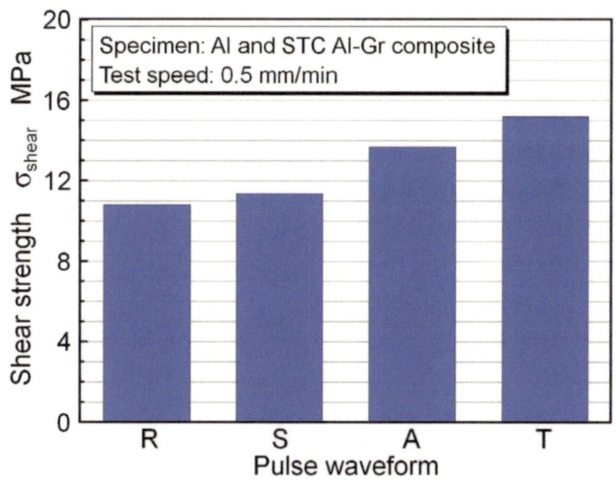

Fig. 22. Shear strength of the welding with and without the control of pulse waveforms (R: Rectangular, S: Spike, A: Annealing, T: Trailing)

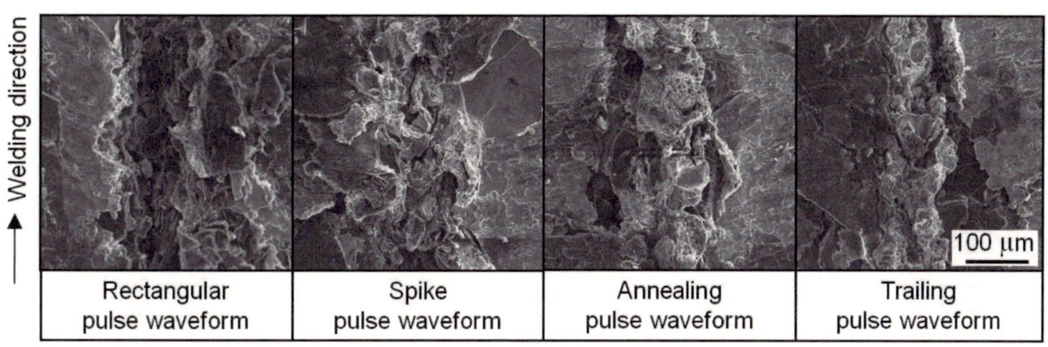

Fig. 23. SEM photographs of fracture on the top of STC Al-Gr composite

Figure 24 shows the side fracture surface of aluminum and STC Al-Gr composite, which were obtained from the shearing test under the welding condition with trailing pulse waveform. It can be seen that the fracture appearances show the brittle fracture. This fracture along the welding interface would deteriorate the weld strength of aluminum and STC Al-Gr composite joint. Figure 24 also shows the distribution map of elements Al and C on the fracture zone, where spot analysis of points 1 to 10 are listed in Table 3. It can not be detected the aluminum carbide Al_4C_3 at the fracture zone, since the solid graphite can not be solidified into liquid aluminum. EDS mapping on the bottom surface of aluminum confirms that after the shearing test, the fine and coarse graphite particles were found sticks on melted aluminum without mixture with solidified aluminum.

In the case of fracture zone on the top surface of STC Al-Gr composite, it also can be seen that the solidified aluminum was squeezed out towards the edge of weld joint region, which showed the aluminum and graphite can not mixed together between both materials.

Furthermore, the formation of the Al_4C_3 during welding process was successfully prevented during the welding process, which could deteriorate the strength of weld joint. In addition, the SEM observation revealed the crack was propagated in the fracture zone, which restricts the further strength of weld joint. However, compared to the bending strength of STC Al-Gr composite which is 7 MPa in the thickness direction, the weld strength between the aluminum and STC Al-Gr composite showed the relatively higher strength. Therefore, it is clearly performed that the controlled laser pulse configurations are effective to produce a higher strength of joint for welding between an aluminum and STC Al-Gr composite.

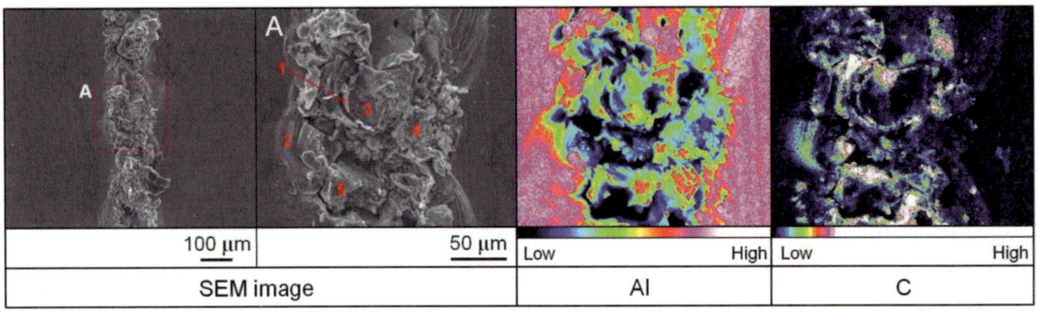

(a) Bottom surface of Al

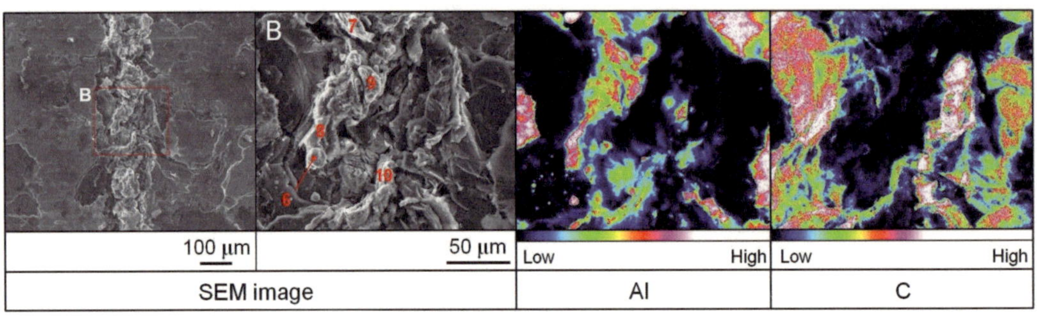

(b) Top surface of STC Al-Gr composite

Fig. 24. SEM photographs and EDS mappings of fracture with trailing pulse waveform on the (a) bottom surface of aluminum and (b) top surface of STC Al-Gr composite

Point No.	1	2	3	4	5	6	7	8	9	10
Al wt. %	49.67	30.94	62.84	41.16	64.45	52.45	35.68	48.66	64.65	48.01
C wt. %	50.33	69.06	37.16	58.84	35.55	47.55	64.32	51.34	35.35	51.99

Table 3. Element composition of points 1-10 in Figure 24

7. Conclusion

The pulsed Nd:YAG laser micro-welding of a super thermal conductive (STC) aluminum-graphite composite was experimentally and numerically investigated. In the bead-on-plate welding of STC Al-Gr composite, the laser power and pulse width had a great influence on the top surface condition of weld bead. Laser power more than 30 W was required to melt the STC Al-Gr composite without evaporation of the graphite element in the composite. The graphite was not mixed with aluminum during welding process to prevent the formation of aluminum carbide, which can degrade the weld joint. The overlap welding of aluminum and STC Al-Gr composite was successfully carried out using an appropriate controlled pulse waveform. Porosity and bump were observed as remarkable weld defects in overlap welding without a control of laser pulse. The proper control of laser power and pulse width could perform a positive result with largely free of weld defects and a relatively small bump. The controlled pulse waveform with slow cooling at the end of laser pulse was essential to relieve internal stress during solidification, since the lack of fusion was observed on the joining zone due to the rapid cooling. The higher shearing strength could be obtained by the control of pulse waveform compared with the uncontrolled rectangular pulse waveform.

8. Acknowledgment

The authors are grateful to Shimane Institute for Industrial Technology for supplying super thermal conductive aluminum-graphite composites.

9. References

Baraker, N.; Tzamtzis, S.; Dhindaw, B.K.; Patel, J.; Babu, N.H. and Fan, Z. (2009). Processing of Aluminum-Graphite Particulate Metal Matrix Composites by Advanced Shear Technology. *Journal of Materials Engineering and Performance,* Vol.18, No.9, pp. 1230-1240, ISSN 1059-9495

Bassani, P.; Capello, E.; Colombo, D.; Previtali, B. and Vedani, B. (2007). Effect of Process Parameters on Bead Properties of A359/SiC MMCs Welded by Laser. *Composites: Part A,* Vol.38, pp. 1089-1098, ISSN 1359-835X

Behler, K.; Berkmanns, J.; Ehrhardt, A. and Frohn, W. (1997). Laser Beam Welding of Low Weight Materials and Structures. *Materials and Design,* Vol.18, No.4/6, pp. 261-267, ISSN 0261-3069

Cao, X.; Wallace, W.; Poon, C. and Immarigeon, J.P. (2003). Research and Progress in Laser Welding of Wrought Aluminum Alloys – Laser Welding Processes. *Materials and Manufacturing Processes,* Vol.18, No.1, pp. 1-22, ISSN 1042-6914

Canilloo, V.; Sola, A.; Barletta, M. and Gisario, A. (2010). Surface Modification of Al-Al_2O_3 Composites by Laser Treatment. *Optics and Lasers in Engineering,* Vol.48, pp. 1266-1277, ISSN 0143-8166

Joseph, M.; Sivakumar, N. and Manoravi, P. (2002). High Temperature Vapour Pressure Studies on Graphite Using Laser Pulse Heating. *Carbon,* Vol.40, No.11, pp. 2031-2034, ISSN 0008-6223

Kannatey-Asibu, E. (2009). *Principles of Laser Materials Processing,* John Wiley & Sons, ISBN 978-0-470-17798-3, United State

Kirillin, A.V. and Kostanovskii, A.V. (2003). Melting Point of Graphite and Liquid Carbon. *Physics-Uspekhi,* Vol.46, No.12, pp. 1295-1303, ISSN 1063-7869

Kutz, M. (2002). *Handbook of Materials Selection,* John Wiley & Sons, Inc., ISBN 0-471-35924-6, Canada

Lienert, T.J.; Brandon, E.D. and Lippold, J.C. (1993). Laser and Electron Beam Welding of SiCp Reinforced Aluminum A-356 Metal Matrix Composite. *Scripta Metallurgica,* Vol.28, pp. 1341-1346, ISSN 0956-716X

Lu, H.; Shen, D.H.; Deng, X.F.; Xue, Q.K.; Froumin, N. and Polak, M. (2001). Study of the Al/Graphite Interface. *Chinese Physics,* Vol.10, No.9, pp. 832-835, ISSN 1009-1963

Massalski, T.B.; Okamoto, H.; Subramanian, P.R. and Kacprzak, L. (1990). *Binary Alloy Phase Diagrams (Vol.1),* ASM International, ISBN 0871704048, United State

Niu, J.; Pan, L.; Wang, M.; Fu, C. and Meng, X. (2006). Research on Laser Welding of Aluminum Matrix Composite SiCw/6061. *Vacuum,* Vol.80, pp. 1396-1399, ISSN 0042-207X

Pan, L.K.; Wang, C.C.; Hsiao, Y.C. and Ho, K.C. (2004). Optimization of Nd:YAG Laser Welding onto Magnesium Alloy via Taguchi Analysis. *Optics and Laser Technology,* Vol.37, pp. 33-42, ISSN 0030-3992

Ueno, T.; Yoshioka, T.; Ogawa, J.; Ozoe, N.; Sato, K. and Yoshino, K. (2009). Highly Thermal Conductive Metal/Carbon Composites by Pulsed Electric Sintering. *Synthetic Metals,* Vol.159, pp. 2170-2172, ISSN 0379-6779

Wang, H.M.; Chen, Y.L. and Yu, L.G. (2000). In-situ Weld-Alloying/Laser Beam Welding of SiCp/6061Al MMC. *Materials Science and Engineering,* Vol.A293, pp. 1-6, ISSN 0921-5093

Yue, T.M.; Xu, J.H. and Man, H.C. (1997). Pulsed Nd:YAG Laser Welding of SiC Particulate Reinforced Aluminium Alloy Composite. *Applied Composite Materials,* Vol.4, pp. 53-64, ISSN 0929-189X

Zhang, J.; Weckman, D.C. and Zhou, Y. (2008). Effect of Temporal Pulse Shaping on Cracking Susceptibility of 6061-T6 Aluminum Nd:YAG Laser Welds. *Welding Journal,* Vol.87, No.1, pp. 18s-30s, ISSN 0043-2296

10

Highly Doped Nd:YAG Laser in Bounce Geometry Under Quasi-Continuous Diode Pumping

Michal Jelínek and Václav Kubeček
Czech Technical University in Prague
Czech Republic

1. Introduction

Nanosecond and picosecond laser pulses play increasing role in a variety of applications, such as microsurgery, micromachining, ranging, optical parametric amplifiers pumping, remote sensing, nonlinear optics, etc. Efficient diode pumped solid state lasers represent widely used sources of such pulses. For applications such as laser ranging and some nonlinear optical effects investigation, the extremely short pulses are not needed and pulses in the range from tens of picoseconds to units of nanoseconds at repetition rates below 1 kHz are optimal. Research and development of such laser systems is the subject of this chapter.

In the diode pumped solid state laser technique several resonator configurations and pumping schemes are utilized. One of interesting setups with unique properties is bounce geometry resonator configuration (Alcock, 1997) when a slab active material is used and the laser beam experiences total internal reflection and amplification on the pumped crystal side. This arrangement has several advantages. The slab crystal shape allows good heat dissipation, because it is relatively thin and can be cooled from one or both sides. As a pump source, high power laser diode single bar or bar stacks can be used without bulky additional optics or radiation delivery fibers. In addition, laser resonator mode can be optimized to match the pumped region profile. Efficient operation of the bounce geometry laser pumped by the laser diode bar requires active medium with a high absorption coefficient for the pump radiation in order to obtain thin region with high gain near the pumping side. Therefore, materials as Nd:$GdVO_4$ and Nd:YVO_4 has challenged Nd:YAG with standard doping concentration of around 1 at.%. When recently Nd:YAG laser crystals with higher doping level (more than 2 at.%) became available (Urata, 2001), efficient operation of these crystals under bounce geometry has been demonstrated operating at 1.06 and 1.3 μm (Sauder, 2006; 2007). Nd:YAG crystals have better thermal and mechanical properties, and better energy storage capacity due to the longer upper state life-time in comparison with vanadates giving potential to generate more energetic pulses in the Q-switched and mode-locked regime.

Highly-energetic pulses in the nanosecond range can be obtained from the oscillator using active or passive Q-switching, which is well-known and widespread technique, mostly used under continuous pumping. In order to obtain more energetic pulses, quasi-continuous (QCW) pumping can be employed. Using the 2 at.% Nd:YAG in bounce geometry under QCW pumping, passively Q-switched operation at 1.06 μm has been reported with maximum output energy of 2.3 mJ in a 12 ns pulse (Sauder, 2006). At 1.3 μm, actively Q-switched operation with maximum pulse energy of 1.8 mJ in 50 ns pulse was produced (Sauder, 2007).

In order to generate pulses in the picosecond and sub-picosecond range, the laser has to be operated in the mode-locking (ML) regime. In the solid state lasers, mostly passive ML techniques are used. These techniques can be divided into several groups. One large group is based on saturable absorbers based on different materials and media. The next group of techniques involves usage of nonlinear effects of different orders. The second order nonlinearity is involved in nonlinear mirror ML (Stankov, 1988; Thomas, 2010) and the third order in Kerr-lens ML (Agnesi, 1994; Spence, 1991). In the case of saturable absorbers, the dye saturable absorber (Schafer, 1976) was widely used in the previous decades, but because of the dye long-term instability, toxicity, and difficult manipulation, it was in the last decade replaced by solid state saturable absorbers. With the semiconductor technique development, the possibility of fabrication of the semiconductor saturable absorber with requested parameters appeared. Such absorber is based on one or more quantum wells grown on the semiconductor substrate. At the wavelength of about 1 μm, the quantum well is usually formed by the InGaAs layer with the thickness of several nanometers surrounded by the GaAs layers. As the InGaAs energy band gap is lower than GaAs band gap, it serves as a quantum well. According to the number of quantum wells and other growing parameters, the resulting characteristics, such as modulation depth, relaxation time constant, and saturation fluence can be adjusted according to the required application and laser operation conditions. Such saturable absorber is placed into the resonator in the transmission mode, usually under the Brewster's angle to avoid refections from the GaAs with high-refractive index. Alternatively, the saturable absorber structure can be grown on the Bragg mirror forming so called semiconductor saturable absorber mirror (SAM), which is placed as a rear mirror in the laser resonator and which has been widely used (Keller, 1996). Use of SAM as the ML technique has also some disadvantages. At first the laser output power is limited by the SAM structure damage threshold. The output pulse duration is also limited by the SAM relaxation time constant. The another approach to the solid-state saturable absorber development utilizes carbon nanotubes or graphene. Mode-locking operation of Nd-doped solid state lasers using carbon nanotubes SA (Yim, 2008) or graphene SA (Lee, 2010; Tan, 2010) was recently demonstrated.

Using semiconductor SAM, the output train of low energy pulses usually under continuous pumping is produced. In order to obtain pulses with higher energy directly from the oscillator, the semiconductor saturable absorber containing multiple quantum wells with higher modulation depth can be used in the laser resonator under QCW pumping. Bounce geometry can be considered as a proper resonator configuration for efficient mode-locking operation in this pumping regime. Efficient operation of quasi-continuously pumped 1 at.% doped Nd:YAG (Kubeček, 2008) and also Nd:YVO_4 and Nd:$GdVO_4$ (Kubeček, 2009a; 2010) using the saturable absorber in the transmission mode was already reported.

For applications requiring more energetic pulses or higher average power, the laser oscillator output can be further amplified using a laser amplifier in various configurations involving regenerative amplifier (Luhrmann, 2009), zigzag slab geometry (Cui, 2009; Fu, 2010), multiple amplification media (Fu, 2009), or hybrid fiber and solid-state system (Wushouer, 2010). These configurations are usually based on Nd:YAG or Nd:YVO_4 active materials with concentrations below 1 at.%. Much simpler amplifier configuration providing high gain involve grazing incidence geometry based on 1 at.% Nd:YVO_4 slab crystal (Agnesi, 2008; 2010).

2. Nd:YAG crystal with doping concentration of 2.4 at.%

High neodymium concentration results in shorter upper-state (fluorescence) lifetime caused by concentration quenching. At the typical Nd concentration of 1 at.% the fluorescence lifetime is about 230 μs (Koechner, 2006). At the concentration of 2 at.% the lifetime shortening to 145 - 180 μs and further at 3 at.% to 120 - 130 μs (Dong, 2005; Urata, 2001) was reported.

In our experiments we have used a Nd:YAG slab crystal with dimensions of 30 $\times$ 5 $\times$ 2 mm grown by Czochralski method with Nd concentration of 2.4 at.%. Two 5 $\times$ 2 mm end faces were angled at 68 degrees. This configuration allows the laser cavity beam to be incident at Brewster's angle to these faces and to experience total internal reflection from the diode pumped face which results in laser operation in the horizontally polarized state. The pumping face 30 $\times$ 2 mm was antireflection coated at the 808 nm pump wavelength.

As a pump source the linear single bar quasi-continuous diode array LD with the fast axis collimation, nominal output power of 180 W and repetition rate up to 100 Hz was used. This diode was mounted on the copper plate with Peltier cooling enabling fine generated radiation wavelength tuning around 808 nm to match the Nd:YAG absorption peak.

2.1 Nd^{3+} lifetime measurement

The integral fluorescence decay from our crystal for excitation at the wavelength of 808 nm was measured using a Si photodiode in the spectral region above 1000 nm and is shown in Fig. 1. Calculated lifetime of 135 μs is in the good agreement with other published results.

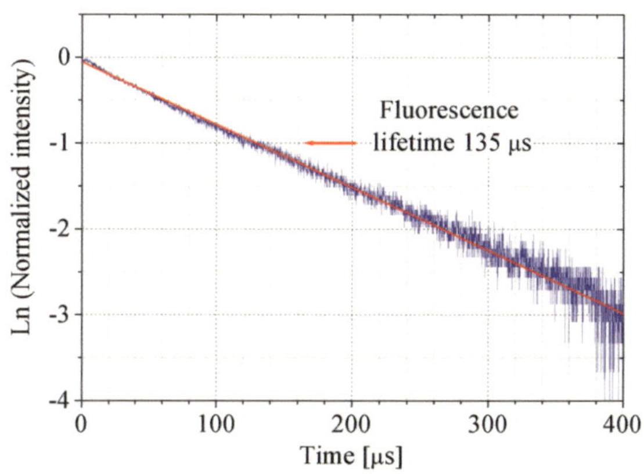

Fig. 1. Fluorescence lifetime measurement of 2.4 at.% Nd:YAG slab crystal.

3. Free-running laser operation at 1.06, 1.3 and 1.4 μm

The laser free-running operation was investigated at the output wavelengths of 1.06, 1.3, and 1.4 μm and is described in details in Kubeček (2009b). Schematic of the laser oscillator is shown in Fig. 2.

Generation at the wavelength of 1.06 μm was studied at first. The 17 cm long laser resonator was formed by a spherical high reflective mirror HR (radius of curvature of 300 mm) and a flat output coupler OC. Various OC reflectivities of 82, 50 and 30 % at 1.06 μm were tested.

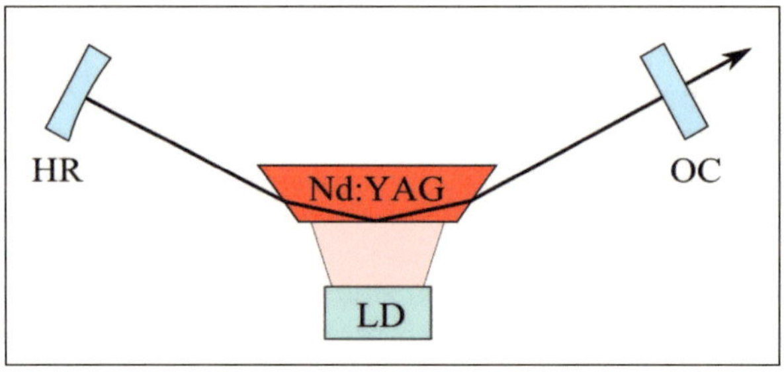

Fig. 2. Schematic of the Nd:YAG laser oscillator in bounce geometry operated in the free-running regime.

The Nd:YAG crystal was inserted approximately in the middle of the resonator. The output pulse energy with respect to the pump pulse energy is shown in Fig. 3. The pump pulse duration was set to 200 μs. For the output coupler with reflectivity of 82 % the slope efficiency of 50 % was obtained. The maximum output energy of 17 mJ for the pump energy of 38 mJ corresponds to the optical to the optical efficiency of 44.6 %. The output beam spatial profile was elliptical in the horizontal axis and is shown in the inset of Fig. 3 (for the output coupler reflectivity of 50 %).

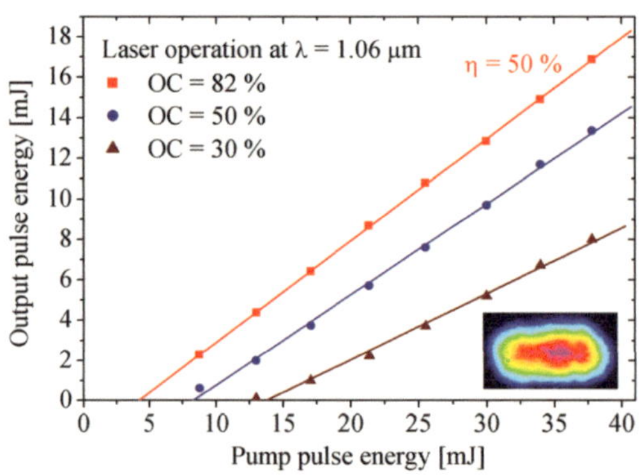

Fig. 3. The laser output characteristics in the free-running regime at 1.06 μm with various output couplers.

The slab shape with uncoated faces allowing the laser beam to be incident on the Nd:YAG at the Brewster's angle has the advantage in possibility of using the same crystal for generation at different wavelengths. We have successfully investigated efficient operation at 1.3 and 1.4 μm and the results were compared with values obtained for the generation at 1.06 μm. For the generation at 1.3 μm the HR mirror (radius of curvature of 150 mm) was highly reflective in this spectral region and had low reflectivity at 1.06 μm. The laser resonator was 95 mm long. For the generation at 1.4 μm the curvature of this mirror was the same, but the mirror was highly reflective at 1.4 μm. The output pulse energy with respect to the pump pulse energy is shown in Fig. 4. The maximum output energy at 1.3 μm was 8.5 mJ for the pump energy of 38 mJ and was obtained using the output coupler with reflectivity of 94 %. This corresponds to the optical to optical efficiency of 22.5 % and slope efficiency of 26 %.

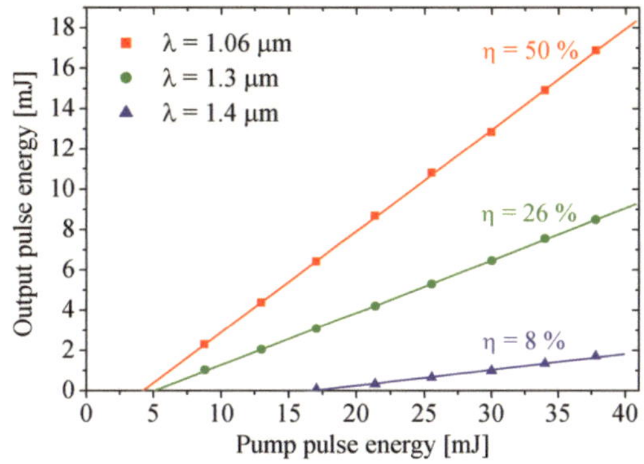

Fig. 4. Comparison of the laser operation at different wavelengths: 1.06, 1.3, and 1.4 μm.

The lowest curve in Fig. 4 shows the output pulse energy at 1.4 μm versus pump pulse energy. The maximum output energy of 1.5 mJ for the pump energy of 38 mJ was obtained using output coupler with reflectivity of 96 %. This corresponds to the optical to optical efficiency of 4.1 % and slope efficiency of 8.4 %.

4. Passively Q-switched laser operation at 1.06 and 1.3 μm

Q-switched laser operation was investigated at the output wavelengths of 1.06 and 1.3 μm and is described in details in Jelínek (2011a).

The laser resonator schematic was similar to that used in the free-running regime (shown in Fig. 2). At 1.06 μm the 80 mm long optical resonator was formed by a spherical high reflective mirror HR (radius of curvature of 300 mm) and a flat output coupler OC with reflectivity of 50 %. In order to obtain Q-switched regime, two 5 mm long anti-reflection (AR) coated Cr:YAG crystals with low signal transmission of 58 % (corresponding to the total low power transmission of 34 %) were introduced into the resonator close to the output coupler OC. Stable Q-switched operation was obtained for the pump energy of 17 mJ. The single 5 ns output pulse with energy of 1.3 mJ was generated directly from the oscillator and its oscillogram is shown in Fig. 5. The output beam spatial profile was Gaussian in both axes and is shown in the inset of the figure.

At 1.3 μm the 100 mm long optical resonator was formed by a spherical high reflective mirror HR (radius of curvature of 300 mm) and a flat output coupler OC with reflectivity of 70 % at 1.3 μm. Both dichroic mirrors have low reflectivity at wavelength of 1064 nm. As a saturable absorber V:YAG crystal with low signal transmission of 60 % and AR coating was used. Because of possible AR coating damage the saturable absorber was placed close to the rear high reflective mirror HR, where the fundamental laser mode radius is larger. Stable Q-switched operation at 1.3 μm was obtained for the pump energy of 30 mJ, which is nearly two times higher than the pump energy for the operation at 1.06 μm, caused by lower gain at 1.3 μm transition. Generation at the single wavelength of 1338 nm was confirmed by the spectrometer (Ocean Optics NIR512). The 13 ns output pulse with energy of 600 μJ was generated directly from the oscillator and its oscillogram is shown in Fig. 6. The output beam spatial profile shown in the inset of the figure was investigated by a standard silicon CCD

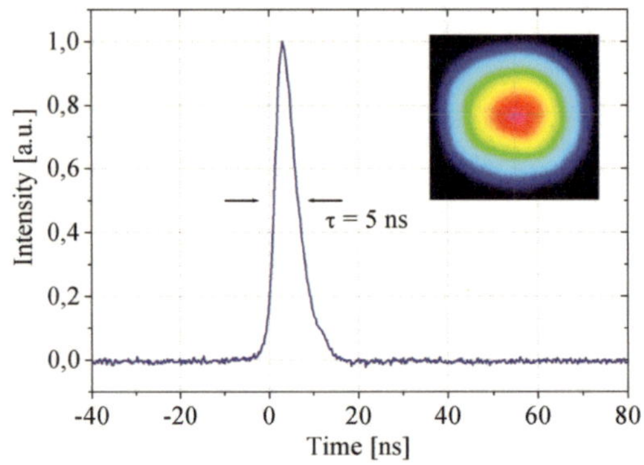

Fig. 5. Oscillogram of the Q-switched output pulse at 1.06 μm.

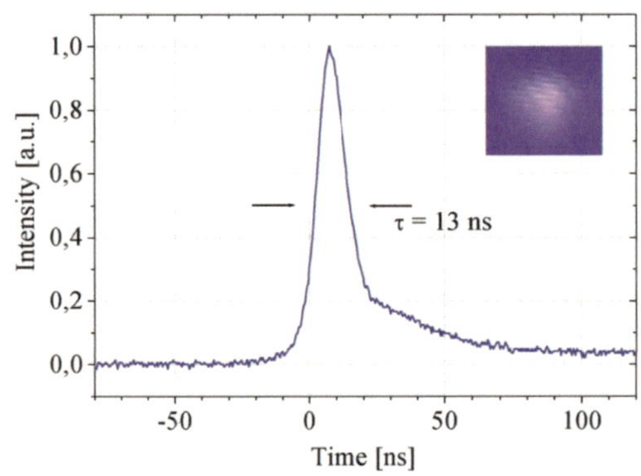

Fig. 6. Oscillogram of the Q-switched output pulse at 1.34 μm.

camera. Because of its low sensitivity in this region the measured signal was quite low but it can be deduced that the beam profile was approximately Gaussian in both axes.

5. Passively mode-locked laser operation at 1.06 μm

In order to obtain mode-locked laser operation, two types of semiconductor saturable absorbers (SA) were tested. In the first case the SA in the transmission mode was used and the output characteristics were compared with the characteristics of the laser mode-locked using the semiconductor saturable absorber mirror (SAM).

5.1 Laser mode-locking by saturable absorber in transmission mode

In the transmission mode two saturable absorbers were investigated in two resonator configurations. The saturable absorbers were grown by molecular beam epitaxy at the Center for High Technology Materials of the University of New Mexico (Diels, 2006; Kubeček, 2009c). The structures were grown on the 400 μm thick GaAs substrates and consisted of 33 or 100 quantum wells (QW) formed by the periods of 9 nm thick $In_{0.275}Ga_{0.725}As$ and 11 nm thick GaAs layers grown at low temperature (around 350° C) in order to decrease the carrier lifetime to the value less than 50 ps (Gupta, 1992). Number of the QWs determines the saturable absorber modulation depth and low-power transmission which were in the case of the SA with 100 QW measured to be ~25 % and ~50 %, respectively. In the case of SA with 33 QW the modulation depth is lower. The GaAs substrate can be used also as a nonlinear element for passive negative feedback by beam defocusing (Corno, 1990; Diels, 2006). The laser resonators differ in length between 25 and 116 cm and also in the ratio of the beam radius inside the Nd:YAG crystal and on the SA. The experimental details are described in Jelínek (2010; 2011b).

5.1.1 Laser operation with short resonator

A schematic diagram of the mode-locked laser oscillator in the configuration (A) is shown in Fig. 7a. The 250 mm long laser resonator was formed by the spherical high reflective mirror HR (radius of curvature of 300 mm) and a flat output coupler OC with reflectivity of 30 %. The Nd:YAG crystal was inserted at a distance of 130 mm from the OC. The fundamental laser beam profile inside the resonator was calculated by software Psst (University of St Andrews, United Kingdom). According to this calculation the beam diameter in the crystal was approximately 600 μm and its diameter on the SA was 400 μm. In order to obtain better spatial overlap between the pump and smaller laser beam, a cylindrical focusing lens L1 (f = 50 mm) was used to focus the pumping radiation.

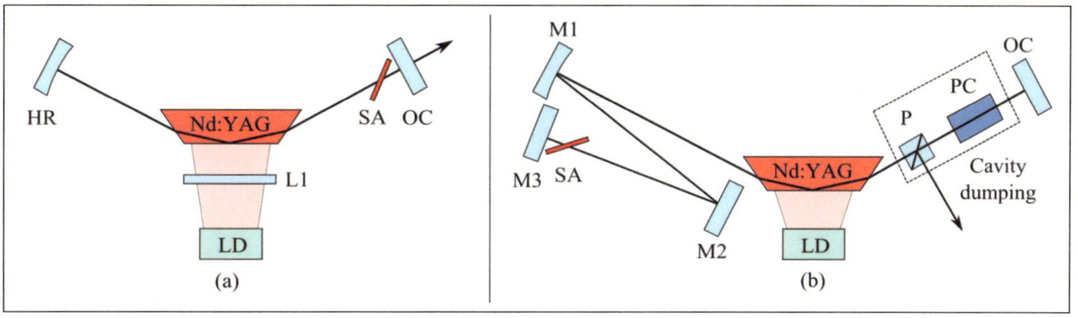

Fig. 7. Schematic of the laser in the mode-locked regime: oscillator (A) 250 mm long and oscillator (B) 1160 mm long.

The laser was passively mode-locked by a multiple quantum well semiconductor saturable absorbers (SA) inserted into the resonator close to the OC under the Brewster's angle in the transmission mode. Two different samples of saturable absorbers, consisted of 33 and 100 quantum wells (QW), were investigated. In addition its GaAs substrate served also as nonlinear element for passive negative feedback by beam defocusing.

Using the SA consisting 33 QW, the mode-locking threshold was achieved for pump energy of 17.5 mJ in 150 μs pulse. Pulse duration evolution along the train was measured by a fast oscilloscope LeCroy SDA 9000 and a PIN photodiode EOT ET-3500 and a typical output pulse train oscillogram together with details of the pulses at the beginning and the end of the train are shown in Fig. 8. The oscilloscope-photodiode system response time was 75 ps.

The measured pulse width was consequently recalculated using the experimentally determined oscilloscope-photodiode instrumental limit and *sum of square* method, which is in details described in Jelínek (2011c); Kubeček (2009d). Although the absolute calculated pulse duration may not be precise, pulse shortening from initial 120 ps to final 35 ps can be clearly observed and in summary is shown in Fig. 9. The whole train consisted of 100 pulses and its

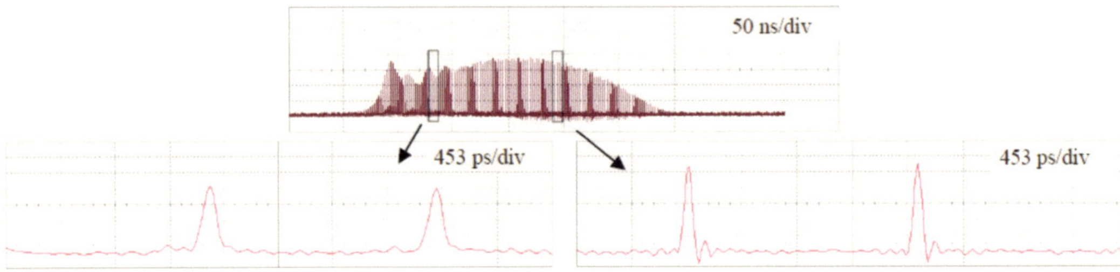

Fig. 8. Oscillogram of the output pulse train using the SA consisted of 33 QWs from the laser resonator (A) (250 mm long) - upper trace, and details of the pulse shapes from the beginning and the end of the train - lower traces.

energy was of 170 μJ. The output beam spatial profile was slightly elliptical in the horizontal axis but it was approximately Gaussian in both axes.

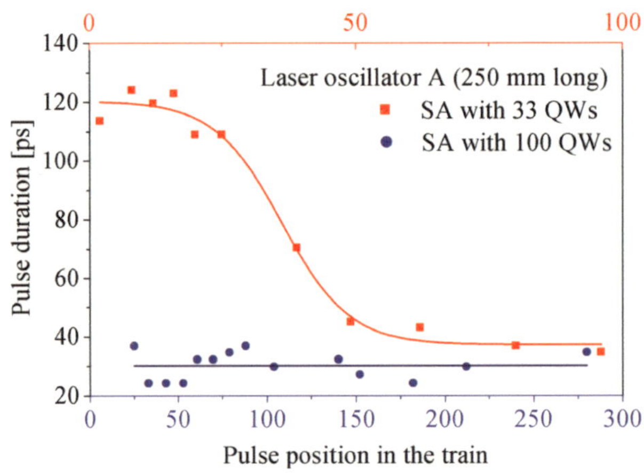

Fig. 9. Pulse duration evolution along the train for the laser resonator (A) mode-locked by the SA consisted of 33 and 100 quantum wells (QW).

Usage of the saturable absorber with 100 quantum wells resulted in higher mode-locking threshold (pump energy of 25 mJ in 180 μs pulse) and also to lengthening of the pulse train. The typical output pulse train oscillogram together with detail of the pulses in the middle of the train is shown in Fig. 10. In this case significant pulse shortening was not observed, the pulse duration of 30 ps remained at approximately constant value from the beginning along the whole train, as can be seen in Fig. 9. The whole train consisted of more than 300 pulses and its energy reached 500 μJ.

Using both saturable absorbers the extended pulse trains typical for passive negative feedback regime of operation were obtained. The corresponding pulse shortening along the train was observed only for pulses with initial pulse duration of 120 ps. With higher depth of modulation the shortening of initially 30 ps long pulses was not observed probably due to the limit of resolution of our detection system.

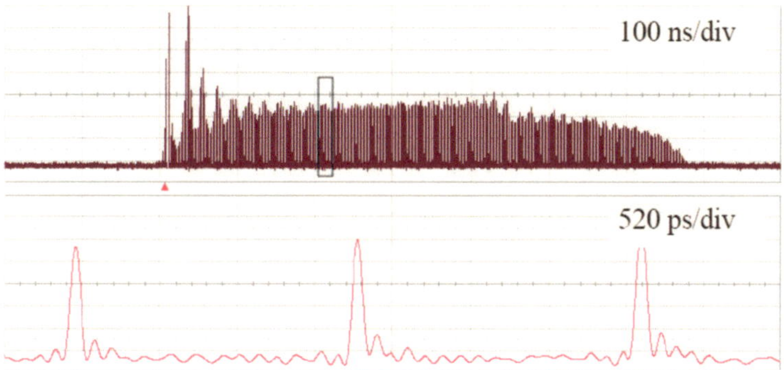

Fig. 10. Oscillogram of the output pulse train using the SA consisted of 100 QWs from the laser resonator A (250 mm long) - upper trace and details of the pulse shapes in the middle of the train - lower traces.

5.1.2 Laser operation with long resonator

In order to obtain shorter pulse trains with longer interval between pulses, which is necessary for cavity dumping, a longer resonator configuration was used. The experimental setup is in details described in Jelínek (2011b).

A schematic diagram of this investigated laser oscillator in configuration B is shown in Fig. 7b. The 116 cm long optical resonator was formed by a flat output coupler OC with the reflectivity of 30% at 1.06 μm, a high reflective folding mirror M1 with the radius of curvature of 1 m, a flat mirror M2, and a flat rear mirror M3. The laser was passively mode-locked by a multiple quantum well saturable absorber SA consisted of 100 quantum wells, inserted into the resonator under the Brewster's angle in transmission mode. The fundamental laser beam profile inside the crystal was approximately 1080 μm and its diameter on the SA was 640 μm.

The mode-locking operation threshold was obtained for the pump energy of 39 mJ in 270 μs long pulse. The output pulse train energy reached 900 μJ and its spatial profile was approximately Gaussian in both axes but slightly elliptical in the horizontal axis and is shown in the inset of Fig. 11. The train consisted of 7 pulses (at half-maximum) and the recorded oscillogram is shown in Fig. 11. Lower graphs show pulse duration histograms of the highest pulse (pulse no. 1) and the pulse no. 4. The Gaussian approximations of the histograms show only small decrease in the pulse duration from initial 76 to 69 ps. Increase in pulse duration stability is also observable, the Gaussian fit FWHM of the histogram decreased from 24 to 19 ps.

5.1.3 Single pulse extraction using cavity dumping

In order to obtain single mode-locked pulse, the cavity dumping technique was used. Into the resonator a KD*P Pockels cell PC and a Glan laser polarizer P were inserted. The radiation transmitted through the output coupler OC was used to trigger a high voltage generator supplying quarter wave step 4.8 kV signal (rise time below 7 ns) for the Pockels cell. The single pulse with the maximum amplitude from the train (pulse no. 1) was extracted with the energy of 30 μJ. In order to obtain maximum energy in the single pulse directly from the oscillator, the 50% output coupler was used. The pump energy of mode-locking operation

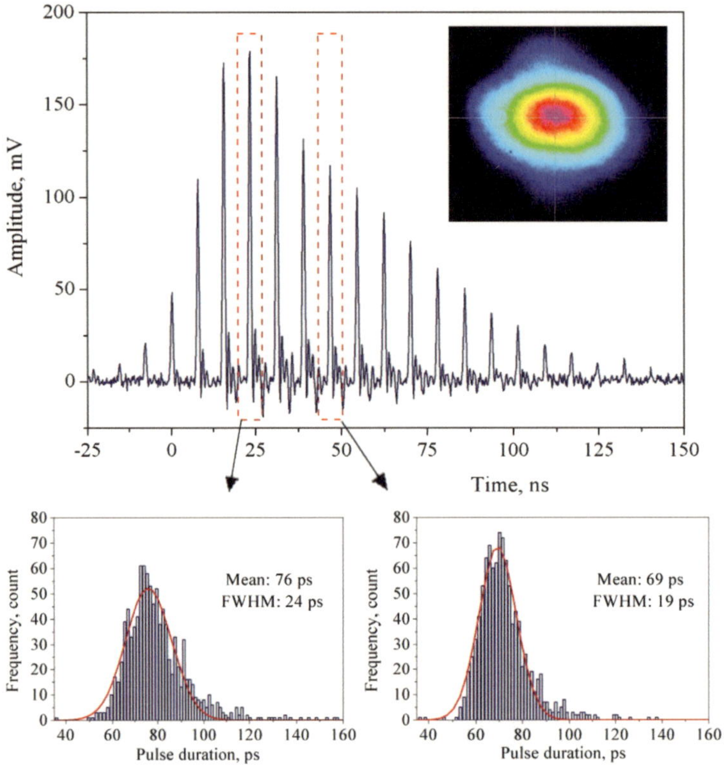

Fig. 11. Upper trace: oscillogram of the oscillator output pulse train. Lower traces: pulse duration histograms of the highest pulse (pulse no. 1) and the pulse no. 4.

threshold decreased to 26 mJ in 180 μs long pulse. The whole output train energy with this output coupler was 500 μJ (measured behind M1 without cavity dumping). The extracted single pulse energy increased to 75±7 μJ and the contrast was better than 10^{-3}. The pulse duration histogram was also measured and is shown in Fig. 12. Its Gaussian approximation gives the mean pulse duration of 113 ps. It can be seen that using the 50% output coupler the pulse duration is longer than using 30 % output coupler which corresponds to the lower saturated gain coefficient in the Nd:YAG.

5.2 Laser mode-locking by saturable absorber mirror

Simpler mode-locked laser configuration was obtained using a saturable absorber mirror (SAM) with high modulation depth. The experimental details and results listed below are in details described in Jelínek (2011d).

The laser oscillator schematic is shown in Fig. 13. The 112 cm long optical resonator was formed by a flat output coupler OC with reflectivity of 30 % at 1.06 μm, a highly reflective folding mirror M1 with the radius of curvature of 1 m, and a flat mirror M2. For passive mode-locking regime a commercially-available semiconductor anti-resonant saturable absorber mirror (SAM) with absorbance of 50 %, modulation depth of 25 %, and relaxation

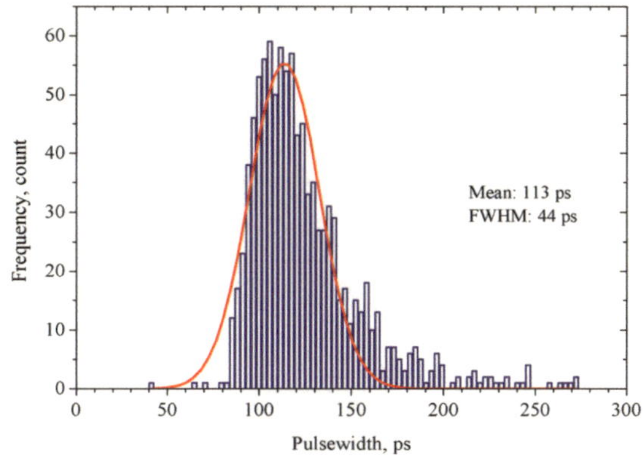

Fig. 12. Extracted pulse duration histogram.

time constant of 5 ps was used. The fundamental laser beam diameter was calculated to be approximately 1 mm in the Nd:YAG crystal and 700 μm on the SAM.

Stable mode-locked laser operation was obtained at the pump energy of 25 mJ in 180 μs long pulse. The 500 μJ output pulse train consisted of $\sim$ 15 pulses (at half-maximum) is shown in Fig. 14.

The output beam spatial profile was close to Gaussian in both axes and is shown in the inset of Fig. 14. Duration of the pulse with the maximum amplitude in the train was measured by the streak camera resulting in the value of 15 ps. The pulse duration stability measured from 10 successive laser shots was better than 2 ps and pulse train stretching was not observed. The duration of the pulse with maximum amplitude was also confirmed by the SHG autocorrelator. Assuming the Gaussian pulse shape the real pulse duration (measured from 1000 laser shots) was calculated to be 16 ps which is in the good agreement with the streak camera measurement. The streak camera and the autocorrelator records can be found in Jelínek (2011d).

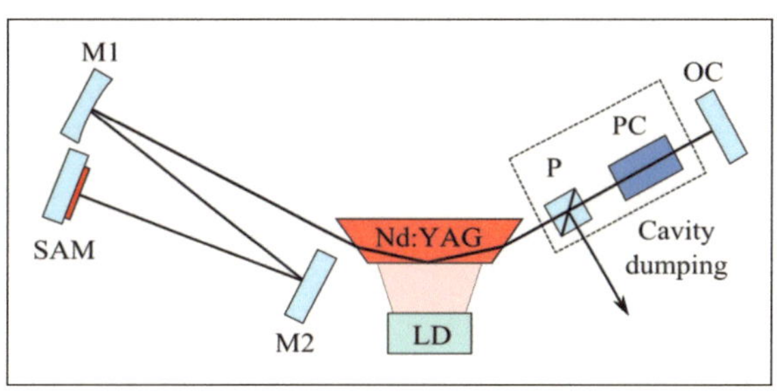

Fig. 13. Schematic of the Nd:YAG laser oscillator in the mode-locked regime using saturable absorber mirror (SAM).

5.2.1 Single pulse extraction using cavity dumping

A single pulse was extracted from the output pulse train by the cavity dumping technique, as it was described in section 5.1.3. In order to obtain high energy single pulse directly from

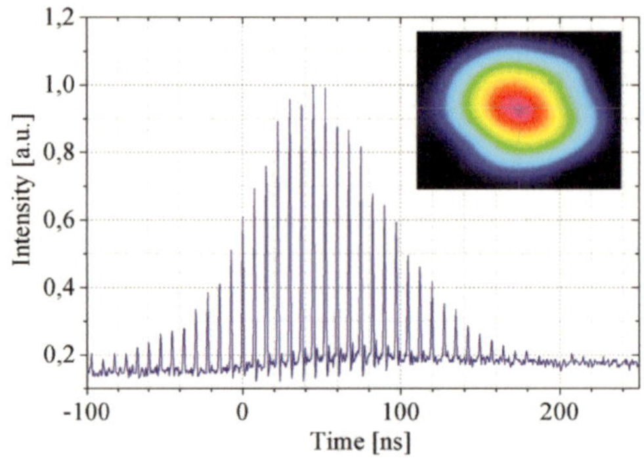

Fig. 14. Output pulse train without cavity dumping.

the oscillator, the output coupler with reflectivity of 70 % was used. Stable mode-locked laser operation was obtained at the pump energy of 19 mJ. Oscillogram of the output pulse train (measured behind the OC) and the extracted single pulse with energy of 25 μJ are shown in Fig. 15. The beam spatial profile was nearly Gaussian in both axes and is shown in the inset of the figure. Duration of the pulse was 17 ps with the stability better than 2 ps. The temporal profile measured by the streak camera is shown in Fig. 16. Inset shows the streak camera record of the pulse and its replica reflected from the 5 mm BK7 plate corresponding to the delay of 50 ps between pulses. In comparison with our previous results described in section 5.1.3, significant (about 5 times) pulse duration shortening and stability improvement was achieved.

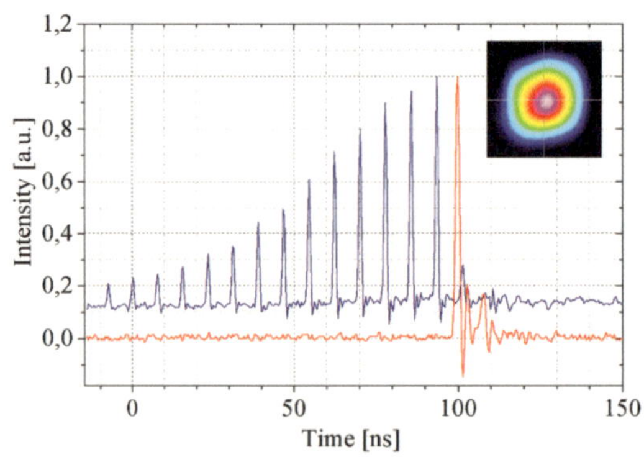

Fig. 15. Oscillogram of the ML Nd:YAG laser output pulse train (blue upper trace) and the extracted pulse (red lower trace).

6. Diode pumped Nd:YAG amplifier in bounce geometry

Performance of the Nd:YAG laser crystal in bounce geometry was also investigated in the single pass amplifier configuration and is in details described in Jelínek (2011a;b). The input pulse diameter was adjusted by a spherical lens to obtain good overlap with the pumped region. Several laser sources generating pulses in the range from 100 ps to 5 ns were used to provide the amplifier input fluence in the range from $\sim 1\,\mu\text{J/cm}^2$ to $3\ \text{J/cm}^2$. According to the laser source performance and certain amplifier setup the pump pulse duration ranging

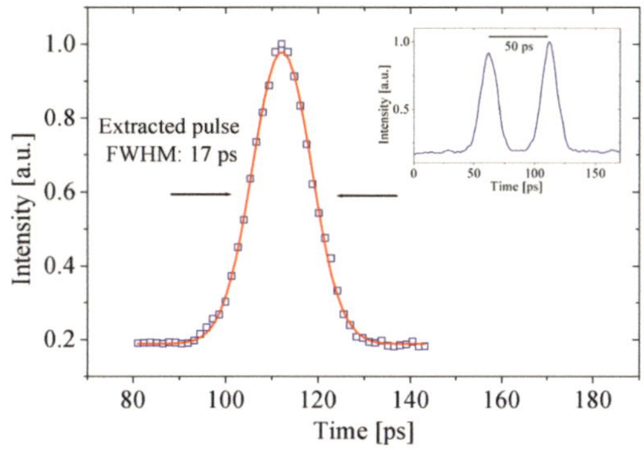

Fig. 16. Streak camera record of the ML Nd:YAG laser extracted pulse temporal profile (dots) and its Gaussian fit (red line).

from 200 to 300 μs with the energy from 29 to 42 mJ was used. Resulting amplification factor is shown in Fig. 17.

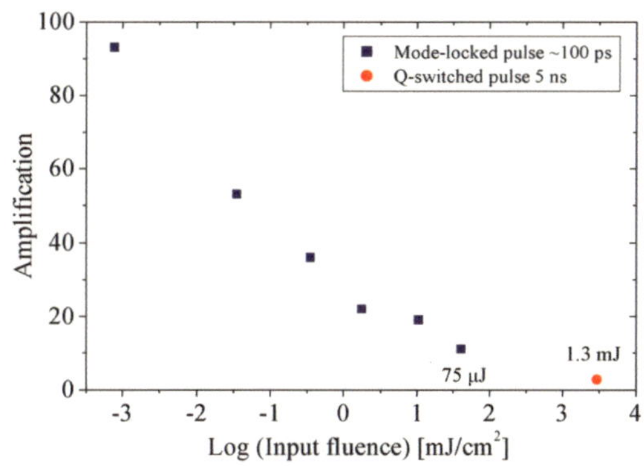

Fig. 17. Nd:YAG laser crystal in bounce geometry amplification under input fluence from 1 μJ/cm^2 to 3 J/cm^2.

It can be seen that in the weak signal regime ($\sim$ 1 μJ/cm^2), amplification up to 93 was obtained. In the strong signal regime (3 J/cm^2), the 1.3 mJ seed pulse was amplified to 3.5 mJ corresponding to the strong signal amplification about 3 times.

7. Conclusion

Operation of the highly 2.4 at.% doped Nd:YAG laser in bounce geometry under quasi-continuous pumping was investigated. The laser was operated in free-running, Q-switched and mode-locked regime. In the free-running regime, efficient operation at the wavelengths of 1.06, 1.3 and 1.4 μm was demonstrated with slope efficiencies of 50, 26, and 8 %, respectively. In the passively Q-switched regime at 1.06 μm, the 5 ns output pulses with the energy of 1.3 mJ were generated. At 1.34 μm, the 13 ns output pulses with energy of 600 μJ were generated.

In order to obtain passively mode-locked regime, two types of saturable absorbers inserted into the resonator either in the transmission or reflection mode were tested. Using the saturable absorber with lower modulation depth (33 quantum wells) in the transmission mode, pulse duration shortening along the extended output pulse train from initial 120 to 35 ps was observed. Whole pulse train energy was 170 μJ. Using the saturable absorber with higher modulation depth (100 quantum wells), the 500 μJ train of 300 pulses with duration of 30 ps was generated without significant shortening effect. In order to be able to extract single pulse from the train, longer laser resonator was constructed. Using the cavity dumping technique, the single 75 μJ pulse with the duration of 113 ± 22 ps was extracted from the resonator.

Simpler mode-locked laser configuration was obtained using a saturable absorber mirror (SAM) with modulation depth of 25 %. The 500 μJ output pulse train consisted of 15 pulses with duration of 15 ps. By the cavity dumping technique, the single 25 μJ pulse with stable duration of 17 ps and Gaussian spatial profile was extracted from the resonator.

Single pass amplification using the Nd:YAG crystal in bounce geometry was also investigated in the weak and also strong signal regime. In the weak signal regime ($\sim 1\,\mu J/cm^2$), amplification up to 93 was obtained. In the strong signal regime ($3\,J/cm^2$), the 1.3 mJ seed pulse was amplified to 3.5 mJ corresponding to the strong signal amplification about 3 times.

8. Acknowledgements

This research has been supported by the Czech Science Foundation under grant No. 102/09/1741, the research projects of the Czech Ministry of Education MSM 6840770022 "Laser Systems, radiation and modern optical applications" and ME 10131 "Picosecond solid state lasers and parametric oscillators for sensors of rotation and other physical quantities."

9. References

A. Agnesi, et al. (1994). Kerr-Lens Modelocking of Solid-State Lasers and Unidirectional Ring Cavities, *IEEE Journal of Quantum Electronics* 30: 1115-1121.

A. Agnesi, et al. (2008). 210-μJ picosecond pulses from a quasi-CW Nd:YVO_4 grazing-incidence two-stage slab amplifier package, *IEEE Journal of Quantum Electronics* 44: 952-957.

A. Agnesi and F. Pirzio (2010). High Gain Solid-State Amplifiers for Picosecond Pulses, in M. Grishin (ed.), *Advances in Solid-State Lasers: Development and Applications*, Intech, Croatia, pp. 213-238.

A. Alcock and J. Bernard (1992). Diode-pumped grazing incidence slab lasers, *IEEE J. Sel. Top. Quantum Electronics* 3: 3-8.

A. D. Corno, et al. (1990). Active-passive mode-locked Nd:YAG laser with passive negative feedback, *Optics Letters* 15: 734-736.

J. Cui, et al. (2009). 500 W Nd:YAG zigzag slab MOPA laser, *Laser Physics* 19: 1974-1976.

J.-C. Diels and W. Rudolph (2006). *Ultrashort Laser Pulse Phenomena*, Elsevier, USA.

J. Dong, et al. (2005). Temperature-dependent stimulated emission cross section and concentration quenching in highly doped Nd^{3+}:YAG crystals, *Phys. stat. sol.* 202: 2565-2573.

X. Fu, et al. (2010). 1 mJ, 500 kHz Nd:YAG/Nd:YVO_4 MOPA laser with a Nd:YAG cavity-dumping seed laser, *Laser Physics* 20: 1707-1711.

X. Fu, et al. (2009). 120W high repetition rate Nd:YVO_4 MOPA laser with a Nd:YAG cavity-dumped seed laser, *Applied Physics B* 95: 63–67.

S. Gupta, et al. (1992). Ultrafast Carrier Dynamics in III-V Semiconductors Grown by Molecular-Beam Epitaxy at Very Low Substrate Temperatures, *IEEE Journal of Quantum Electronics* 28: 2464-2472.

M. Jelínek, et al. (2010). Passively mode locked quasi-continuously pumped 2.4% doped crystalline Nd:YAG laser in a bounce geometry, *Proc. SPIE* 7721: 772115.

M. Jelínek, et al. (2011a). Passively Q-switched quasi-continuously pumped 2.4% Nd:YAG laser in a bounce geometry, *Proc. SPIE* 7912: 791221.

M. Jelínek, et al. (2011b). 0.8 mJ quasi-continuously pumped sub-nanosecond highly doped Nd:YAG oscillator-amplifier laser system in bounce geometry, *Laser Physics Letters* 8: 205-208.

M. Jelínek, et al. (2011c). Single shot diagnostics of quasi-continuously pumped picosecond lasers using fast photodiode and digital oscilloscope, in J.-W. Shi (ed.), *Photodiodes - Communications, Bio-Sensings, Measurements and High-Energy Physics*, Intech, Croatia.

M. Jelínek and V. Kubeček (2011d), 15 ps quasi-continuously pumped passively mode-locked highly doped Nd:YAG laser in bounce geometry, *Laser Physics Letters* 8: 657-660.

U. Keller, et al. (1996). Semiconductor saturable absorber mirrors (SESAMS) for femtosecond to nanosecond pulse generation in solid-state lasers, *IEEE J. Sel. Top. Quantum Electronics* 2: 435-453.

W. Koechner (2006). *Solid-state laser engineering*, Springer, USA.

V. Kubeček, et al. (2008). Side pumped Nd:YAG slab laser mode-locked using multiple quantum well saturable absorbers, *Laser Phys. Letters* 5: 29-33.

V. Kubeček, et al. (2009a). Quasi-continuously pumped passively mode-locked operation of a Nd:$GdVO_4$ and Nd:YVO_4 laser in a bounce geometry, *Laser Physics* 19: 396-399.

V. Kubeček, et al. (2009b). Quasi-continuously pumped operation of 2.4% doped crystalline Nd:YAG in a bounce geometry, *Proc. SPIE* 7193: 719320.

V. Kubeček, et al. (2009c). Quasi-Continuously Pumped Passively Mode-Locked Operation of a Nd:$GdVO_4$ and Nd:YVO_4 Laser in a Bounce Geometry, *Laser Physics* 19: 396-399.

V. Kubeček, et al. (2009d). Pulse shortening by passive negative feedback in mode-locked train from highly-doped Nd:YAG in a bounce geometry, *Proc. SPIE* 7354: 73540R.

V. Kubeček, et al. (2010). 0.4 mJ quasi-continuously pumped picosecond Nd:$GdVO_4$ laser with selectable pulse duration, *Laser Physics Letters* 7: 130-134.

C.-C. Lee, et al. (2010) Ultra-short optical pulse generation with single-layer graphene, *Journal of Nonlinear Optical Physics and Materials* 19: 767-771.

M. Luhrmann, et al. (2009). High energy cw-diode pumped Nd:YVO_4 regenerative amplifier with efficient second harmonic generation, *Optics Express* 17: 22761-22766.

D. Sauder, et al. (2006). High efficiency laser operation of 2at.% doped crystalline Nd:YAG in a bounce geometry, *Opt. Express* 14: 1079-1085.

D. Sauder, et al. (2007). Laser operation at 1.3 μm of 2at.% doped crystalline Nd:YAG in a bounce geometry, *Opt. Express* 15: 3230-3235.

F. P. Schafer, et. al. (1976). Organic dyes in laser technology, *Topics in Current Chemistry* 61.

K. A. Stankov (1988). A Mirror with an Intensity-Dependent Reflection Coefficient, *Appl. Phys. B* 45: 191-195.

D. E. Spence, et. al. (1991). 60-fsec pulse generation from a self-mode-locked Ti:sapphire laser, *Optics Letters* 16: 42-44.

W. D. Tan, et. al. (2010). Mode locking of ceramic Nd:yttrium aluminum garnet with graphene as a saturable absorber, *Applied Physics Letters* 96: 031106.

G. M. Thomas, et al. (2010). Nonlinear mirror modelocking of a bounce geometry laser, *Optics Express* 18: 12663-12668.

Y. Urata, et al. (2001). Laser performance of highly neodymium-doped yttrium aluminum garnet crystals, *Opt. Letters*, 26: 801–803.

X. Wushouer, et al. (2010). High peak power picosecond hybrid fiber and solid-state amplifier system, *Laser Phys. Letters* 7: 644-649.

J. H. Yim, et al. (2008). Fabrication and characterization of ultrafast carbon nanotube saturable absorbers for solid-state laser mode locking near 1 μm, *Applied Physics Letters*, 93: 161106.

11

Application of Nd:YAG Laser in Semiconductors' Nanotechnology

Artur Medvid', Aleksandr Mycko, Pavels Onufrijevs and Edvins Dauksta
Riga Technical University
Latvia

1. Introduction

The main tendency in development of modern electronics and optoelectronics is the use of functional objects of small size. Unique properties of the nanoobjects are mainly determined by their atomic and electronic processes occurring in the structure, which has a quantum character.

Surface nanostructures and their unique properties play a significant role in such objects as highly dispersed systems - adsorbents and catalysts, fillers, composite materials, film and membrane systems (Crommie et al., 1993; Beton et al., 1995; Junno et al., 1995). Formation on the surfaces of ordered and disordered ensembles of nanoparticles allow creating materials with new unique physical properties. Nanostructuring of the surface leads to improvement of optical, electrical, thermal, mechanical and field electron emission properties of materials, for example, reducing of the work function of electron from silicon (Evtukh et al., 2010), enhancing biocompatibility with implants in living tissue and prosthetic devices used in orthopaedics and dentistry. Such materials find application in selective nanocatalyse, microelectronics, nanophotonics, photovoltaic, spectroscopy, and optics. On their base devices are created for recording and storing information with ultra-high density, as well as light-emitting devices (Vu et al., 2010).

Study of different mechanisms of nanostructured materials' formation is required due to the need to develop new and effective methods of forming low-dimensional nanostructures when at least one dimension of crystal is less or equal Bohr' radius of electron, hole or exciton. At this condition the quantum confinement effect manifests in the system, and it leads to the essential change of material physical properties.

One promising direction is to use laser radiation to form nanostructures on solid surfaces.

Lasers have been used for more than 50 years in diverse fields of application starting from simple laser micromachining processes for micro-electro-mechanical systems (MEMS), such as, cutting, drilling, to more smarter ones, e.g.: restoration of art works (laser stone cleaning), pulse laser deposition of coatings and films; local defect annealing after ion implantation; formation the precipitation areas of impurities in Si in medicine, medical diagnosis, treatment, or therapy etc.

Among solid state lasers, Nd:YAG laser has an important role due to its high efficiency, possibility to tune it in different wavelengths from infrared (λ=1064 nm) till ultraviolet (λ=213 nm) and change pulse duration from milliseconds down to picoseconds.

Nowadays, lasers are used for formation of the following semiconductors' structures: p-n junctions, heterostructures Ge/SiGe, CdTe/CdZnTe, and graded band gap (Medvid' & Fedorenko 1999; Medvid', Mychko et al., 2007; Medvid', I.Dmytruk et al., 2007; Medvid', P.Onufrijevs et al., 2008; Medvid', Mychko et al., 2008; Medvid', Onufrijevs et al., 2010; Medvid', Mychko et al., 2010).

Moreover, the most actual application of lasers could be in nanotechnology. We have elaborated a new laser method for nanocones' formation on a surface of semiconductors by laser radiation. The proposed model is characterized by two stages - Laser Redistribution of Atoms (LRA) and Selective Laser Annealing (SLA).

In the case of $Si_{0.7}Ge_{0.3}$ solid solutions the first stage, LRA, is characterized by formation of hetero-structures, such as: Ge/Si due to drift of Ge atoms toward the irradiated surface of the sample in the gradient of temperature, the so-called Thermogradient effect (Medvid', 2002). This process is characterized by positive feedback: after every laser pulse the gradient of temperature increases due to increase of Ge atoms' concentration at the irradiated surface. New Ge phase is formed at the end of the process. Ge atoms are localized at the surface of Si like a thin film. As a result, LRA stage gradually transits to SLA stage.

The second stage, SLA, is characterized by formation of nanocones on the irradiated surface of a semiconductor by selective laser heating of the top layer with a following mechanical plastic deformation of the layer as a result of relaxation of the mechanical compressive stress arising between these layers due to mismatch of their crystal lattices and selective laser heating. SLA occurs due to higher absorption of the laser radiation by the top layer than the buried layer.

In the case of the elementary semiconductors, at the first stage of the process a thin top layer with mechanical compressive stress due to separation and redistribution of interstitials and vacancies in gradient temperature field (Medvid', 2002) on the irradiated surface of the semiconductors is formed. As a result, interstitials are concentrated at the irradiated surface of the semiconductor, forming the top layer. Vacancies are concentrated under the top layer forming a buried layer with mechanical tension due to absence of atoms. Sometimes vacancies form nanocavities (Medvid, 2009). At the second stage of the process nanocones are formed on the irradiated surface of the semiconductors due to plastic deformation of the top layer in the same way as in the previous case with semiconductor solid solutions.

A study of the nanocones' optical, mechanical and electrical properties which are used for construction of third generation solar cells, Si white light emitting diode, photondetector with selective or "bolometer" type spectral sensitivity, Si tip for field electron emitting with low work function of electron is conducted in the chapter.

2. Applications of laser redistribution of atoms

2.1 P-n junction formation in i-Ge crystal by laser radiation

P-n junction is the main component of many semiconductor devices such as diodes, transistors, microchips etc (Makino, Kato, et al. 2008). Thermodiffusion, ion implantation

(Benda, Gowar , et al. 1999) and molecular beam epitaxy (Tomas, Jennings, et al., 2007) are only a few methods to form a p-n junction. The main drawback for these methods is a high cost per p-n junction, since the equipment for these methods is expensive.

A possibility of p-n junction formation by laser radiation was shown in several p- and n-type semiconductors: p-Si (Mada & Ione, et at., 1986; Blums & Medvid', et al., 1995), p-CdTe (Medvid' , Litovchenko et al., 2001), p-InSb (Fujisawa, 1980; Medvid', Fedorenko, et al., 1998), p-InAs (Kurbatov & Stojanova, 1983), p-PbSe (Tovstyuk, Placko, et al., 1984), p-Ge (Kiyak & Savitsky, 1984) and n-HgCdTe (Dumanski, Bester, et al., 2006) due to inversion of conductivity type. Different mechanisms have been proposed to explain the nature of inversion of conductivity type, for example, impurities' segregation, defects' generation (Tovstyuk, Placko, et al., 1984), amorphization (Ljamichev, Litvak, et al. 1976) and oxygen related donor generation (Mada & Ione, et at., 1986). However, n-type impurities in Si irradiated by laser cannot be oxygen atoms, according to paper (Blums & Medvid', et al., 1995). Several authors have tried to explain p-n junction formation in n-type HgCdTe by defects generation, based on a model of defects formation related to interstitial mercury diffusion (Dumanski, Bester, et al., 2006). On the other hand, the authors of those papers did not take into account the effect of temperature gradient on the diffusion of atoms in solid solution. For example, diffusion of Hg interstitials toward the bulk of the semiconductor is impeded due to gradient of temperature directed to the irradiated surface of the semiconductor. Moreover, it is theoretically shown, that the p-n junction can be formed by redistribution of impurities in co-doped Si in gradient temperature field (Medvid' & Kaupuzs, 1994). In this case the atoms with bigger effective radii than Si atom radius drift toward the irradiated surface of the Si crystal, but those with radii smaller than Si atom radius - to the opposite direction.

Characteristics of p-n junction, formed by laser radiation in semiconductors, are comparable to the commercial ones, that is why laser technology has a promising future. The main advantages of this technology are low cost and the possibility to locate p-n junction fast and precisely.

Unfortunately, the mechanism of p-n junction formation by laser radiation has not been clear until now. In this paper a new mechanism of p-n junction formation in intrinsic semiconductor by laser radiation is proposed.

According to the model, p-n junction is formed in a semiconductor by strongly absorbed laser radiation due to generation and redistribution of intrinsic point defects (interstitial atoms and vacancies) in temperature gradient field, so called Thermo gradient effect (Medvid', 2002).

For this purpose i-Ge crystal is irradiated by Nd:YAG laser with different energy of quantum. I-type Ge crystal was used in the experiments as a model material because the concentration of impurities in this material is lower than the concentration of intrinsic point defects at room temperature.

In the experiments i-Ge single crystals with $N_a = 7.4 \times 10^{11}$ cm^{-3}, $N_d = 6.8 \times 10^{11}$ cm^{-3} ,where N_a and N_d are acceptors' and donors' concentration, and slab dimensions of 16.0 × 3.5 × 2.0 mm^3 were used. Samples were mechanically polished with diamond paste and chemically treated with H_2O_2 and CP-4 ($HF:HNO_3:CH_3COOH$ in volume ratio 3:5:3), therefore all surfaces of samples are characterised by minimum surface recombination velocity of S_{min} = 100 cm/s (Pozela, 1980).

Different intensities and energies of laser radiation quanta: $h\nu_1$=1.16 eV (λ_1 = 1064 nm), $h\nu_2$=2.32 eV (λ_2 = 532 nm), and $h\nu_4$=4.66 eV (λ_4 = 266 nm), where ν_1- fundamental frequency, ν_2- second harmonics, ν_4- fourth harmonics and λ- wavelength, of nanosecond Nd:YAG laser with pulse duration τ_1 = 6.0 ns, τ_2 =4.0 ns and τ_4 =3.0 ns, were used to irradiate the samples. Current-voltage characteristics were measured for the non-irradiated and the irradiated samples. Measurements of current-voltage characteristics were done by soldering tin electrical contacts directly on the irradiated surface of i-Ge and on the opposite side. Also, current-voltage characteristics were measured at different number of laser pulses in order to determine the mechanism of the effect. Maximum electric field applied to the samples was 100 V/cm. Measurements of current-voltage characteristics were done at room temperature (T) and atmospheric pressure. Rectification ratio was calculated at constant 30 V voltages.

Current-voltage characteristics of i-Ge samples before and after irradiation by Nd:YAG laser with energy of laser radiation quantum $h\nu_4$=4.66 eV and different intensities, are shown in Fig. 1. The current-voltage characteristic of the non-irradiated sample is linear. It means, current-voltage characteristics obey Ohm's law and therefore there are no potential barriers between the electric contacts and the sample. But after irradiation by the laser current-voltage characteristics becomes diode-like. Moreover, this process takes place in threshold manner - resistance of the sample increases by 10 times and rectification effect appears at a certain intensity of the laser radiation.

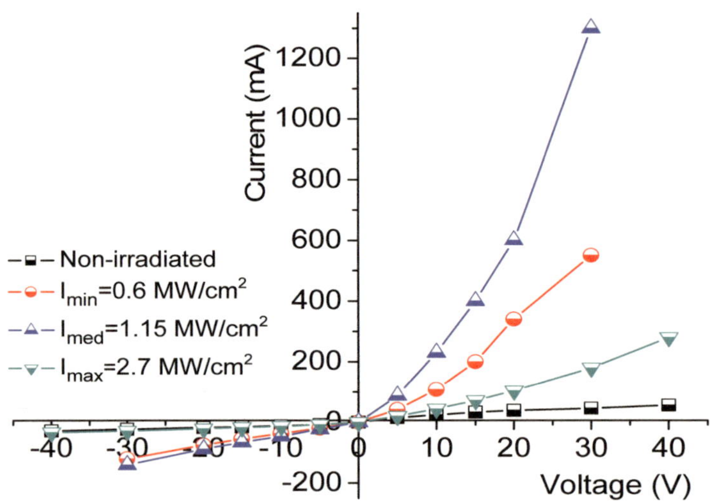

Fig. 1. Current-voltage characteristics of a non-irradiated and an irradiated i-Ge sample by Nd:YAG laser with different intensities of quantum energy $h\nu_4$=4.66 eV and 350 laser pulses.

Threshold intensity (I_{th}) decreases with increase of energy of laser radiation quantum, as seen in Fig. 2. The rectification effect of current-voltage characteristics, quantitively determined by rectification ratio, is increased with intensity and energy of laser radiation quantum, as shown in Fig. 1. and 2. Threshold intensities are observed at the fundamental frequency I_{th1}= 5.5 MW/cm^2, the second harmonic I_{th2}= 1.5 MW/cm^2 and the fourth harmonic I_{th4}= 1.15 MW/cm^2.

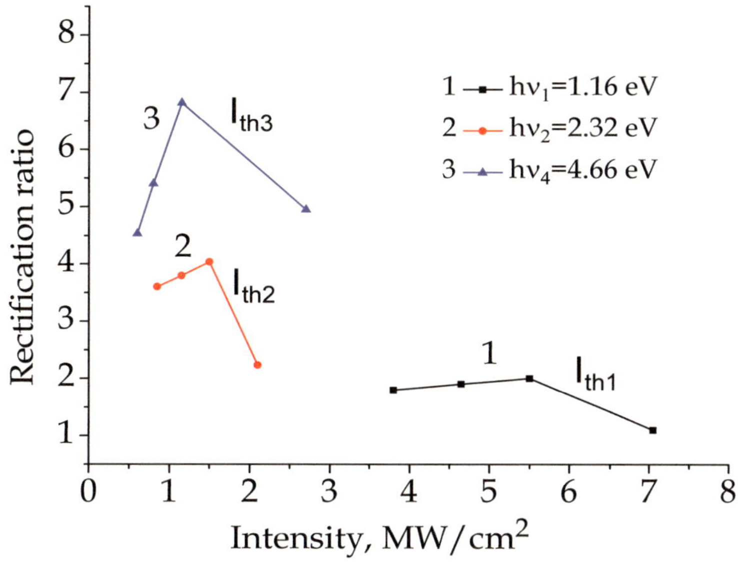

Fig. 2. Rectification ratio as a function of irradiation intensity of Nd:YAG laser for different energies of the laser radiation quantum and 350 laser pulses.

The rectification ratio of 9.3 at 350 laser pulses was observed after irradiation with the highest quantum energy $h\nu_4$=4.66 eV, but the highest rectification ratio of 400 and potential barrier height 0,472 eV were observed at 650 laser pulses for the second harmonics $h\nu_2$=2.32 eV as can be seen in Fig. 3.

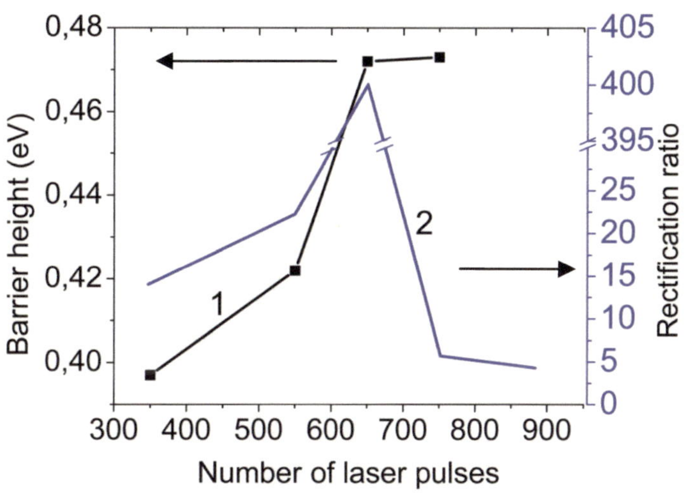

Fig. 3. Barrier height and rectification ratio as a function of number of laser pulses with energy of laser radiation quantum $h\nu_2$=2.32 eV.

Current-voltage characteristics of samples irradiated with energies of laser radiation quanta $h\nu_1$=1.16 eV and $h\nu_2$=2.32 eV are not shown here because they are similar to current-voltage characteristics in Fig. 1. Decrease of the threshold intensity with increase of energy of laser radiation quantum and appearance of current-voltage characteristics rectification effect we explain in the following way: irradiation of the sample by strongly absorbed laser radiation leads to transformation of the light energy to the thermal one. Heating up the sample by laser radiation increases additional generation of intrinsic defects in crystal - interstitials and vacancies which are quenched in crystal lattice after the end of the laser pulse. The formation of a potential barrier takes place due to separation of vacancies and interstitials in temperature gradient field (Medvid, 2002). Calculations of T_{max} on the surface of the sample at threshold intensity of laser radiation using the heat balance equation in approach of adiabatic process show that T_{max1} in the case of fundamental frequency is lower than T_{max2} for the second and T_{max4} for the fourth harmonics.

Energy of laser radiation quantum, eV	Wavelength, nm	Absorption coefficient, x10^8 m^{-1}	Absorption depth, x10^{-9} m	Rectification ratio	Barrier height, eV
4,66	266	1.60	6.25	9.29	0.36
2,32	532	0.45	22.2	4.16	0.34
1,16	1064	0.05	200	2.14	0.32

Table 1. i-Ge and p-n junction parameters

It means that gradient of temperature has the main role in formation of the potential barrier, because gradient of T is proportional to the light absorption coefficient in i-Ge crystal which increases with energy of laser radiation quantum, as seen in Table 1.

Decrease of the threshold intensity of laser radiation with increase of laser radiation quantum, as can be seen in Fig. 2, is an evidence of this suggestion. Concentration of interstitials at the irradiated surface and vacancies in the buried layer of the sample leads to formation of n-p junction because interstitials are donors and vacancies are acceptors in Ge (Shaw, 1973). Schematic illustration of n-p-i structure formed by the laser in i-Ge is shown in Fig. 4.

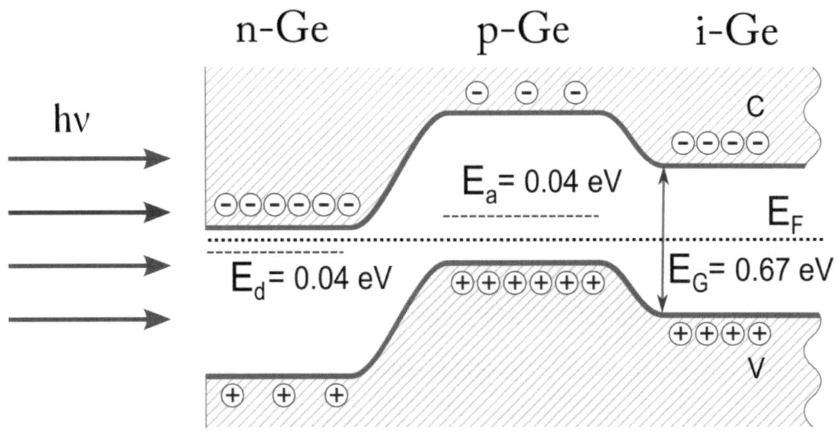

Fig. 4. Schematic illustration of n-p-i structure after irradiation of i-Ge sample. Interstitial atoms are known to be of n-type forming E_d level 0.04 eV below the conduction band, but vacancies are known to be of p-type forming E_a level 0.2 eV above the valence band (Claeys & Simon, 2007).

Nonmonotonous dependence of rectification ratio as function of the laser intensity and decrease of rectification ratio at maximal intensities of laser irradiation are explained by formation of nanocones on the irradiated surfaces of the samples (Medvid', Onufrijevs, et al., 2011) and therefore a partial destruction of the n-type layer on the irradiated surface of the samples. The current-voltage characteristics measured at a different number of the laser pulses (N) from 350 to 880 at irradiation of the samples by the second laser harmonic and laser intensity 1.5 MW/cm² are shown in Fig. 3. We can see that rectification ratio of the current-voltage characteristics is a non-monotonous function of N with maximum at 650 pulses, as can be seen in Fig.3. At the initial stage of the function rectification ratio increases with N, but at 650 it sharply decreases. Increase of rectification ratio with N (Fig. 3.) we explain by accumulation effect: after every laser pulse concentration of interstitials at the irradiated surface and concentration of vacancies in the buried layer of the sample increases and, of course, n-p junction barrier height increases, too, as shown in Fig. 3, curve 1.

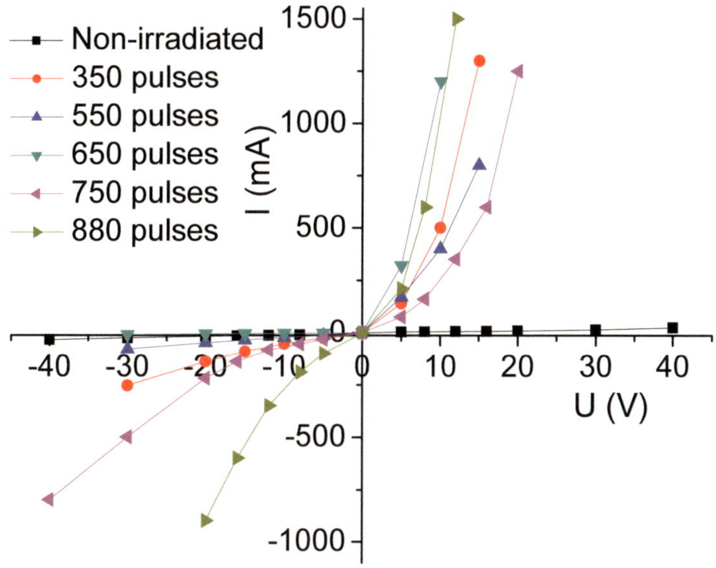

Fig. 5. Current-voltage characteristics of i-Ge samples before and after Nd:YAG laser irradiation at intensity 1.5 MW/cm² and quantum energy $h\nu_2$=2.32 for different number of laser pulses.

To calculate barrier height equation 1 was rearranged to the express barrier height (Gupta, Yakuphanoglu, et al., 2011):

$$I_s = AA^*T^2 \exp\left(-\frac{\phi_b}{kT}\right), \tag{1}$$

where A is the irradiated area (0.3 × 0.3 cm²) of i-Ge, A^* is the Richardson constant, which is 40.8 A cm⁻² K⁻² for i-Ge (Chua, Nikolai, et al., 2003), k is the Boltzmann constant, ϕ_b - the barrier height.

A sharp decrease of rectification ratio at N more than 650 pulses is explained by nanostructure formation on the irradiated surface of the sample due to relaxation of mechanic compressive stress, as shown in Fig. 6. It arises in the structure due to a high concentration of interstitials in the top layer and vacancies in the buried layer.

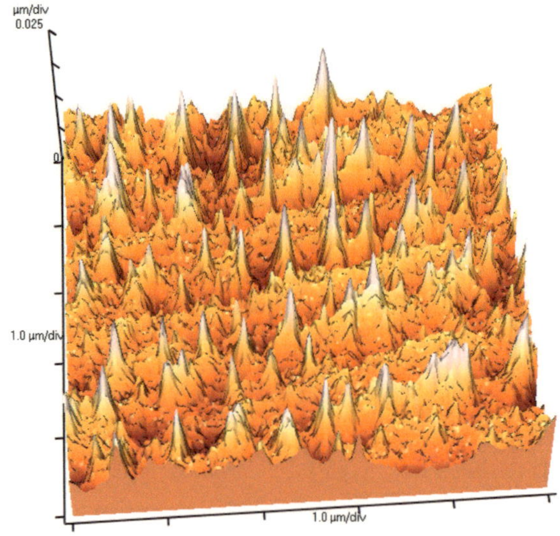

Fig. 6. AFM image of i-Ge surface irradiated by the Nd:YAG laser at intensity 7.0 MW/cm^2.

Conclusion

- For the first time we have proved that the mechanism of p-n junction formation in intrinsic semiconductor is caused by generation and redistribution of intrinsic point defects in temperature gradient field.
- Increase of rectification ratio and the potential barrier of p-n junction with the increase of laser radiation intensity and number of the laser pulses are explained by Thermogradient effect, therefore this effect has the main role in formation of p-n junction.

2.2 X- and Y-ray detector with increased radiation hardness

It was noted that radiation damage occurs in semiconductor radiation detectors during their operation, while measuring ionizing radiation, which impairs the ability of the device (Lindstrom, 2003).The main expressions of radiation damage are the following: the increase of leakage current in a semiconductor detector, the need to increase the bias voltage, reduction of the efficiency of collecting the charge created by ionization. The main objectives of research are not only the elucidation of the causes of degradation, but also the development of methods for increasing the operation time of devices.

The preliminary irradiation with neutrons or charged particles with subsequent annealing also allows increasing radiation hardness of material.

One of the promising materials for the use in radiation detectors is CdTe and solid solutions based on this material (Fougeres et al., 1999; Niraula et. al., 1999). Due to the wide band gap, detectors based on CdTe can operate at room temperature. One of the important tasks for application of radiation detectors and semiconductor devices is a study of the possibility to reduce the influence of ionizing radiation on their parameters.

Investigation of influence of γ radiation on the work of CdTe detectors is studied in papers (Franks et. al., 1999; Taguchi et. al.,1978; Cavallini et.al., 2000; Cavallini et.al., 2002). It was noted that CdTe possesses a high enough radiation hardness due to high mass of elements in the compound material (Shapiro, (2002).

In this work the possibility to increase radiation hardness of CdZnTe crystals using laser radiation has been studied.

Single crystals of $Cd_{1-x}Zn_xTe$ with x= 0.1 and sizes 10.0 mm × 10.0 mm × 2.0 mm grown by High-Pressure Vertical Zone Melting method were used in our experiments. After irradiation by pulsed Nd:YAG laser samples were γ-irradiated by a ^{60}Co source (E_γ photons = 1.2 MeV) at room temperature with a dose rate of 5×10^5 Rad = 5.0 KGy. Nd:YAG laser with the following parameters: wavelength λ = 0.532 μm, pulse duration τ = 10.0 ns, power P = 1.0 MW and intensity range I = 0.48 - 1.80 MW/cm^2 was used. The irradiated surface of crystal was covered with a thin layer of SiO_2 in order to avoid material evaporation during laser irradiation. The thickness of SiO_2 layer was 0.3 μm and it was transparent for laser radiation. To assess the damage of the crystalline lattice after irradiation by γ-ray, photoluminescence method was used. Photoluminescence spectra were measured at 5K using 632.8 nm line of He-Ne laser with excitation power of less than 200 mW. It is known that γ-radiation mainly causes intrinsic defect generation in a semiconductor.

Photoluminescence spectra of irradiated and non-irradiated $Cd_{1-x}Zn_xTe$ crystal by Nd:YAG laser are shown in Figure 7. There are typical photoluminescence spectra of p-type $Cd_{0.9}Zn_{0.1}Te$ containing an intense band at 1.645 eV (A^0X) ascribed to excitons bound to shallow acceptors (Cd vacancies - V_{Cd}), longitudinal optical phonon replicas at 1.624 eV (A^0X-LO) and a weak band at 1.657 eV (D°X) of excitons bound to shallow donors (Cd interstitial - Cd_I). The PL band around 1.54 eV is caused by recombination of donor-acceptor pairs (D-A) with vacancy impurity complexes (V_{Cd}-D_{Cd}, D=group III or VII elements) (Yang et. al., 2006; Suzuki et. al., 2001).

This non-monotonic change of the intensity of the A^0X line in photoluminescence can be attributed to the effects of the temperature gradient induced by strong absorption of laser radiation (Thermogradient effect). This effect causes intense generation of point defects and their redistribution in the irradiated area (Medvid', 2002).

Point defects and vacancies are generated under the influence of the temperature gradient in the surface layer during irradiation by laser: Cd (V_{Cd}) vacancies and interstitial atoms of Cd (Cd_I). The increase in V_{Cd} concentration causes an increase in the intensity of the A^0X line in the photoluminescence spectrum. The D^0X line in the photoluminescence spectrum is not shown, since this semiconductor is initially p-type and therefore the concentration of donors is very small (Figure 7 curves 2 and 3).

After irradiation at high laser intensity generation of point defects in the irradiated area increases. In the irradiated area of semiconductor processes of generation and recombination and simultaneous redistribution of defects in the temperature gradient field take place. Due to the temperature gradient Cd vacancies move in to the bulk of semiconductor in the region of lower temperature, but interstitial atoms of Cd - in the opposite direction, to the surface of semiconductor. When the laser irradiation intensity reaches 1.2 MW/cm^2 gradient of temperature partially compensates vacancies of Cd atoms by Zn atoms (Figure 7 curve 4). This is due to the greater bonding energy of Zn and Te atoms in comparison to Cd and Te atoms. At lower intensities of laser irradiation, this process is less likely to happen.

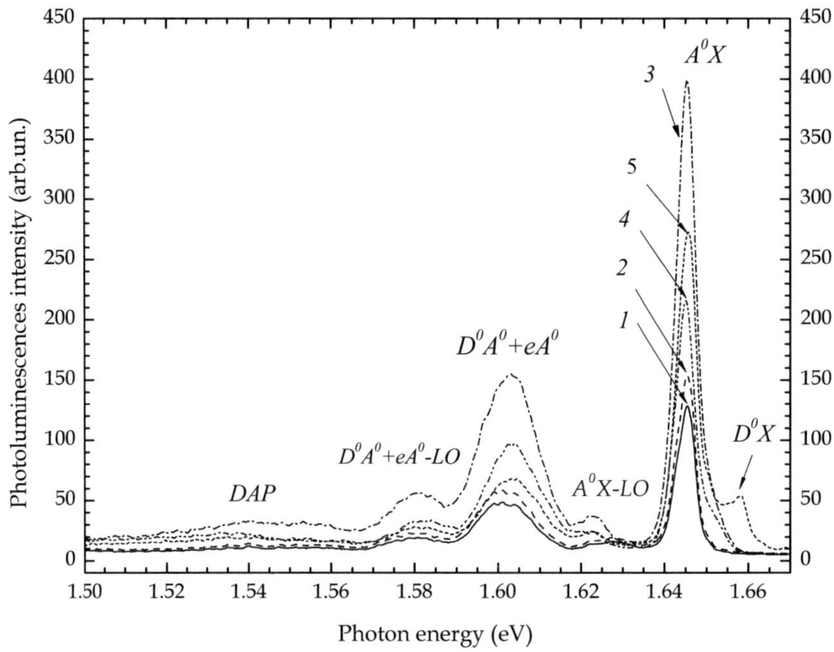

Fig. 7. The photoluminescence intensity as a function of photon energy for $Cd_{1-x}Zn_xTe$ before and after irradiation by Nd:YAG laser: curve 1 – non irradiated; curve 2 – irradiated with intensity 0.48 MW/cm^2; curve 3 - irradiated with intensity 0.84 MW/cm^2; curve 4 – irradiated with intensity 1.20 MW/cm^2; curve 5 - irradiated with intensity 1.80 MW/cm^2.

As a result, the A^0X line intensity decreases due compensation of vacancies by Zn atoms as well as localization of V_{Cd} in the bulk of semiconductor. Intensive generation of interstitial atoms of Cd and increase of their concentration, and their localization at the surface can be proved by appearance of a the D^0X line in the photoluminescence spectrum (Figure 7 curve 5 and Figure 8 curve 2). Thus, in the surface layer of irradiated semiconductor a region highly enriched by point defects V_{Cd} and Cd_I is formed.

Irradiation of $Cd_{1-x}Zn_xTe$ crystal by γ-ray with a dose rate of 5×10^5 Rad = 5.0 KGy leads to a strong increase of the A^0X band intensity in photoluminescence spectra of $Cd_{1-x}Zn_xTe$ crystal, as shown in Figure 8. For example, intensity of the A^0X band is increased 10 times for curve 3 in comparison with intensity for curve 1. At the same time, the D^0X band in photoluminescence spectrum of $Cd_{1-x}Zn_xTe$ crystal disappears fully, as shown in Figure 9, curve 2 and Figure 10, curve 2.

We explain this effect by Cd vacancies' generation and localization in the excited luminescence thin layer after γ-irradiation of $Cd_{1-x}Zn_xTe$ crystal.

The main effect observed in the study is suppression of V_{Cd} generation and /or localization by γ-radiation at the irradiated surface of $Cd_{1-x}Zn_xTe$ crystal if the crystal is preliminary irradiated by the laser.

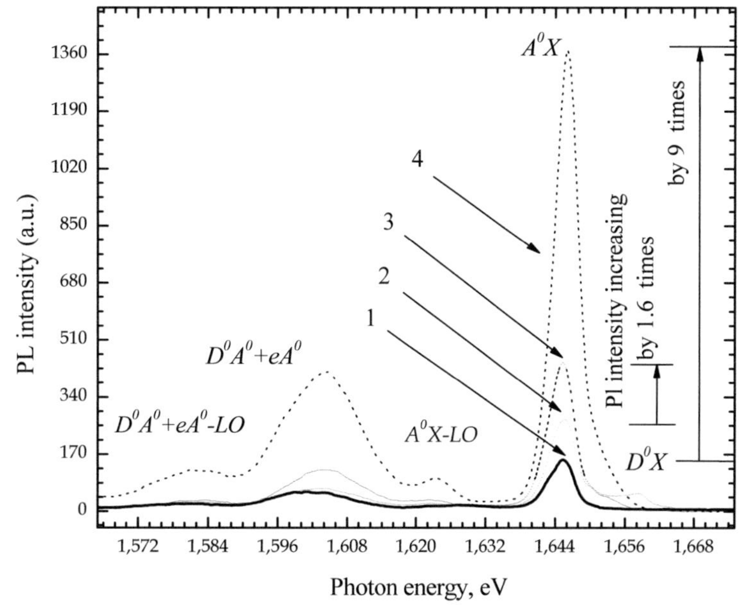

Fig. 8. The photoluminescence spectra of $Cd_{1-x}Zn_xTe$ crystal before (curve 1) and after irradiation by Nd:YAG laser: curve 2 - irradiated by Nd:YAG laser with intensity 1.80 MW/cm²; curve 4 – irradiated by γ- ray with a dose rate of 5.0 KGy; curve 3 – previously irradiated by Nd:YAG laser with intensity 1.80 MW/cm_2 and subsequently by γ- ray with a dose rate of 5.0 KGy

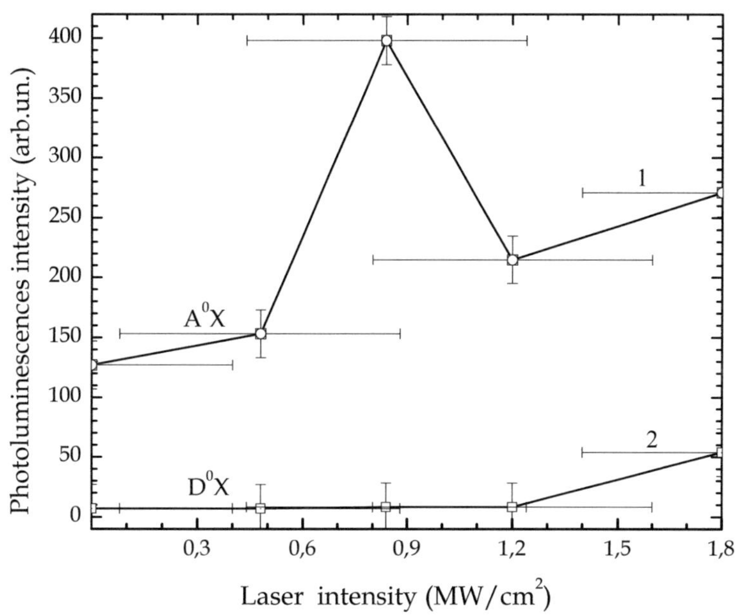

Fig. 9. The photoluminescence intensity of the A^0X (curve 1) and D^0X (curve 2) bands of CdZnTe crystal irradiated by the laser as a function of Nd:YAG laser intensity.

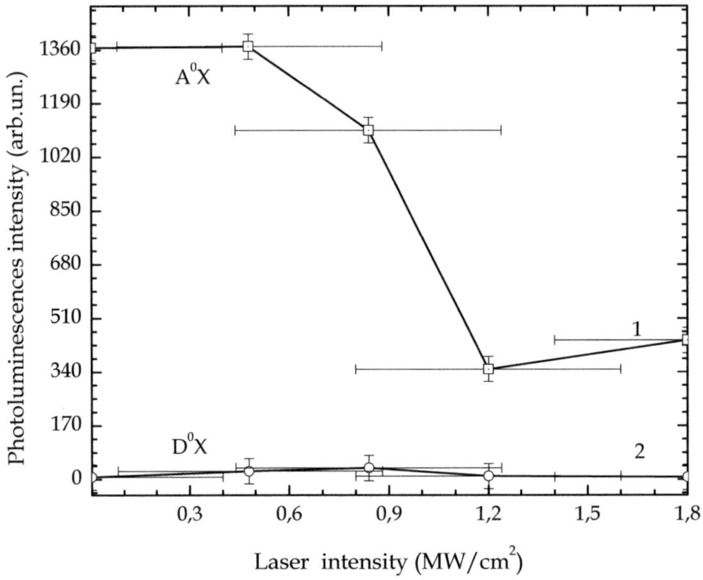

Fig. 10. The photoluminescence intensity of A^0X (curve 1) and the D^0X (curve 2) bands of CdZnTe crystal irradiated by the laser and after subsequent irradiation by γ-rays with a dose rate of 5.0 KGy as a function of Nd:YAG laser intensity.

The effect increases with intensity of the laser in region of the laser intensity up to 0.50 - 2.0 MW/cm^2, as shown in Figure 8, curve 3. This effect is clearly observed in Fig. 3 and Fig.4. Intensity of the A^0X band in photoluminescence spectrum of $Cd_{1-x}Zn_xTe$ crystal increases only 1.7 times (for comparison, non-irradiated by the laser: 9.3 times) after γ-radiation if the crystal preliminary was irradiated by the laser at intensity 1.2 WM/cm^2.

Gamma radiation causes the generation of additional Cd vacancies and interstitial Cd atoms, which causes the growth of the intensity of the A^0X line in the photoluminescence spectrum. Generation and recombination process takes place simultaneously, when the existing interstitial atoms or atoms generated under the influence of gamma radiation move and fill the existing vacancies or newly generated vacancies. As a result, the rate of increase in the concentration of vacancies and interstitial atoms is slowed down because of their partial recombination. These processes can be explained by a decrease of the A^0X line intensity growth, which also means an increase in radiation resistance of semiconductor.

It should be noted that such mechanism of "defect healing" is valid only up to certain values of radiation doses at which the number of donors formed by laser irradiation may still restrain the growth of the number of vacancies due to the effects of gamma radiation.

3. Application of selective laser annealing

3.1 Third generation solar cells on the base of ITO/Si/Al structure

Indium-tin-oxide (ITO) thin films are widely used as a transparent conductive oxide in optoelectronics devices such as solar cells (Bruk *et al.*, 2009), liquid crystal displays and plasma display panels. This material has high transmittance in the visible region of spectra (Adurodija *et al.*, 2002), surface uniformity and process compatibility (Blasundaraprabhu *et al.*, 2009). ITO is a perspective material for elaboration of new generation solar cells using nanotechnology.

We investigated the ITO/p-Si/Al structure irradiated by Nd:YAG laser with the aim to form a p-n junction and to grow nanocones on the interface of ITO/Si.

ITO/p-Si/Al structures, with ITO top layer thickness of 70 nm, Si layer thicknes of 500 μm and an Al back layer thickness of 100 nm in the experiments were studied. The structures were irradiated from the ITO side by Nd:YAG laser second harmonic with a wavelength of λ = 512 nm and a pulse duration of τ = 10 ns. The diameter of the laser beam spot was 0.9

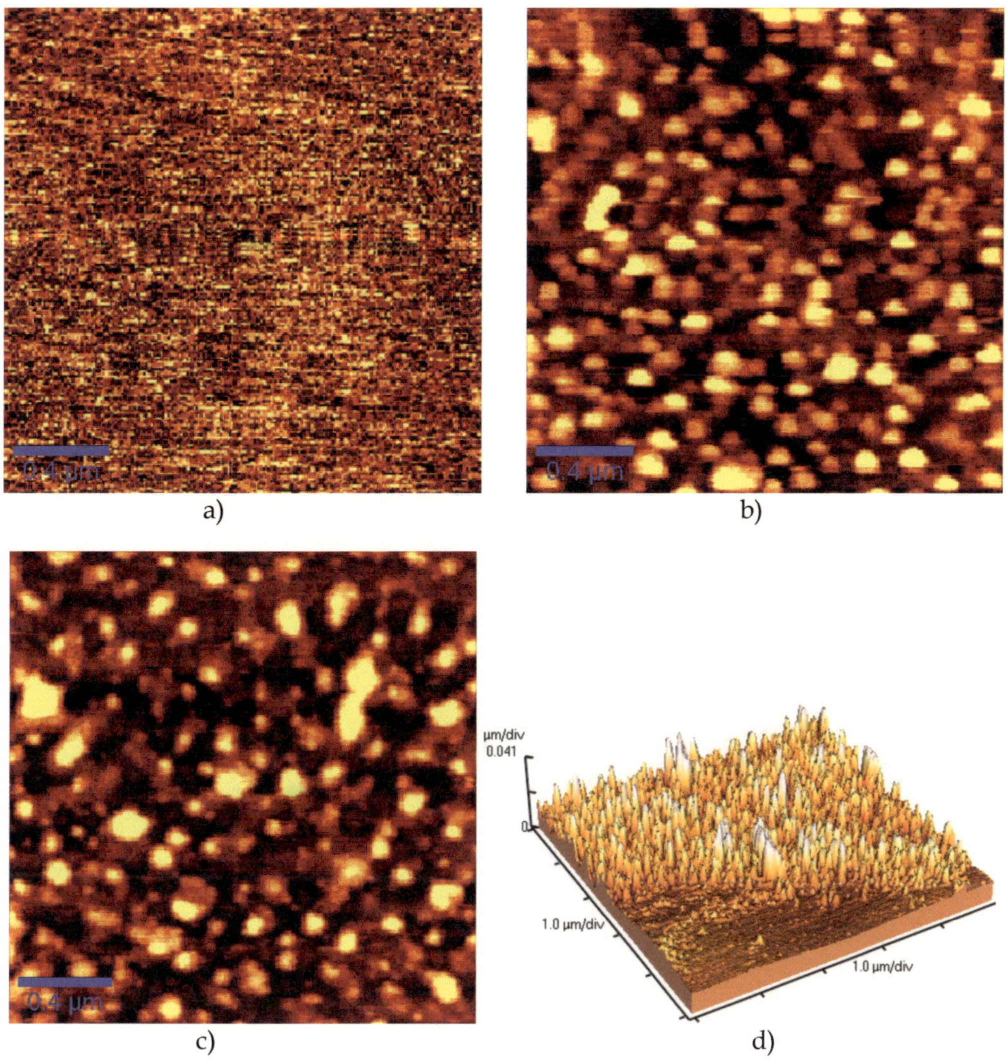

Fig. 11. Atomic force microscope images of ITO/p-Si structure: a) non-irradiated, irradiated by Nd:YAG laser at intensities of b) 1.13 MW/cm^2; c) 2.83 MW/cm^2 and d) boarder of non-irradiated (smooth surface) and irradiated by Nd:YAG laser at intensity 2.83 MW/cm^2 (surface with nanocones).

mm. Irradiation of the samples in the scanning regime using a two coordinate manipulator with 20 μm step was carried out. Experiments in ambient atmosphere at pressure of 1 atm, T = 20^{0}C, and 60% humidity were performed.

The irradiated structures were studied using atomic force microscope and photoluminescence spectroscopy method. The measurements of current - voltage characteristic to observe the changes of electrical parameters for ITO/p-Si/Al structure after irradiation by laser were carried out. The n-type Si layer was formed on p-type Si, due to drift of interstitial Si atoms towards the irradiated surface, as a result of huge gradient of temperature induced by laser radiation, that so called Thermogradient effect (Medvid', 2002).

The two-dimensional surface topography of ITO/p-Si structure was studied by atomic force microscope, non-irradiated and irradiated by the Nd:YAG laser at intensities of 1.13 MW/cm^2 and 2.83 MW/cm^2 (figures 11a, 11b, 11c, respectively). Nanocones were observed on the irradiated Si surface with the average height 12 nm and 26 nm formed by laser radiation at the intensity of 1.13 MW/cm^2, and 2.83 MW/cm^2, respectively (e.g. see figures 11d).

Photoluminescence spectra of the ITO/p-Si structures with the maxima at 575 nm and 490 nm obtained after laser irradiation at intensities of 1.13 MW/cm^2 and 2.83 MW/cm^2 are shown in figure 12. Position of the observed photoluminescence maximum compared with the bulk Si shows a significant "blue shift". The maxima of the photoluminescence band at 575nm and 490 nm are explained by presence of the Quantum confinement effect (Brus, 1984) on the top of nanocones.

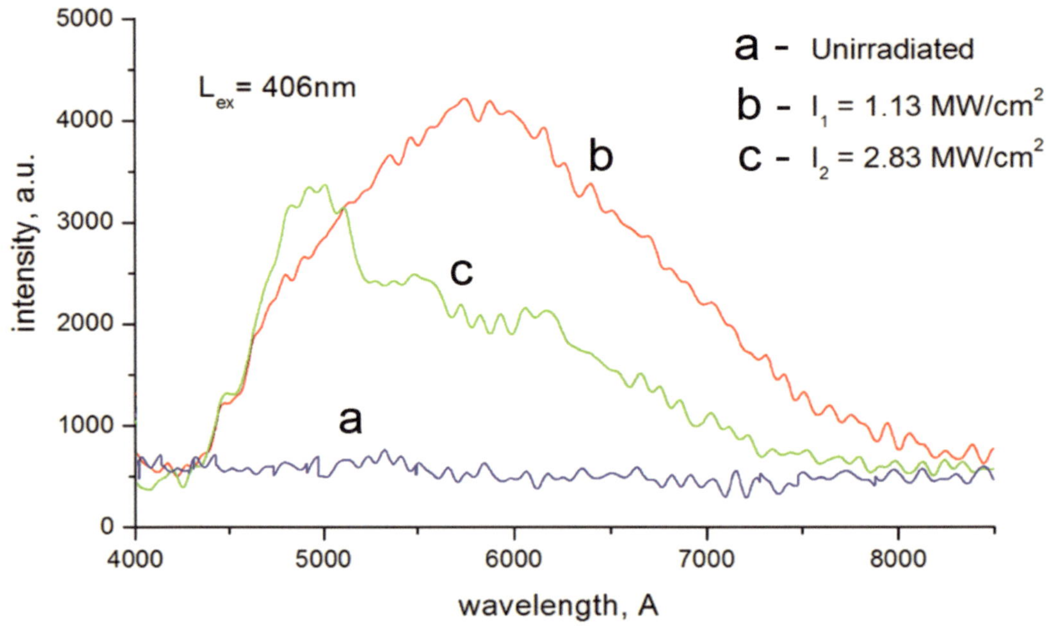

Fig. 12. Photoluminescence spectra of ITO/p-Si structure: before and after irradiation by the Nd:YAG laser.

After irradiation of ITO/p-Si/Al structure by the laser, photocurrent - voltage characteristic measurements (see figure 13) have shown, that electrical power output increased by two times in comparison to the nonirradiated structure. This effect is explained by increase of absorption coefficient due to quantum confinement effect in nanocones. That means Si band gap is increasing with the increase of laser radiation intensity.

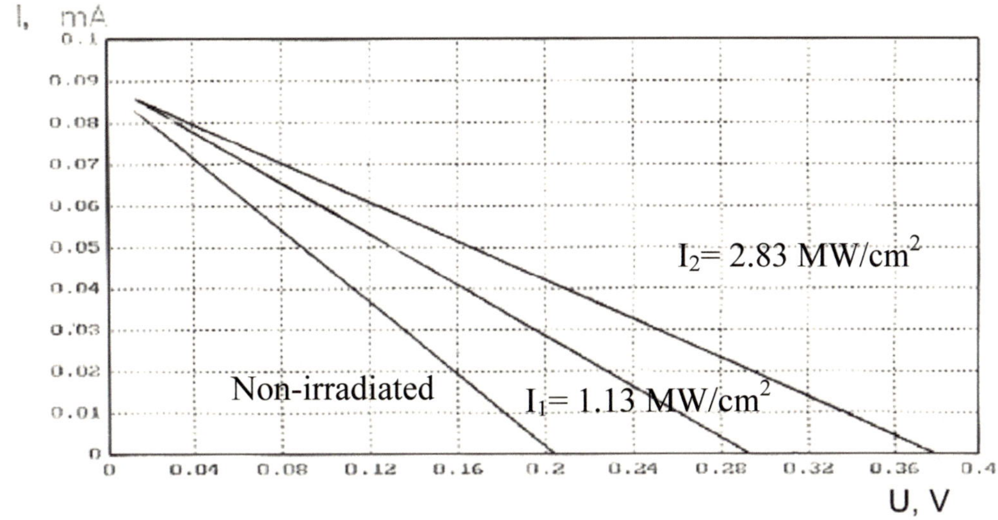

Fig. 13. Photocurrent-voltage characteristic of ITO/p-Si/Al structure before and after irradiation by laser with different intensities.

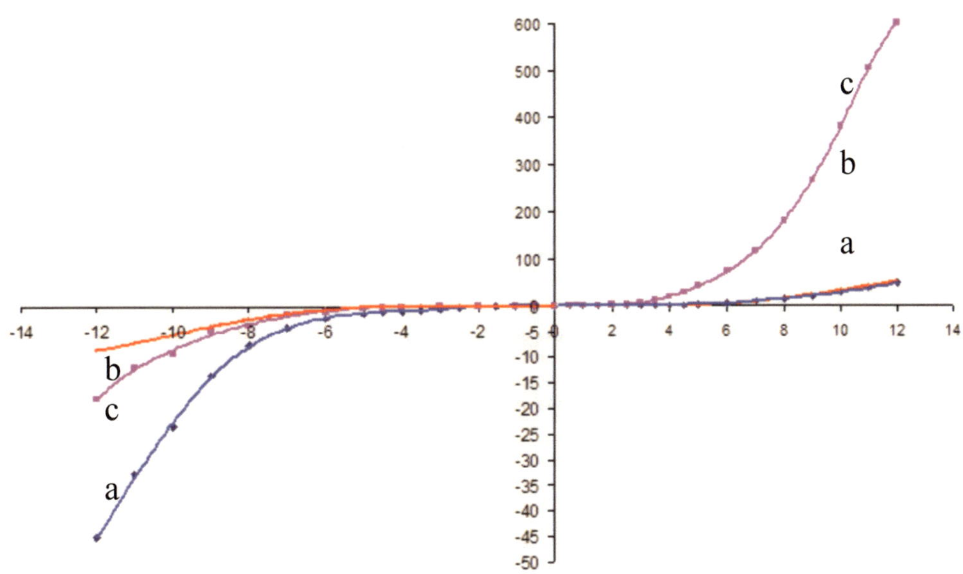

Fig. 14. Dark current-voltage characteristic of ITO/p-Si/Al structure: a) non-irradiated; irradiated by Nd:YAG laser at intensities of b) 1.13 MW/cm^2; c) 2.83 MW/cm^2.

It is found that the dark current-voltage characteristic becomes diode-like with rectification coefficient K = 10^5 at 5 V. The result is caused by the laser irradiation with intensity I_2 = 2.83

MW/cm2 (see figure 14). The improvements of ITO/p-Si/Al structure as a solar cell can be explained by the increase of potential barrier between ITO an p-Si layers.

Conclusion

- The possibility of p-n junction and Si nanocones' formation by laser irradiation in ITO/p-Si/Al structure was shown.
- The photoluminescence spectra from irradiated ITO/p-Si/Al structure by laser irradiation have been found in visible range of spectra and explained by presence of quantum confinement effect in nanocones with graded band gap.
- Study of the dark current-voltage characteristics showed diode-like character with rectification coefficient $K = 10^5$ at 5 V caused by laser irradiation with intensity $I = 2.83$ MW/cm2.
- After irradiation of ITO/p-Si/Al structure by laser, the solar cell electrical power output increased by two times in comparison to the nonirradiated structure.

3.2 Electron field emitter based on Si nanocones

Silicon electron field emitters may be attractive for numerous applications, mostly due to their compatibility with the dominating Si-based solid state micro- and nanoelectronics.

Nanocones have been formed on the surface of p-Si crystal after irradiation by the second harmonic of Nd:YAG laser with an intensity of 2.0 MW/cm2 (Evtukh et al., 2010). The measurements of electron field emissions were performed with a flat diode configuration with a glass spacer (figure 15). Distance between the cathode (silicon wafer) and the anode was 0.8 mm. The applied field emission setup allows achieving a vacuum of 1×10^{-5} Pa. The applied voltage varied between 1500-3600 V.

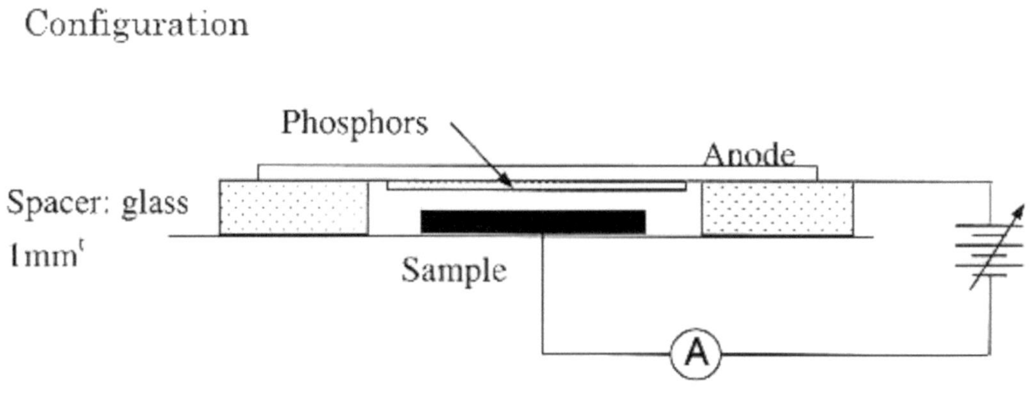

Fig. 15. Cell for electron field emission measurements.

Nanocones with a nanosphere on top of each cone were formed after irradiation by the laser (figure 16). A decrease in the nanosphere diameter from 600 nm to 20 nm with an increase of intensity of laser pulse from 2.0MW/cm2 to 20.0 MW/cm2 was observed. Nanocones have been formed on the irradiated surface of p-Si crystal by second harmonic of Nd:YAG laser with intensity 2.0 MW/cm2.

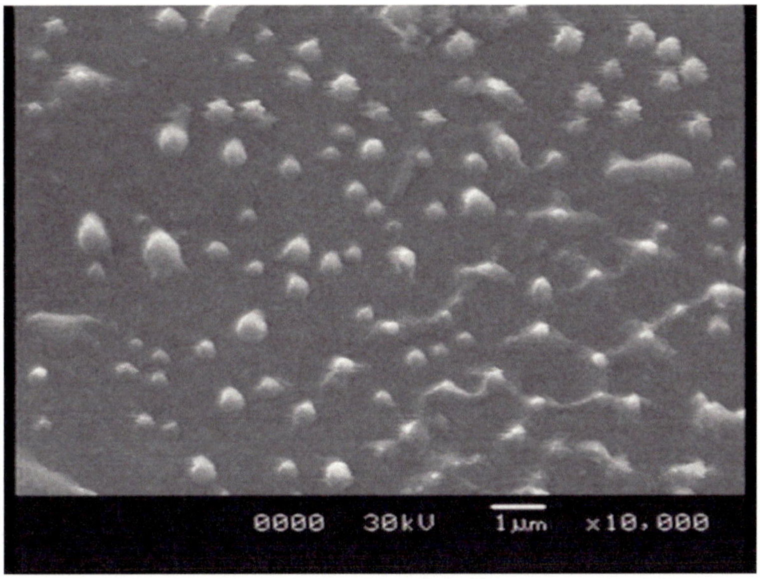

Fig. 16. Scanning electron microscope image of Si nanocones created by laser irradiation.

The electron field emission from such nanocones has some peculiarities, namely: (i) decrease of the threshold field from E_{th}=4×10^4 V/cm at the first measurement to E_{th}=3.5×10^4 V/cm in subsequent measurements, (ii) two slopes of Fowler-Nordheim curves (higher slope at low fields and lower slope at high fields) (figure 17). Analysis of the scanning electron microscope micrographs and electron field emission curves allows us to estimate (i) the electron field enhancement coefficient, β≈100, (ii) work functions (Φ_1=6.8 eV at the first

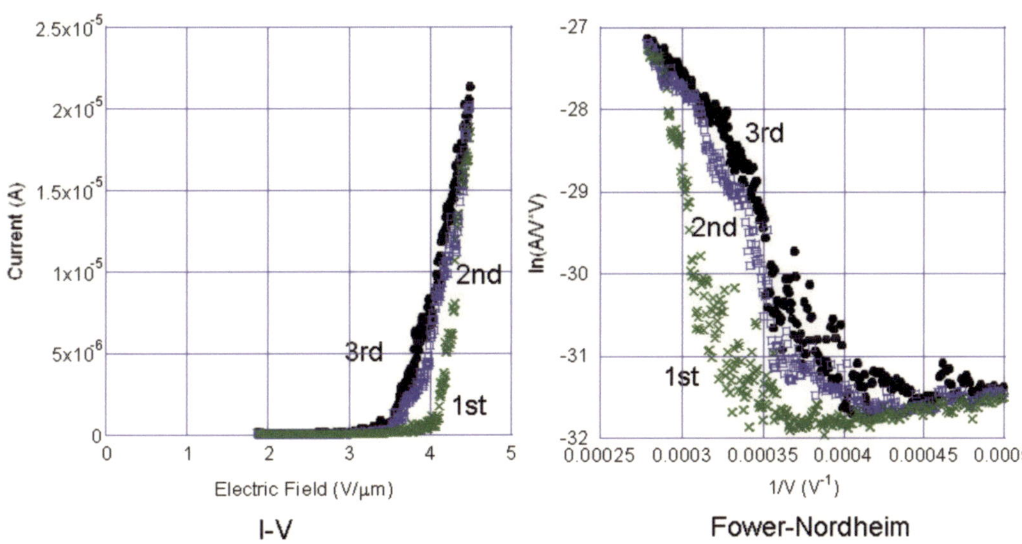

Fig. 17. Current-voltage characteristics (a) and corresponding Fowler-Nordheim plots (b) of Si nanocones electron field emission. Nanocones have been formed by second harmonic of Nd:YAG laser with intensity 2.0 MW/cm^2.

measurement and Φ_2=3.9 eV, Φ_3=2.38 eV from the two slopes in subsequent measurements), (iii) effective emission area, $\alpha=3\times10^{-8}$-1.8×10^{-5} cm^2. The lower work function in relation to the known value for high doped n-type silicon Φ_0=4.15 eV we explain by increase of Si band gap on the top of nanocones.

3.3 High intensity Si source of white light and photodetector with selective or "bolometer" type of spectral sensitivity on the base of nanocones

A result of investigation of photoluminescence spectra of Si and Ge_xSi_{1-x} crystals has shown (figures 18 and 19), that nanocone radiate wide spectrum of light like sun or glow lamp. At the same time, this structure can be used as photon detector with "bolometer" type of spectral photo sensitivity (Capassoa Federico, 1987), as shown in figure 20, curve 1, if irradiation of the structure takes place from the wide band gap part of semiconductor with gradient band gap structure, it means from top of cones, or as photon detector with selective type of spectral photo sensitivity, if irradiation of the structure takes place from narrow band gap part of graded band gap structure, as shown in figure 20, curve 2.

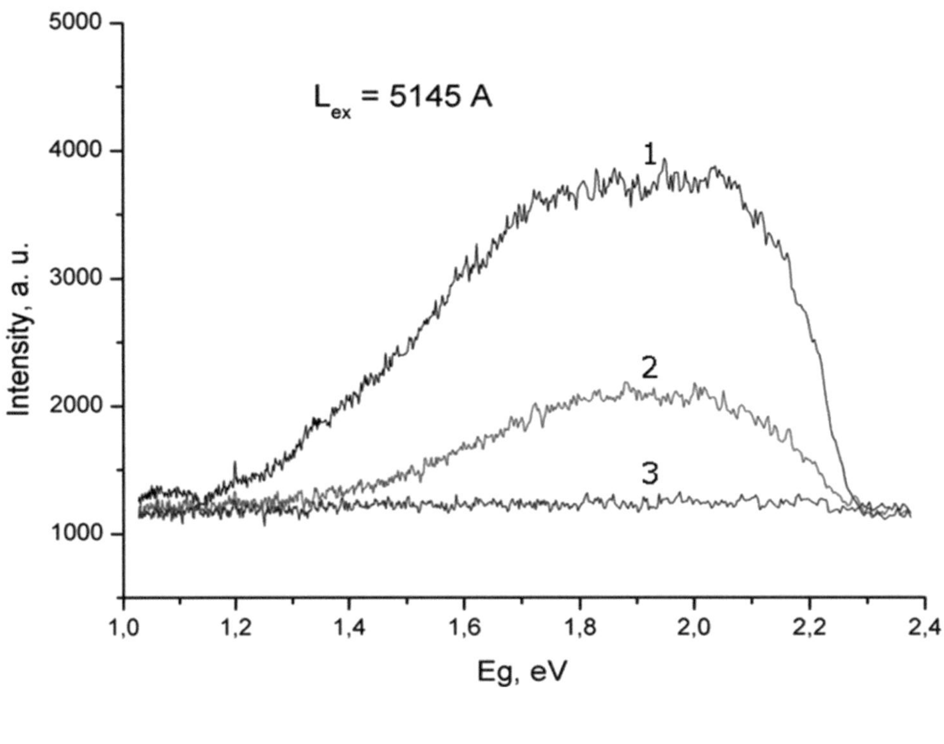

Fig. 18. Photoluminescence spectra of the SiO_2/Si structure irradiated by the laser at intensity 2.0 MW/cm^2 (2. and 3. curves), after removing of SiO_2 layer by chemical etching in HF acid (3. curve). 1. curve corresponds to photoluminescence of the non-irradiated surface.

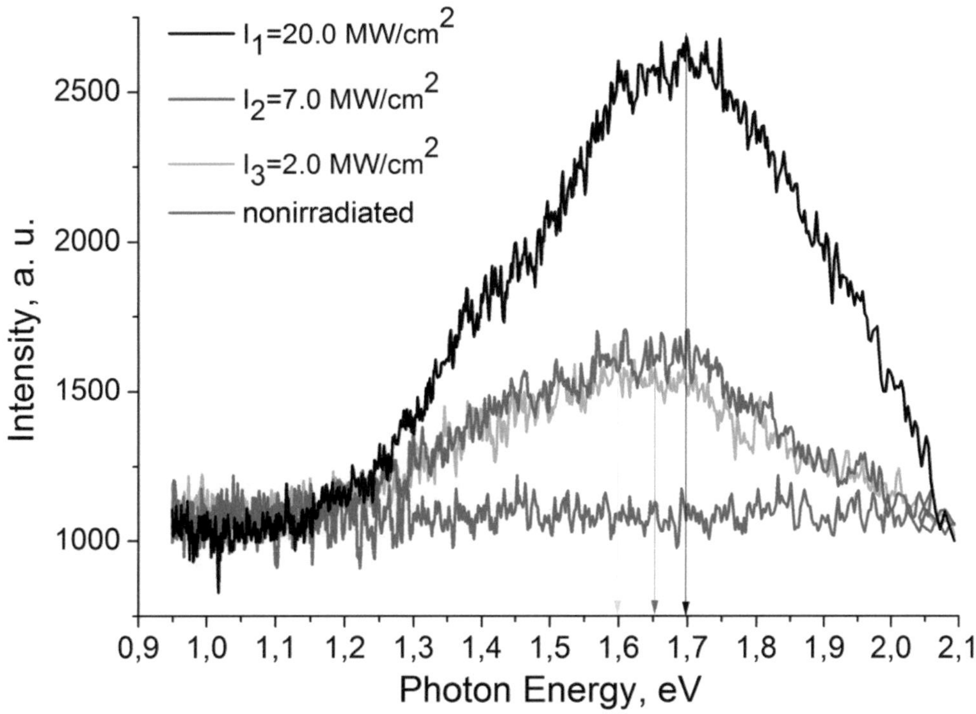

Fig. 19. Photoluminescence spectra of $Si_{0.7}Ge_{0.3}$/Si heteroepitaxial structures before and after irradiation by Nd:YAG laser.

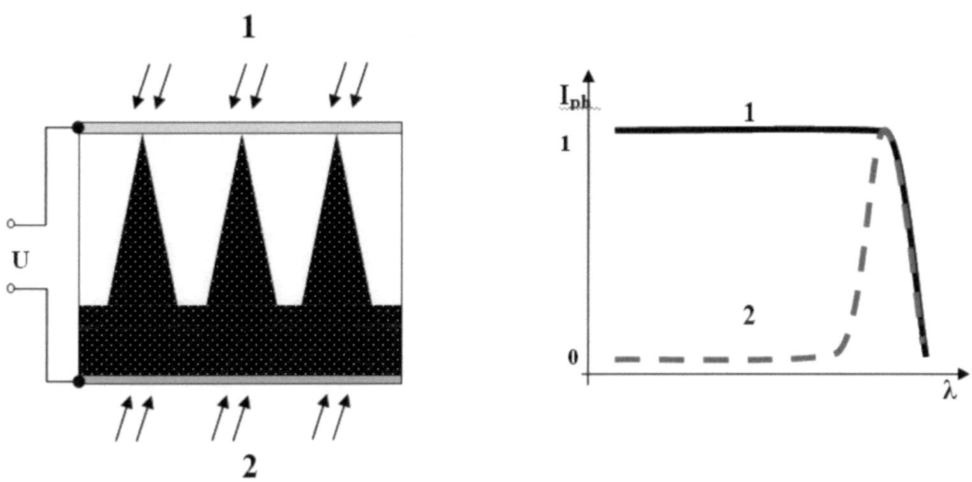

Fig. 20. Scheme of photodetector with graded band gap structure and photoconductivity spectra of photodetector with "bolometric" type of photo sensitivity - curve 1(detector irradiated from nanocones' side 1) and photodetector with selective type photo sensitivity – curve 2 (detector irradiated from base side 2).

4. Summary

Two-stage mechanism of nanocones' formation on a surface of semiconductor by Nd:YAG laser radiation consists of Laser Redistribution of Atoms and Selective Laser Annealing. This model was successfully applied for explaining p-n junction formation in i-Ge, increase of radiation hardness, formation of nanocones by laser radiation.

Power output of ITO/p-Si/Al solar cell structure has been increased by two times after irradiation by Nd:YAG laser second harmonic with intensity 2.83 MW/cm^2 due to formation of Si nanocones with p-n junction on their top. Decrease by two times of work function of electrons from Si nanocones formed by Nd:YAG laser second harmonic with intensity 2.0 MW/cm^2 is explained by increase of Si band gap on the top of nanocones due to quantum confinement effect.

5. Acknowledgment

The author gratefully acknowledges financial support in part by Europe Project in the framework FR7-218000 "COCAE", European Regional Development Fund within the project "Sol-gel and laser technologies for the development of nanostructures and barrier structures", the ESF Projects No. 1DP/1.1.1.2.0/09/ APIA/VIAA/142 and «Support for the implementation of doctoral studies at Riga Technical University».

6. References

Adurodija F.O.; Izumi H.; Ishihara T.; Yoshioka H.; Motoyama M.& Murai K. (2002). Effect of laser irradiation on the properties of indium tin oxide films deposited at room temperature by pulsed laser deposition,*Vacuum*, Vol. 67, pp. 209-216, ISSN 0042-207X

Balasundaraprabhu, R; Monakhov, E.; Muthukumarasamy, N.; Nilsen, O & Svensson, B. G. (2009). Effect of heat treatment on ITO film properties and ITO/p-Si interface, *Materials Chemistry and Physics*, Vol. 114, pp. 425- 429, ISSN 0254-0584

Baojun, Li.; Chua, S. J.; Yakovlev, N.; Lianshan, W.; & Eng-Kee, S. (2003). Properties of Schottky contact of titanium on low doped p-type SiGeC alloy by rapid thermal annealing, *Solid-State Electronics*, Vol. 47, No. 4, pp. 601-605, ISSN 0038-1101

Benda V.; Gowar J. & Grant D. (1999) *Power Semiconductor Devices: Theory and Applications*, Wiley, ISBN 047197644X, Chichester, England

Beton, P.H.; Dunn, A.W. & Moriarty P. (1995). Manipulation of C molecules on a Si surface, *Applied Physics Letters*, Vol. 67, pp. 1075 – 1077, ISSN 0003-6951

Blums J. & Medvid' A. (1995). The Generation of Donor Centres Using Double Frequency of YAG:Nd Laser, *Physics Status Solidi (a)*,Vol. 147, pp. K91-K94, ISSN 1862-6319

Bruk L.; Fedorov V.; Sherban D.; Simashkevich A.; Usatii I.; Bobeico E. & Morvillo P. (2009). Isotype bifacial silicon solar cells obtained by ITO spray pyrolysis, *Materials Science and Engineering* B, Vol. 159-160, pp. 282-285, ISSN 0921-5107

Brus L. E. (1984). Electron–electron and electron-hole interactions in small semiconductor crystallites: The size dependence of the lowest excited electronic state, *Journal of Chemical Physics*, Vol. 80, pp. 4403- 4409, ISSN 0021-9606

Capassoa F. (1987). Semiconductors and Semimetals, Chapter 6, Graded-Gap and Superlattice Devices by Bandgap Engineering , Volume 24, 319-395, At&T Bell Laboratories Murray Hill, New Jersey

Cavallini, A. ; Fraboni, B. & Chirco, P. (2000). Electronic properties of traps induced by γ-irradiation in CdTe and CdZnTe detectors, *Nuclear Instruments and Methods in Physics Research Section A: Accelerators, Spectrometers, Detectors and Associated Equipment*, Vol. 448, pp. 558-566, ISSN 0168-9002

Cavallini, A.; Fraboni, B. & Dusi, W. (2002). Radiation effects on II-VI compound-based detectors, *Nuclear Instruments and Methods in Physics Research Section A: Accelerators, Spectrometers, Detectors and Associated Equipment*, Vol. 476, pp. 770-778, ISSN 0168-9002

Claeys, C. & Simoen, E. (2007). Germanium-Based Technologies From Materials to Devices, *Elsevier*, ISBN 978-0-08-044953-1, London

Crommie, M.F.; Plutz, C. & Eigler, D.M. (1993). Confinement of electrons to quantum corrals on a metal-surface, *Science*, Vol. 262. pp. 218-220, ISSN 0036-8075

Dumanski L.; Bester M.; Virt I.S. & Kuzma M. (2006). The p–n junction formation in Hg1−xCdxTe by laser annealing method, *Applied Surface Science*, Vol. 252, pp. 4481–4485, ISSN 0169-4332

Evtukh, A.; Medvid, A.; Onufrijevs, P.; Okada, M. & Mimura, H. (2010). Electron field emission from the Si nanostructures formed by laser irradiation, *Journal of Vacuum Science and Technology B: Microelectronics and Nanometer Structures*, Vol. 28., No. 2, pp. C2B11- C2B11, ISSN 1071-1023

Fougeres, P.; Hage-Ali, M. & Koebel, J.M. (1999). CdTe and Cd1−xZnxTe for nuclear detectors: facts and fictions, *Journal of Crystal Growth*, Vol.428, pp. 38-44, ISSN 0022-0248

Franks, L.A.; Brunett, B.A. & Olsen, R.W. (1999). Radiation damage measurements in room-temperature semiconductor radiation detectors, *Nuclear Instruments and Methods in Physics Research Section A: Accelerators, Spectrometers, Detectors and Associated Equipment*, Vol. 428, pp. 95-105, ISSN 0168-9002

Fujisawa, I. (1980). Type Conversion of InSb from p to n by Ion bombardment and laser Irradiation. *Japanese Journal of Applied Physics*, Vol. 19, pp. 2137-2140, ISSN 0021-4922

Gupta R. K.; Yakuphanoglu F.; Ghosh K.& Kahol P. K. (2011). Fabrication and characterization of p-n junctions based on ZnO and CuPc Microelectronic Engineering, *Microelectronic Engineering*, Vol. 88, No. 10, pp. 3067-3069, ISSN 0167-9317

Junno, T.; Depent, K.; Montelius L. & Samuelson L. (1995). Controlled manipulation of nanoparticles with an atomic-force microscope, *Applied Physics Letters*, Vol. 66., pp. 3627-3629, ISSN 0003-6951

Kaupuzs J. & Medvid' A. (1994) New Conception in Transistor Technology Using Nonhomogeneous Temperature Field. - Vol. 2335-36, *SPIE Proceedings Microelectronic Manufacturing '94 Conference*, Austin, Texas, USA, October 18-22, pp. 134-145, 1994

Kiyak S.G. & Savitsky G.V., (1984). Formation of p-n junction on p-type Ge by millisecond laser pulses. *Physics and techniques of semiconductors,* Vol. 18, pp. 1958 - 1964

Kurbatov, L., Stojanova I., Trohimchuk,P.P. & Trohin,A.S. (1983), Laser annealing of AIIIBV compound. Rep.Acad.Sc.USSR, Vol.268, pp.594- 597, (in Russian);

Li B.; Chua S.J.; Nikolai Y.; Wang L. & Sia E.K (2003) Properties of Schottky contact of titanium on low doped p-type SiGeC alloy by rapid, *Solid State Electronics*, Vol. 47, pp. 601-605, ISSN 0038-1101

Lindstrom, G. (2003). Radiation damage in silicon detectors. *Nuclear Instruments and Methods in Physics Research Section A: Accelerators, Spectrometers, Detectors and Associated Equipment*, Vol. 512. pp. 30-43, ISSN 0168-9002

Ljamichev, I.J.; Litvak, I.I. & N.A. Osvhepkov, Devises on amorphous semiconductors and their application. Moskva, Sov. Radio, 1976.

Mada, Y.& Ione, N. (1986). p-n junction formation using laser induced donors in silicon, *Applied Physics Letters,* Vol.48, pp.1205-1207, ISSN 0003-6951

Makino T.; Kato H.; Ri S.G.; Yamasaki S. & Okushi H. (2008). Homoepitaxial diamond p–n+ junction with low specific on-resistance and ideal built-in potential *Diamond and Related Materials*, Vol. 17, pp. 782-785, ISSN 0925-9635

Medvid', A. Fedorenko L. & Frishfelds V. (1998). Electrical properties of donor defects at the surface of InSb after laser irradiation, *Vacuum*, 51, pp. 245 - 249, ISSN 0042-207X

Medvid', A. & Fedorenko, L. (1999). *Generation of Donor Centers in p-InSb by Laser Radiation,* Materials Science Forum, Vol. 297-298, pp. 311-314, ISSN 0255-5476

Medvid', A., Litovchenko, V.G., Korbutjak, D., Krilyk, S.G., Fedorenko, L.L.& Hatanaka ,Y. (2001). Influence of laser radiation on photoluminescence of CdTe, *Radiation Measurements*, Vol. 33, pp. 725-730, ISSN 1350-4487

Medvid', A. (2002). Redistribution of the Point Defects in Crystalline Lattice of Semiconductor in Nonhomogeneous Temperature Field, *Defects and Diffusion Forum*, Vol. 210- 212 , pp. 89-101, ISSN 1662-9507

Medvid', A.; Mychko, A.; Fedorenko, L. & Korbutjak, D. (2007). Formation of graded band-gap in CdZnTe byYAG:Nd laser radiation, *Radiation Measurements*, Vol. 42, No. 4-5, pp. 701-703, ISSN 1350-4487

Medvid', A.; Dmytruk, I.; Onufrijevs, P. & Pundyk, I. (2007). Quantum Confinement Effect in Nanohills Formed on a Surface of Ge by Laser Radiation, *Physica Status Solidi (c),* Vol. 4, pp. 3066-1069, ISSN 1610-1642

Medvid', A.; Onufrijevs, P.; Dmitruk, I. & Pundyk, I. (2008). Properties of Nanostructure Formed on SiO_2/Si Interface by Laser Radiation, *Solid State Phenomena*, Vol. 131-133, pp. 559-562, ISSN 1012-0394

Medvid', A.; Mychko, A.; Strilchyuk, O.; Litovchenko, N.; Naseka, Yu.; Onufrijev, P. & Pludonis, A. (2008). Exciton quantum confinement effect in nanostructures formed by laser radiation on the the surface of CdZnTe ternary compound, *Physica Status Solidi (c),* Vol. 6, pp. 209-212, ISSN 1610-1642

Medvid A.; Onufrijevs P.; Dauksta E.; Barloti J.; Grabovskis D. & Ulyashin A. (2009). Dynamics of Nanostructure Formation using Point Defects on Semiconductors by Laser Radiation, *Physica Status Solidi C,* Vol. 6, pp. 1927-1928, ISSN 1610-1642

Medvid', A.; Onufrijevs, P.; Lyutovich, K.; Oehme, M.; Kasper, E.; Dmitruk, N.; Kondratenko, O.; Dmitruk, I. & Pundyk I. (2010). Self-assembly of nanohills in SixGe1-x/Si by laser radiation, *Journal of Nanoscience and Nanotechnology*, Vol. 10, pp. 1094-1098, ISSN 1533-4880

Medvid', A.; Mychko, A.; Dauksta, E.; Naseka, Y.; Crocco, J. & Dieguez E. Increased Radiation Hardness of CdZnTe by Laser Radiation, *2010 IEEE Nuclear Science Symposium and Medical Imaging Conference*, ISBN 978-1-4244-9106-3, USA, October 2010

Medvid, A.; Onufrievs, P.; Dauksta, E.; Barloti, J.; Ulyashin, A.; Dmytruk, I. & Pundyk I. (2011). P-n junction formation in ITO/p-Si structure by powerful laser radiation for solar cells applications, *Advanced Materials Research*, Vol. 222, pp. 225-228, ISSN 1022-6680

Niraula, M.; Mochizuki, D.; Aoki, T.; Hatanaka, Y.; Tomita Y. & Nihashi, T. (1999). Improved spectrometric performance of CdTe radiation detectors in a p-i-n design, *Applied Physics Letters*, Vol.75, pp. 2322-2325, ISSN 0003-6951

Philipp, H.P. & Taft, E.A. (1959). Optical constants of germanium in the region 1 to10 eV, *Physical Review*, Vol. 113, No. 4, pp. 1002-1005

Pozela J. (1980) *Semiconductor Transducers*, Mokslas, Vilnius

Romanov, V. & Serdega, B. (1975). Bipolar electric conductivity of germanium in a high-frequency electric field, *Physics and technics of semiconductors*, Vol. 9, No.4, pp. 665 - 668.

Shapiro, J. (2002). *Radiation protection: a guide for scientists, regulators, and physicians*, 4th ed., Harvard college

Shaw, D. (1973). *Atomic Diffusion in Semiconductors, Plenum Press*, London and New York

Suzuki, K.; Seto, S.; Savada, T.; Imai, K.; Adachi, M. & Inabe, K. (2001). Photoluminescence measurements on undoped CdZnTe grown by the high-pressure Bridgman method, *Journal of Electronic Materials*, Vol.30 pp. 603-607, ISSN 0361-5235

Taguchi, T.; Shirafuji, J. & Inuishi, Y. (1978). Gamma and electron radiation effects in CdTe, *Nuclear Instruments and Methods in Physics Research Section A: Accelerators, Spectrometers, Detectors and Associated Equipment*, Vol.150, pp.43-48, ISSN 0168-9002

Tomas A.P.; Jennings M.R.; Davis M.; Shah V.; Grasby T. & Covington J.A. (2007). High doped MBE Si p–n and n–n heterojunction diodes on 4H-SiC, *Microelectronics Journal*, Vol. 38, No. 12, pp. 1233-1237, ISSN 0026-2692

Toshiharu M.; Hiromitsu K.; Sung-Gi Ria, Satoshi Y. & Hideyo, O. (2008). Homoepitaxial diamond p-n+ junction with low specific on-resistance and ideal built-in potential, *Diamond & Related materials*, Vol. 17, pp. 782 - 785, ISSN 09259635

Tovstyuk, K.D.; Placko, G.V.; Orletskiy, V.D.; Kiyak, S.G. & Babytskiy, J. V. (1976). The formation of p-n and n-p-junction in semiconductor by laser radiation, *Ukrainian Journal of Physics*, Vol. 21, 1918 - 1920.

Vu, V. T.; Nguyen, D. C.; Pham, H. D.; Chu, A. T. & Pham, T. H. (2010). Fabrication of a silicon nanostructure-based light emitting device, *Advances in natural sciences: nanoscience and nanotechnology*, Vol. 1, pp. 025006-025010, ISSN 2043-6254

Yang, G.; Jie, W. & Zhang, Q. (2006). Photoluminescence investigation of CdZnTe:In single crystals annealed in CdZn vapors *Japanese Journal of Applied Physics*, Vol.21 pp.1807-1809, ISSN 0021-4922

12

Applications of Polidocanol in Varicose Vein Treatment Assisted by Exposure to Nd:YAG Laser Radiation

Adriana Smarandache[1,*], Javier Moreno Moraga[2], Angela Staicu[1], Mario Trelles[3] and Mihail-Lucian Pascu[1]
[1]*National Institute for Lasers, Plasma and Radiation Physics, Bucharest*
[2]*Instituto Medico Laser, Madrid*
[3]*Instituto Médico Vilafortuny/Fundacion Antoni de Gimbernat, Cambrils*
[1]*Romania*
[2,3]*Spain*

1. Introduction

The understanding of the interaction between Polidocanol (POL) and the target veins tissues is important in utilizing it in varicose veins diseases treatment. Generally, the development of new drug delivery routes may represent methods to improve the efficacy and/or safety of the active pharmaceutical ingredients. With this respect, the treatment involving POL administration as foam has gained widespread use (Cavezzi and Tessari, 2009). Although the main approach in the treatment of small diameter veins, in venulectasias and reticular veins of less than 4 mm in diameter (class I/II and III) is sclerotherapy (Alos et al., 2006; Nijsten et. al., 2009; Parsi et al., 2007; Railan et al., 2006), lasers, especially the Nd:YAG laser, have shown interesting and non-negligible capabilities in treating these cases (Redondo and Cabrera, 2005; Santos et al., 2008; Trelles et al., 2005). Clinical experimental results prove that the exposure of the tissues impregnated with POL to laser radiation emitted at 1.06 µm improves the efficiency of the treatment (Moreno Moraga, n.d.).

The aim of this chapter is to present an extensive study concerning the optical properties of commercially available Polidocanol (Aethoxysklerol 2%, Kreussler & Co. GmbH, Germany) and the possible modifications induced at molecular level in this medicine as supplied by the manufacturer, by exposing it to Nd:YAG laser radiation.

Almost 60% of the adult population of Europe and the USA present varicose or spider veins in the lower limbs and this is the seventh most common chronic vascular disorder, nine times more frequent than arterial diseases (Tibbs, 1997). In recent years, a significant, growing number of patients are consulting in order to correct the symptoms originated in this vascular disorder, as well as the aesthetic implications that cause this ailment.

* Corresponding Author

The lower extremities contain both superficial and deep veins. The superficial venous system consists of the greater saphenous vein draining most of the leg, the lesser saphenous vein draining the posterior and lateral lower leg, and the lateral (subdermal) venous system draining the lateral leg above and below the knee. Clinically, this last system may be observed as visible veins. Dilatations of the most superficial veins are called telangiectasias. These are usually bright red and measure 0.03 mm to 0.3 mm in diameter. Slightly larger are postcapillary venulectasias, which measure 0.4 mm to 2 mm, and may be red or blue depending on their oxygenation. Telangiectasias and venulectasias are commonly referred to as spider veins. These connect to larger reticular veins measuring 2 mm to 4 mm that are typically blue because of the Tyndall effect and their deeper location within the dermis (Kunishige et al., 2007).

The treatment of varicose veins reduces the symptoms and the complication rate of venous insufficiency and increases the patient's health related quality of life. It can roughly be divided into four groups: compression therapy, surgical treatment, sclerotherapy and endovenous thermal ablation. To improve efficacy and treatment satisfaction and to reduce serious side effects, costs, and postoperative pain, new minimally invasive techniques, such as ultrasound-guided foam sclerotherapy (UGFS), endovenous laser therapy (EVLT), and radiofrequency ablation (RFA), have been introduced in the last decade (Nijsten et. al., 2009). Also, sclerotherapy (mainly foam sclerotherapy) combined with 1.06 μm Nd:YAG laser beam therapy showed very encouraging clinical results. This new approach is ideal for treating varicose veins originating in the lateral subdermal plexus of the lower limbs or reticular veins of other locations. Veins of up to 4 mm calibre located at a depth between 2 mm and 4 mm can be treated. Telangiectasias or spider-veins, located between 1.2 mm and 1.7 mm below the surface of the skin, either associated with or independent from reticular veins in any location are also ideal for combined treatment. Surgery or EVLT continue to be recommended for the great saphenous veins (GSV) of a calibre greater than 8 mm. It is possible to treat truncal veins with combined POL and laser, but this requires particular expertise. In these cases ultrasound monitoring is well advised, both when puncturing and injecting the microfoam. When POL enters the veins it is detected by the ultrasound, so vein spasms, which occur after the injection, can be well controlled as well as their duration (Moreno Moraga, n.d.; Trelles et al., 2011).

Sclerotherapy is the targeted elimination of varicose veins by injection of a sclerosing substance into the vein lumen. Sclerosing agents cause a chemical irritation of the venous intimate that produces an inflammation of the endothelial lining of the vessel. Subsequently, a secondary, wall-attached local thrombus is generated and, in long term, the vein will be transformed into a fibrous cord (Redondo and Cabrera, 2005). Each class and individual type of sclerosant within the same class produce this effect in different ways associated with diverse and highly variable patterns of efficacy, potency, and complications (Artemi, 2007). According to their potency, sclerosing agents can be classified as major (alcohol, iodine, sodium tetradecyl sulfate), intermediate (sodium salicylate, POL), or minor (chromated glycerin).

Modern sclerotherapy is performed using sclerosing detergents such as POL. With their introduction in the 1920s and 1930s, detergent sclerosants, also known as fatty acids and fatty alcohols, soon became and are still the most popular sclerosant types used worldwide for the treatment of varicose veins. Detergent sclerosants, with increasingly favorable risk-to-benefit ratios, rapidly replaced older sclerosants thought to be ineffective if they could not produce tissue necrosis or significant thrombosis at vasodestructive concentrations. After intravenous injection, liquid detergent sclerosants become protein bound and inactivated when mixed with blood. When the foam sclerosant is injected it fills the vein completely; it also displaces blood from the vein and ensures that all endothelium is exposed to the sclerosant agent and destroyed.

Medical observations prove that foam sclerotherapy is preferable instead of the use of liquid sclerosing substances (Ouvry et al., 2008). Detergent-like sclerosing agents can be transformed into fine-bubbled foam by special techniques. Orbach was the first who described the use of froth in sclerotherapy (Orbach, 1950). In 1997 new methods of transforming the sclerosing liquid into foam were proposed (Cabrera et al., 1997a, 1997b; Henriet, 1997; Monfreux, 1997) and Cavezzi and Frullini in 1999 reported their 13 month experience of duplex-guided sclerotherapy with sclerosing foam prepared by Monfreux's method (Cavezzi and Frullini, 1999). Sadoun and Benigni (Sadoun and Benigni, 1998) and Garcia-Mingo (Garcia-Mingo, 1999) suggested new ways of manufacturing sclerosing foam. In December 1999 Tessari described a safe and easy method to generate a fairly stable and compact foam (made of a mixture of micro-bubbles of detergent drug and air) using two plastic syringes and a three-way tap (Tessari, 2000). Subsequently Frullini (Frullini, 2000) and Gachet (Gachet, 2001) suggested other ways to produce a sclerosing foam. The ability to be agitated and foamed increases the potency of detergents from 2 to 4 times by mechanically displacing blood and thus maximizing the surface area and the contact time with the endothelium. This process has the advantage of using small and presumably less allergenic or less tissue-toxic concentrations and volumes of sclerosant, although greater risks may occur if foam passes in the ocular or cerebral circulation (Duffy, 2010).

The precise procedure to prepare the microfoam is the subject of a confidential agreement with a third party who is the proprietary. It is, however, based on the procedure described in granted European and US patents EP 656 203 and US 5 676 962, respectively (Cabrera and Cabrera Jr., 1997b). The Polidocanol microfoam produced by Provensis has successfully undergone phase III clinical trials in Europe and is currently in the process of obtaining approval from the Food and Drug Administration (BTG International Ltd., 2011; Redondo and Cabrera, 2005).

Dermatologic application of laser technology was initiated by Leon Goldman. His work was further elucidated by some authors (Anderson and Parrish, 1983), who described the principle of selective photothermolysis. The application of this principle shows how laser energy is preferentially absorbed by the intended target tissue or chromophore, resulting in their controlled destruction while minimizing collateral thermal damage to surrounding normal tissue. To achieve selective photothermolysis, a wavelength that is preferentially absorbed by the intended target tissue (in our case the vein tissue) or chromophore is selected. The pulse duration of the irradiating beam is also important for selectivity. If the pulse duration is too long, heat absorbed by the veins will diffuse out in the surrounding tissues so that the veins will be not selectively heated to the necessary degree in order to be destroyed. If the pulse duration is too short, the light absorbing chemical species such as hemoglobin in blood will be heated causing vaporization. This effect can cause purpura. Theory dictates that the proper pulse width should match the thermal diffuse time of the targeted structure. For smaller vessels, for example, this thermal diffusion time can be of the order of hundreds of microseconds to several milliseconds, while for larger leg veins it could be 5-100 ms.

In the last decade, the treatment technique has been improved so that predictable, reproducible results can be achieved. Several investigations (Mordon et al., 2003) have enabled us to better understand the role played by the transformation of oxyhemoglobin (OxiHb) into methemoglobin (MetHb) in order to obtain a more marked, more selective laser light-blood chromophore interaction.

Long-pulsed (as mentioned above) Nd:YAG lasers offer great advantages for treating varicose veins of the lower limbs. They can penetrate to a depth of 4 to 6 mm and may be used to treat more deeply seated vessels (Rogachefsky et al., 2002; Ross and Domankevitz, 2005). Also, they are less absorbed by melanin (which means that Nd:YAG lasers can be used to treat dark phototypes). On the other hand, in order to produce penetration, high fluence is required and hence treatment is painful and requires cooling and even topical anaesthesia (Anderson and Parrish, 1983).

The light absorption of whole blood is different depending on the osmolarity, hematocrit, hemodynamics and red cell morphology variables. The MetHb and protoporphyrin IX (PpIX) have an absorption coefficient at 1,064 nm four times greater than Hb. It is shown that one pulse of the Nd:YAG laser causes substantial transformation of Hb into MetHb, followed by a drastic decrease in blood circulation speed (Mordon et al., 2003). Then, a second Nd:YAG laser pulse, shortly after the first one, would be four times more effective. Following this theory, a system was manufactured with two different lasers built into the same console. The energy of the two lasers is sequentially pulsed. First, a pulse of a dye laser emitted at 585 nm produces an increase in the rate of intraerythrocytic MetHb, changing the coefficient of absorption of the Nd:YAG laser beam which is delivered after the first sequence. The long pulse emitted by the 1,064 nm Nd:YAG laser induces vessel damage at lower fluence. In reality, the pulsed dye laser has low penetration and so its efficiency is limited to the most superficial dermal plexus, since the vessels located at a greater depth than 1.5 mm are not reached by these laser pulses. Hence, significant rates of MetHb cannot be formed in the deepest veins (Trelles et al., 2010).

Clinical experimental results have proved that the exposure of tissues impregnated with POL to laser radiation emitted at 1.06 µm improves the efficacy of the treatment (Moreno Moraga, n.d.; Trelles et al., 2011). It combines the effect of sclerotherpy with that of the laser therapy and more than this, with the capabilities of drug induced modifications following the exposure to the laser radiation. Sometimes, the exposure of a medicine to laser radiation yields a product that is more biologically active than the un-exposed parent (Pascu et al., 2011; Wainwright, 2008).

In this respect, the data reported in this chapter represent a step to better understand the interaction of POL with the target tissues in the context of its exposure to pulsed laser beams emitted at 1,064 nm by the Nd:YAG laser.

2. Description of Polidocanol

Detergent sclerosants produce endothelial damage by multiple mechanisms associated with a decrease in endothelial cell surface tension, interference with cell surface lipids, disruption of intercellular cement and extraction of cell surface proteins. Detergents are most effective in the form of micelles (molecular aggregates) drawn in Fig. 1.

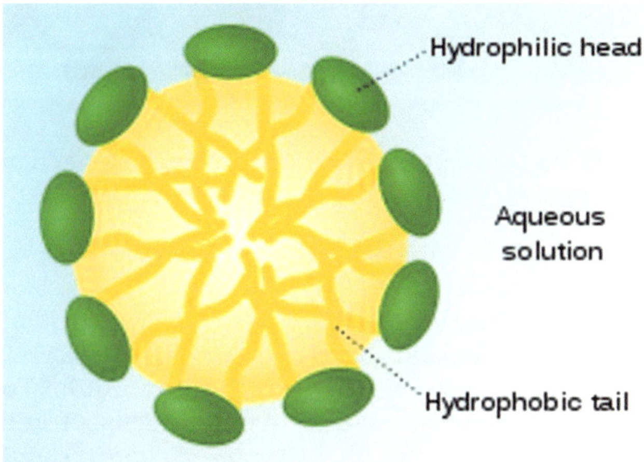

Fig. 1. Scheme of a micelle

At low concentrations and temperatures most detergent molecules are individually dissolved in solution, micelles are not formed and toxicity to endothelium is considered minimal. By increasing sclerosant concentrations at temperatures encountered in living tissue, a predictable threshold for endothelial damage occurs with a commensurate increase in sclerosing potency. The degree of damage can be controlled varying the detergent concentration and the time length of the sclerosant contact with the vessel wall. Unfortunately, due to the dilution with blood, it is not possible to determine the quantity of the sclerosing agent in intimate contact with the endothelial surface and the quantity which remains in circulation. Thus, the effects obtained are not always in direct proportion to their concentration; ''downstream'' effects with unintended damage to interconnecting vessels can always occur (Duffy, 2010).

2.1 Introduction of the compound

Polidocanol or Lauromacrogol 400, $C_{14}H_{30}O_2$ (CID 24750, molecular weight 230.3868 [g/mol]), is the pharmaceutical active ingredient of commercially available drugs listed in Table 1. It is a polyethylene glycol ether of Lauryl alcohol, where the average value of polymer is 9 (PubChem, 2011). This chemical compound is a viscous liquid at room temperature, having a melting point of 15–21°C. It is miscible in water, has a pH of 6.0–8.0 and has a density of 0.97 g/cm^3 at room temperature, closed to that of the water.

The 3D image of this molecule is shown in Fig. 2, where the carbon atoms are in dark grey colour, the hydrogen atoms are light grey and the oxygen atoms are red.

Polidocanol laurel macro gel 400 laureth-9 was first synthesized in 1936 and marketed as a topical and local anesthetic under the trade name SCH-600 (Chemische Fabrik Kreussler & Co. GmbH, Germany). This urethane local anesthetic differs from classic ester and amide anesthetic agents because it lacks an aromatic ring.

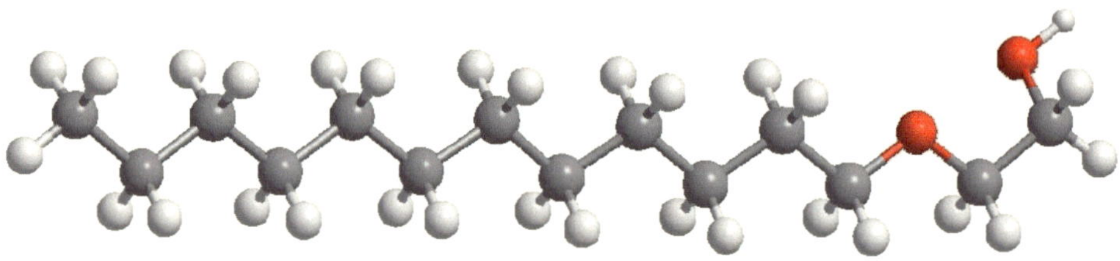

Fig. 2. 3D image of Polidocanol molecule, where the carbon atoms are figured in dark grey, the hydrogen atoms are light grey and the oxygen are in red color.

It is used as a topical anesthetic in ointments and lotions for skin irritation, burns, and insect bites and as an epidural anesthetic. The ability of POL to sclerose blood vessels without significant damage to surrounding tissues led to its use as a sclerosant in the 1960s. By 1967, it was registered in the Federal Republic of Germany as Aethoxysklerol, and it is now the only sclerosant approved for use in Germany (Duffy, 2010). Today, the leading sclerotherapy treatment in Europe is based on Aethoxysklerol. It is the only sclerosant approved by the Japanese Ministry of Health, Labor and Welfare (Eckmann, 2009). Recently - March 2010, the U.S. Food and Drug Administration approved Asclera (POL) injection (produced by Chemische Fabrik Kreussler & Co. GmbH, Germany, too) for the treatment of varicose veins in USA (U.S. Department of Health & Human Services, FDA., 2010).

In Table 1 are synthesized the main pharmaceutical companies, researchers, developers, manufacturers, distributors and suppliers to provide POL.

Companies	Product name
□ Chemische Fabrik Kreussler & Co., Germany	Aethoxysklerol, Asclera(USA)
□ Berlin Pharmaceutical Industry, Germany	Aethoxysklerol
□ Cem Farma Ilac, Turkey	Aethoxysklerol
□ Codali, Belgium	Aethoxysklerol
□ Felo, Denmark	Aethoxysklerol
□ Globopharm, Switzerland	Aethoxysklerol
□ IBI International, Czech Republic	Aethoxysklerol
□ Institute of Pharmaceutical Research and Technology, Grece	Aethoxysklerol
□ Inverdia, Sweden	Aethoxysklerol
□ Lomapharm Rudolf Lohmann, Germany	Aethoxysklerol
□ Nycomed, Switzerland	Aethoxysklerol
□ Repharma, Azerbaijan	Aethoxysklerol
Resinag, Switzerland	Sclerovein
□ Sigma-Tau, Italy	Aethoxysklerol
□ Tamro Distribution, Sweden	Aethoxysklerol

Table 1. The main pharmaceutical companies, researchers, developers, manufacturers, distributors and suppliers of POL

POL should be stored at room temperature (15-25^{0}C). No special precautions are necessary when disposing of this material. POL at concentrations of 0.5% to 1% is stable for 3 years and should be stored between 15⁰C and 30⁰C (59–86⁰F). There are no special precautions for the disposal of unused POL (product insert).

The common pharmaceutical presentation form of Aethoxysklerol is as ampoule of 2ml injection solution whose composition is given in Table 2.

Composition of injected solution in concentration of:	0.5%	2%	3%
Lauromacrogol 400 (active)	10mg	40mg	60mg
Ethanol (inactive)	0.10ml	0.10ml	0.10ml
Pure water (inactive)	2ml	2ml	2ml
Sodium hydrogen phosphate (buffer - inactive)			
Potassium dihydrogen phosphate (buffer - inactive)			

Table 2. Composition of Aethoxysklerol injected solution

2.2 Mechanism of the sclerosing action

Polidocanol exerts its sclerosant effects by causing concentration dependent differential cell injury (Eckmann and Kobayashi, 2005). Cellular calcium signaling and nitric oxide pathways become activated by the administration of the sclerosant, followed by cell death. The timing of endothelial cell death is predictable based on sclerosant concentration during exposure. This ultimately results in endothelial cell lysis but that can potentially also involve erythrocytes, platelets, and lead to platelet-derived microparticle formation (Parsi et al., 2008). Although hemolysis occurs experimentally in whole blood samples at Polidocanol concentrations greater than 0.45%, erythrocyte lysis, platelet lysis and platelet-derived microparticle formation have not been a significant concern reported in any clinical trials of sclerosant therapy.

2.3 Pharmacokinetics

POL: 12 hours after intravenous application, about 90% of the POL administered would have been eliminated from the blood. No accumulation is to be expected even after repeated doses at intervals which are normally used for sclerotherapy.

In a study made by Artemi and reported in 2007, the following values were determined after a single intravenous dose: protein binding 64%, terminal elimination half-life 4 hours, volume of distribution 24.5 l, total clearance 11.7 l/h, renal clearance 2.43 l/h and biliary clearance 3.14 l/h.

Ethanol: No data are available for the level of the ethanol absorption rate. The volume of distribution of ethanol is 0.68 l/kg for a man and 0.55 l/kg for a woman, and is reached very quickly. Ethanol enters the foetus and mother's milk. Ethyl alcohol is oxidised by alcohol dehydrogenase in the liver to form acetaldehyde, and acetaldehyde is in turn broken down by acetaldehyde dehydrogenase to form acetic acid. This metabolic breakdown method accounts for 90-96% in man. The rate of elimination is independent of concentration and is 0.1 g/kg/h for a man and 0.085 g/kg/h for a woman (hourly breakdown c. 0.15‰). Insignificant amounts are eliminated via the lungs (2-3%) and kidneys (1-2%).

2.4 Foaming process

There are many works providing good results about the use of POL foam sclerotherapy (Alos et al., 2006; Cabrera et al., 2000; Cavezzi and Frullini, 1999; Frullini, 2000; Gachet, 2001; Garcia-Mingo, 1999; Hsu and Weiss, 2003; Ouvry et al., 2008; Rabe et al., 2008; Redondo and Cabrera, 2005; Santos, 2008; Tessari, 2000). The ability to be agitated and foamed increases the potency of detergents from 2 to 4 times by mechanically displacing blood and thus maximizing surface area and time in contact with endothelium. This process has the advantage of using small and presumably less allergenic or less tissue-toxic concentrations and volumes of sclerosant, although greater risks may occur if foam passes in the ocular or cerebral circulation (Duffy, 2010). Since foam displaces the intravascular blood and is not diluted in it, as in the case of liquid sclerosant injection, the concentration of the sclerosing agent in the vessel is known and controlled. Foam instillation leads to a homogeneous distribution of the sclerosant within the vessel lumen, except in very large veins where gravitational effects maintain better contact between the upper venous wall and the foam, which is far less dense than blood. Foam can be produced to be sufficiently stable from breakdown so being possible to provide adequate therapeutic effects from relatively short times in which it is kept in contact with the lumenal surface of the vein, as reported (Rao and Goldman, 2005).

Generally, the sclerosing foam is defined as a mixture of gas and liquid sclerosing solution with tension-active properties. Besides being composed of different specific ingredients, foams can differ in their internal cohesion, which is related to the size of the air bubbles. Macrofoam contains bubbles larger than 500 μm, minifoam contains bubbles between (250 – 500) μm, and microfoam is composed of bubbles smaller than 250 μm (Frullini, 2011; Hsu and Weiss, 2003).

A given volume of liquid can be used to produce 4 or 5 times its volume in foam, depending on the utilized foaming method. This volume allows the use of a lower total dose of the sclerosant to achieve the desired effect. Sclerosing foam is characterized by variables such as: the type and concentration of the tension-active sclerosing agent, the gas type, the liquid-to-gas ratio, the preparation conditions (temperature, pH), the time between preparation and use, and the bubble size.

The foam homogeneity is an important prerequisite for its flow behavior (viscosity) and stability. The ideal foam would have uniform bubble diameter and inner gas pressure. The durability of foam is related to the bubble size, the tension-activity (surface tension properties) of the liquid solution and the conditions under which the foam is formed and kept. Foam should be sufficiently durable to avoid its separation into gas and liquid components during injection but sufficiently ephemeral to break down once injected. The gas must be physiologically tolerated at therapeutic doses. The mean bubble size of the foam should be considerably less than 100 μm (Redondo and Cabrera, 2005).

There are many methods and mechanisms that have been applied to mix and agitate liquid sclerosants with gas admissions to create foams for clinical use. With the exception of one self-contained formulation, POL microfoam for endovenous use (Varisolve®), foamed sclerosants are typically produced by cyclical mechanical agitation of the liquid agent in the presence of a gas to generate the froth used for intravascular injection. Such "home made" foams commonly employ ratios of gas to liquid ranging from 1:1 to 8:1, producing foams of varying densities and rheological properties. In any event, the result is a froth containing 79% nitrogen and 21% oxygen and having a characteristic wide bubble size distribution (Cavezzi and Tessari, 2009).

Orbach was the first who described the use of froth in sclerotherapy (Orbach, 1950). After 1995 new methods of transforming the sclerosing liquid into foam were described (Cabrera et al., 1997a; Cavezzi and Frullini, 1999; Garcia-Mingo, 1999; Henriet, 1997; Monfreux, 1997; Sadoun and Benigni, 1998). In December 1999 Tessari described a safe and easy method to generate a fairly stable and compact foam (made of micro-bubbles of detergent drug and air) using two plastic syringes and a three-way tap (Tessari, 2000). Nowadays this is the main method used in foam sclerotherapy. Subsequently, Frullini and Gachet suggested other ways to produce sclerosing foam (Frullini, 2000; Gachet, 2001).

In the Monfreux technique, negative pressure is generated by drawing back the plunger of a glass syringe, the tip of which is tightly closed. The resulting influx of air produces large-bubbled, fairly fluid foam. In the Tessari technique, the turbulent mixture of liquid and air in two syringes connected using a three-way stopcock produces the foam. It is fine-bubbled and fluid at low concentrations and rather viscous at high concentrations. The mixing ratio for sclerosant to air is 1:3 to 1:4. The double syringe system technique involves the turbulent mixing of POL with air in a sclerosant to air ratio of 1:4 in two syringes linked using a connector. The resulting product is a fine-bubbled, viscous foam (Rabe and Pannier, 2010).

Some authors described in 2008 a method to obtain standardized POL foam prepared by using the EASY-FOAM® kit (two 10 ml silicone-reduced syringes, connected by a two-way valve connector, with one syringe prefilled with 7.4 ml sterile air (Laboratoire Kreussler Pharma, Paris, France). After aspiration of 1.6 ml POL into the other syringe (1:5.6) standardized movements of the syringe plungers were achieved by using the Turbofoam® machine, with a controlled number of movements, speed, and power (Hamel-Desnos et al., 2005; Rabe et al., 2008).

Lately, BTG International (United Kingdom) has developed Varisolve® (polidocanol endovenous microfoam, PEM) as a first line treatment for incompetent great saphenous veins (GSV) and associated varicosities, above and below the knee and for use alongside endovenous thermal ablation. PEM has a controlled density, consistent bubble sizes, and proprietary gas mix that make it a simple and comprehensive treatment for symptomatic and aesthetic varicose veins.

PEM is a pharmaceutical form of micro-foam that emulates the foam originally produced and studied by Cabrera (Cabrera et al., 2000, 2003, 2004) in a standardized way. It is generated and dispensed using a pressurized canister mechanism depicted in Fig. 3.

An European Phase III clinical trial showed that 90% of patients treated with PEM had no reflux in the GSV at 3 months and fewer than 10% of patients had recurrence at 1 year. Patients can generally return to work or their usual activities the same day they are treated, and cosmetic results after PEM treatment are apparent at 6 weeks. PEM is progressing through three US Phase III trials to explore its safety and efficacy as a treatment for moderate to severe varicose veins. All studies are expected to be completed by the end of 2011 and potential approval in 2013 (BTG International Ltd., 2011).

The system contains the liquid agent and a gas mixture of oxygen and carbon dioxide with only trace (0.01–0.08%) nitrogen present. Passage of the gas and liquid under pressure through a microfoam producing system yields micro-structurally consistent 1% POL microfoam having reproducible rheological properties. The bubble size for Varisolve® foam is appreciably smaller than that resulting from manual foam production techniques, and the absence of nitrogen facilitates more rapid absorption of bubbles within the body. Both these considerations are important to the safety profile, given the possibility for gas embolism to occur with any type of foam sclerosant therapy (Eckmann, 2009).

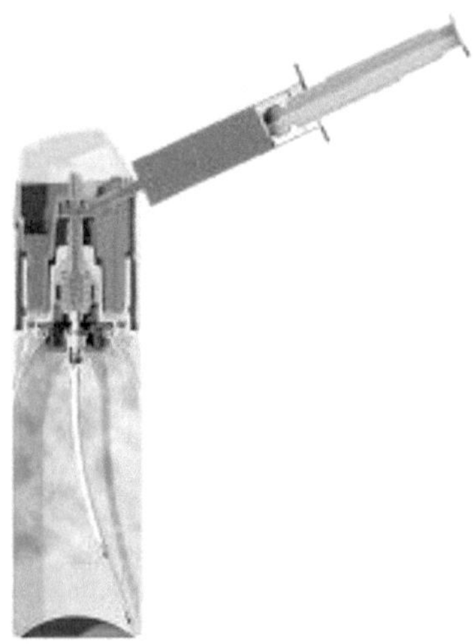

Fig. 3. The Varisolve® PEM generation and dispensing canister.

3. Laser sources description

For laboratory measurements regarding the photophysical properties of POL, we used a Q-switched Nd:YAG laser, while for clinical applications long pulsed laser types with the same active medium were employed.

The difference between the two types of lasers is the fact that for long pulsed medical lasers the emission of the laser resonator is obtained from a simple laser cavity where elements characteristic to Q-switch mode are not present. In this way the laser pulse duration is given by the long life time of the upper laser level (of the order of milliseconds) and lamps emission modulation.

Our Q-switched laser is a Continuum Surelite II. The head of the laser consists in a 115 mm length rod of Nd^{3+} doped YAG pumped by a linear flash lamp. The high pumping efficiency is achieved through a closed coupled configuration surrounded by a high brilliance magnesium oxide diffuser. The flashlamps have Xe as discharge gas with a pressure of (1-3) atm.

The active Q-switch is performed by the combination of a polarizer, Pockels cell and $\lambda/4$ waveplate. The Q-switch allows the accumulation of high energy in the laser resonator till it opens and allows the cavity to oscillate. By this, fast and high peak power laser pulses are produced. The typical laser pulses at 1,064 nm emission wavelength for our Surelite system have 5 ns pulse duration, 10 Hz repetition rate, and maximum peak energy of 685 mJ.

The fundamental laser beam emitted at 1,064 nm can be doubled, tripled, and quadrupled by using high harmonic generation BBO crystals. The characteristics of the beams obtained are: 532 nm, maximum energy per pulse 340 mJ; 355 nm, maximum energy per pulse 180 mJ; 266 nm, maximum energy per pulse 120 mJ.

The Nd:YAG lasers used for medical applications in leg veins treatment are long pulsed lasers, not having a Q-switch operation. Two different commercially available, long-pulsed 1,064 nm Nd:YAG laser systems were engaged for the clinical results presented here, namely:

- The Laserscope Lyra "i" (Laserscope, San Jose, CA, USA). To protect the epidermis from any adverse effects, Lyra-i system combines the long available wavelength with a unique flat beam profile to maximize patient safety and efficacy. The patients are protected by continuous parallel contact cooling that protects the epidermis. The cooling is carried out with a continuous flow of air (Zimmer Cryo 5® system, Zimmer Elektromedizin, Neu Ulm, Germany). The depth of penetration for 1,064 nm is due to its low scattering and epidermal transparency allowing the effective delivery of energy to deeply seated targets, such as blue and purple vessels while safely avoiding epidermal pigment. Computer adjusted fluence, delivered in adjustable pulse widths coupled with 1 mm to 5 mm continuously adjustable spot size is ideal for treating vessels from 0.5 mm to 4 mm.

The adjustable spot size and laser fluence allow the appropriate treatment procedure for several classes of veins. A spot size of 2 mm and fluence of 200 J/cm^2 are selected for spider (class I) and reticular (class II) veins, while 5 mm and 60 J/cm^2 are the laser system parameters for truncal veins (class III) treatment. The pulses delivered are 25 to 35 ms at a frequency of 5 Hz, for class I and II varicose veins, and 70 ms and 2 Hz, for class III varicose veins.

- A CoolTouch Varia® (CoolTouch Inc., Roseville, CA, USA) laser which has an adjustable pulsed cryogen cooling. The system is foreseen with an attached gas cooling canister that enables cooling the epidermis before and after laser pulses, at the same time interval between pulses. The thermal feedback is performed by precisely adjustment of the laser power to safely reach target temperature. Fluences used are similar, except for class I and II veins for which less than 150 J/cm^2 are programmed due to the fact that the minimum spot size is 3 mm and not 2 mm as in the Laserscope Lyra "i" system.

3.1 Laser tissue interaction

Basic requirements for a laser or a light source to treat leg veins are a wavelength that is proportionately better absorbed by the target (hemoglobin) than surrounding chromophores and penetration to the full depth of the target blood vessel. Sufficient energy must be delivered to damage the vessel without damaging the overlying skin, and this must be delivered over an exposure time long enough to slowly coagulate the vessel and its lining without damaging surrounding tissue.

Usually, optical radiation in the near-infrared wavelength range (from about 700 - 2,000 nm) is used, though when appropriate chromophores are available, visible wavelengths (e.g. green) can also be used. Photons launched into tissue are submitted to at least four optical processes: refraction, scattering, absorption, or output from the tissue.

In the context of this chapter, the light scattering in the tissues plays an important role since it regulates in a sense the penetration depth of the laser radiation in the tissue. There are different types of light scattering which may be produced in a tissue by its components, such as the Rayleigh scattering produced by very small particles, the Gans and Debye scattering produced by spherical and cylindrical macromolecules, globular proteins etc. (Grossweiner, 2005). Actually, the presence in the tissue of macromolecules, cells, water/liquid volumes, structures that have constituent layers (such as veins and arteries, cornea) enhance the light scattering within the tissue.

When photons are absorbed, the energy from the photon is converted into inter- and intra-molecular energy and results in generation of heat within the tissue. At the same time the good absorption in tissue limits the size of the lesion created by the laser irradiation. So, a compromise between good penetration and good absorption has to be found. After the initial absorption the temperature generated spreads through the tissue and enlarges the lesion somewhat, depending on the perfusion of the tissue. Large vessels will transport the heat away from the site and the effectively achieved temperature is reduced. For vascular structures the thermal relaxation time is of the order of milliseconds, so that lasers with pulse durations less than 1 ms are likely to produce little thermal injury.

Following Altshuler's expanded theory of selective photothermolysis, the chromophore is the Hb of the circulating blood. This captures the laser energy and turns it into heat. The chromophore must heat up sufficiently for thermal damage to occur in the vessel and especially on the vascular wall (Altshuler et al., 2001). The chromophore must heat intensely enough for heat to propagate to the wall of the vessel, but not to the extent that this becomes carbonized, which would involve the end of heat transfer in the vein structure. These data are highly important when deciding upon laser pulse dosimetry.

When the parameters (fluence and pulse width) are well chosen, intraparietal bleeding and rupture of the elastic fibers of the vein wall are observed immediately. Then, eosinophil appears in the muscle fibers of the treated vein. With the appropriate histology staining, there is a notable increase in the presence of beta-2-type tissue growth factor (TGF-beta 2). This intermediary acts on the fibroblasts increasing the secretion of proteins in response to thermal damage (heat shock protein 70), which are responsible for accelerating collagen and elastin fiber synthesis in the rough endoplasmic reticulum of the fibroblast (Black and Barton, 2004; Sadick et al., 2001). The first sequence is to induce vessel fibrosis and its re-absorption.

Optical absorption in the near-infrared spectral range is generally due to combination of overtone bands of fundamental molecular stretches. For wavelengths near 1,000 nm, water is a primary absorber of optical energy.

The choice of wavelength and pulse duration is related to the type and the size of the target vessel. Deeper vessels require a longer wavelength to allow penetration to their depth. Pulse duration must be matched to vessel size; the larger the vessel diameter, the longer the pulse duration required to effectively damage the vessel thermally. To be most effective, thermal injury must encompass the full thickness and circumference of the vein wall endothelium, rather than just the most superficial aspect of the vein wall. The relative importance of the hemoglobin absorption peaks in green (541 nm) and red to infrared (800-1,000 nm) shifts as the depth and the size of the blood vessel changes. Absorption by hemoglobin in the long-visible to near-infrared range appears to become more important for vessels more than 0.5 mm in diameter and at least 0.5 mm below the skin surface.

This approach of using pulsed lasers to limit thermal effects can also be used to ablate tissue. The limiting pulse length would be determined by the thermal relaxation time of the material. This is the time for heat to diffuse out of the irradiated volume and it is determined by the thermal diffusivity of the tissue and the dimensions of the volume. When laser energy is deposited in pulses shorter than the thermal relaxation time, heat accumulates and high temperatures are achieved. The ablative event can then occur before the heat diffuses out of irradiated volume. This confinement of heat can reduce the thermal damage incurred by adjacent tissue.

4. Experimental devices

First of all, we measured the optical properties of Aethoxysklerol 2% (Chemische Fabrik Kreussler & Co., Germany) solution. The absorption spectra measurements in spectrophotometer cells of 10 mm optical length are performed in UV-VIS-NIR spectral ranges using a Perkin Elmer Lambda 950 spectrophotometer (PerkinElmer, Massachusetts, USA) that has UV/VIS resolution ≤ 0.05 nm, NIR resolution ≤ 0.20 nm, with an error limit for absorbance of ± 0.004 %. This high-performance spectrophotometer features a double beam, double monocromator and ratio-recording optical system. As for the IR spectral range we used a FTIR Spectrophotometer Nicolet Magna 550 (Nicolet Instruments Inc., Madison, USA), working at (4000 - 400) cm^{-1} spectral domain and resolution 0.125 cm^{-1}.

Secondly, we exposed medicine solution samples to laser radiation in order to detect the possible molecular modifications induced in drug by the laser beam. The irradiation experimental set-up is shown in Fig. 4. We exposed Aethoxysklerol solution samples at laser beam emitted by a pulsed Nd:YAG laser (Continuum, Santa Clara, USA,, Excel Technology, Surelite II Frequency 10 Hz, FTW 5 ns) at 1.06 μm, the laser radiation having the following characteristics: repetition rate 10 pps, beam energy on the sample 50 mJ.

The exposure time was made on samples in bulk, between 2 min and 30 min; the sample was introduced in spectrophotometer cells of 10 mm (C - Fig. 4) and the corresponding irradiation dose varied from 60 J/cm^2 to respectively 900 J/cm^2; 8-10 % of the laser radiation was directed through a beam splitter (BS) unto a powermeter (P).

The effect of the laser light may be enhanced if the POL is used as foam since the light scattering in the tissue becomes more important and the effective absorption of the laser beam becomes larger. So, one might have in this case a larger number of modified POL molecules and/or a larger volume of tissue which is exposed to laser radiation.

Using the Tessari method - two disposable 5 ml syringes (Luer Slip) connected to a three-way stopcock, we produced foam from a mixture of Aethoxysklerol solution and atmospheric air; the mixture ratio was 1:4 (Fig. 5). This batch was passed between the two syringes about 40 times and the resulting hand-made foam was stable for 5-6 min.

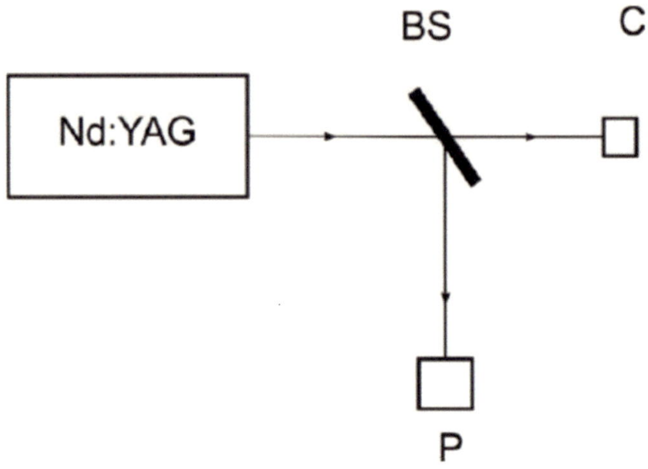

Fig. 4. Experimental set-up for laser irradiation of POL samples (P- powermeter, C- sample cell, BS- beam splitter)

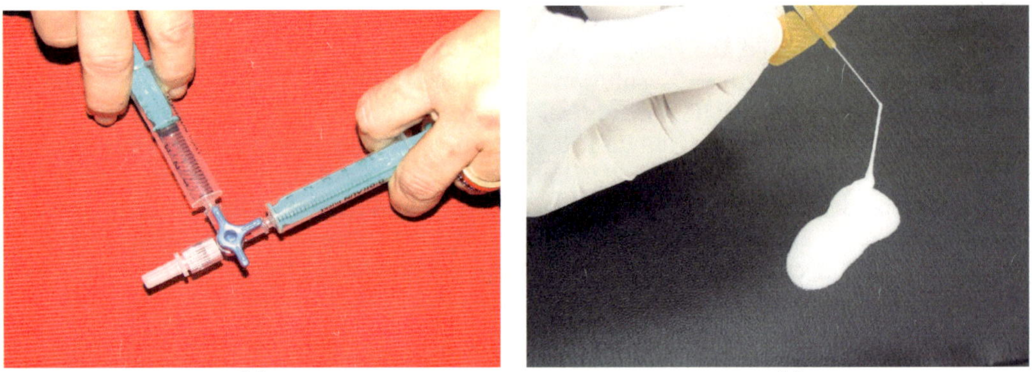

(a) (b)

Fig. 5. The Tessari foaming method (a) and foam consistency (b)

Foam and solution samples of POL were investigated by Raman spectroscopy as shown in the experimental set-up in Fig. 6. The laser radiation used to excite the Raman emission is the second harmonic of the same pulsed Nd:YAG laser emitting pulsed radiation at 10 Hz repetition rate, pulse duration 5 ns, and 250 mJ energy at 532 nm. The complete description of the laser source was made in Section 5. The detection and analysis of Raman signal are made by a spectrograph (SpectraPro 2750, Acton Research) and ICCD camera (PI-MAX 1024RB, Princeton Instruments).

The SpectraPro 2750 is a 750-mm, f/9.7-aperture, triple-grating monochromator/ spectrograph that features a versatile multiport optical system, 0.0025 nm drive-step size, built-in computer compatibility, and a wide scanning mechanical range up to 1,400 nm). As a monochromator, it offers built-in stepping-motor scanning and 0.023 nm resolution, plus easy integration into automated spectral-data acquisition systems. As a spectrograph, the SpectraPro 750 provides 1.1 nm/mm dispersion, a large 14 mm high by 27 mm wide focal plane, and interchangeable triple grating turrets. The grating types are as follows:

holographic UV, 2,400 lines/mm, for (185-375) nm; holographic VIS, 2,400 lines/mm, for (300-800) nm, blaze 1 μm and 600 lines/mm, for (650-1,500) nm.

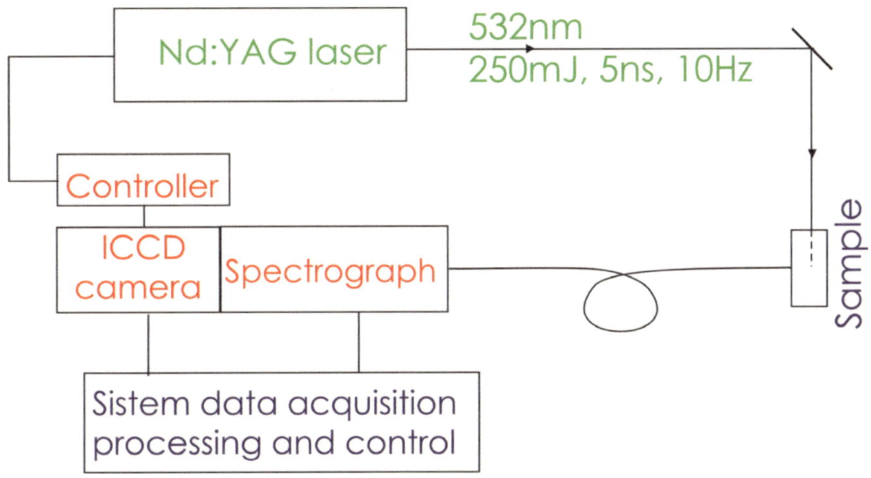

Fig. 6. The Raman spectroscopy system

The PI-MAX 1024RB from Princeton Instruments is a high performance intensified camera system featuring a spectroscopy format CCD. The imaging array is 1,024 x 256 and the spectral range (185-900) nm. It is fiber optically coupled to Gen II intensifiers with wide spectral coverage, quantum efficiency between (10-30)% in spectral range (200- 550) nm and 54 to 64 lp/mm resolution. Sub-nanosecond gating capability and an integrated programmable timing generator (PTG) make these ICCD cameras suitable for time-resolved spectroscopy applications.

5. Target patients groups and specific materials and method

From the clinical point of view and to illustrate this chapter we present the evaluation of leg telangiectasias and reticular veins (classes I to III) which were treated with the Nd:YAG laser following POL micro-foam injections. From our large casuistry we have selected a representative sample that underwent a prolonged control and follow-up.

Materials and method

This review corresponds to 200 patients, ages from 21 to 76 years (mean 37.2 y.o.a), treated with POL and Nd:YAG laser (Table 3). All patients were followed up and assessed periodically up to two years after treatment. Permission was granted to check medical files from the Administrative Council of the Instituto Médico Láser, Vilafortuny. The patients were females presenting leg varicosities that were treated with the same protocol. No patient has been exposed to sunlight or artificial tanning with UV-B during the 2 months prior to treatment; neither had they used any oral contraceptives for at least 12 weeks before the treatment. Exclusion criteria included: patients less than 18 years of age, pregnant women, lactation, scars or skin infections in the treatment area, the use of anticoagulants, a history of photosensitivity, keloids and/or hypertrophic scars and repeated herpes infections. None of the patients has been previously treated with laser for varicose veins.

60 patients suffered red venulectasias of less than 0.5 mm in diameter (class I); 72 of blue venulectasias of 0.5 to 1.5 mm (class II), and 68 patients had reticular veins 2 to 4 mm in diameter (class III).

	CLASS I VEINS Total no. pat. 60	CLASS II VEINS Total no. pat. 72	CLASS III VEINS Total no. pat. 68
AGE			
20-35 years	15	6	4
36-50 years	27	23	27
51-over	18	43	37
VEIN COLOR			
Red	41	39	12
Blue	6	5	42
Red/blue	13	18	14

Table 3. Patients demographics

All patients signed an informed consent for treatment and prior to treatment underwent a Duplex Ultrasound test (Soniline 050, Siemens, Issaqua, Japan) to screen for presence of reflux in the deep system, saphenous axes and in the perforant veins.

The micro-foam was obtained using two 10 ml Omnifix syringes with a Luer-Lock connection, a 3-way stopcock to connect the syringes and a 15G load needle with an air micro-filter. Two ml of POL (Aetoxysclerol® tamponné/lauromacrogol 400 (Kreussler Pharma, Germany) were used at 0.3% and 8 ml of air. Pumping and passing from one syringe to the other 15 times, produces stable micro-foam following the Tessari technique (Cavezzi et al., 2002; Frullini, 2000; Hamel-Desnos et al., 2003; Tessari, 2001; Tessari et al. 2001; Wollmann, 2004). The resulting foam is rich in nitrogen, with an irregular bubble size and a highly internal cohesion (Redondo and Cabrera , 2005).

The 200 patients received two treatments. The second treatment was given six weeks after the first. Parameters and methodology were the same for both treatments. The laser used was a 1,064 nm long pulse Nd-YAG laser, *Laserscope Lyra* "i"™ (Laserscope, San Jose CA, USA). The laser had a variable spot diameter, from 1 mm to 5 mm. The laser energy per pulse was delivered according to the treatment program at the same time that constant cooling of skin surface occurred because of the glass chamber that is adapted to the tip of the hand piece nozzle. The coolant is constantly circulated to cool the skin. In addition, a continuous flow of cold air that is capable of giving up to 1,000 l/m was used.

The device offers various programs to select the air flow. Program #5 was used which gave 600 l/m. The nozzle was pointed, at all times towards the place where the laser beam was directed. The temperature of the cold air was of 4°C. This was measured by pointing the thermometer under the air flow coming out of the hose nozzle. The approximate time of the air flow focusing on the treatment area was of one second; according to the external temperature detected with an IR thermometer (Laser Infrared Thermometer Center 350®, Center Technology Corp., Taiwan) the skin temperature was reduced to 22°C from the basal temperature of 33°C. The cooling device (Cryo 5, Zimmer Elektromedizin, Neu Ulm, Germany) followed the vein path treated with the laser pulses.

For class I and II veins, a 2 mm laser beam spot size was used, whereas for class III veins a 5 mm beam was used. Fluences were 200 J/cm^2 for class I and II and 60 J/cm^2 for class III. The laser pulse width was 30 ms for class I and II, and 70 ms for class III with a pulse repetition rate of 5 Hz for class I and II and 2 Hz for class III veins. According to the utilized protocol, all patients received treatment of the whole limb in the same session. For this, areas of 30x30 cm were selected continuing until the whole leg was treated. Disinfection of the areas for treatment was carried out using hydrogen peroxide. No anesthesia (i.e. any type of anesthesia) was used. Then, the POL micro-foam injection was injected using a 2 ml Omnifix syringe with a 30G needle. The sclerosant concentration was 0.8% in all cases and the maximum amount injected was 30 ml per session which proved to be enough in all cases. The average amount of sclerosant injected in each of the two scheduled treatments was around 20 ml.

Injection caused vessels to blanch and then a delay time of (1-3) min was necessary for the vessels to recover a soft pink color. Following the injection, laser was applied on the whole length of the vessel in the 30 x 30 cm area until it was totally covered with laser pulses. Because the total varicosity that the leg presented were treated in one session (with a repetition session after six weeks), the average number of laser pulses per session varied from 600 to 900.

After 6 weeks, the treatment was repeated with the same parameters. The patients returned for control at 18 weeks after the first treatment and the final assessment was at 2 years, taking as a reference the last (second) treatment carried out. At the 18-week control, complications were evaluated and further follow-up of their evolution was carried out at the 2 year assessment.

At this final assessment, all patients answered questionnaires to score in percentages their subjective impression regarding results. The percentages scored were: >90%, Very Good Results; 70 to 89%, Good Results; 50 to 59%, Fair Results and < 50%, Poor Results. Patients' satisfaction (PS) was determined by adding up the Very Good and Good results.

To objectively determine vessel clearance, photographs were taken, at a distance of 18 cm, using a Canon EOS 400D photo camera, equipped with a macro-lens (Tokina ATX Pro 100 f 2.8 Macro, Sea&Sea Flash Macro DRF 14, Tokyo, Japan). Photographs were taken before first treatment and at 2 years after the second treatment and they were used as objective references for evaluation by two independent Doctors, familiar with leg vein treatment with laser.

6. Physical and clinical results and discussions

It is important to provide a thorough understanding of the basic physical principles that underlie the use of coherent light in therapeutic applications. In this way, with respect to the new method to treat the varicose veins by combining POL microfoam and pulsed 1,064 nm Nd:YAG laser beam we must consider three factors: the drug sclerosis abilities, the action of the laser beam on the surrounding tissues and the possible increasing of the drug's potency under the laser light exposure. Of course, in practice it is important to adjust each of the parameters involved in the exposure to laser radiation in order to obtain the best results.

6.1 Optical results and discussions

Lasers used for the percutaneous treatment of varicose veins of the lower limbs must fulfil the principles of selective photothermolysis: a wavelength that is preferably more absorbed by Hb than by other similar chromophores; sufficient penetration as to achieve the depth at which the vessel for treatment is found; the appropriate energy to damage the vessel without damaging the skin; and release of energy slowly so that the heat reaches the vein wall without damaging the adjacent tissues (Suthamjariya et al., 2004). In order to achieve the above, the following variables must be taken into account: laser beam spot size, fluence, pulse width, pulse repetition rate and the technique used to apply the treatment.

Our laser irradiation and Raman spectroscopic measurements are made with experimental devices that involved time width pulses, energies, and irradiation doses that are different from those used in medical practice. However, they are made in order to better understand the physical processes specific to the interaction of the laser light with the tissues of medical interest in the treatment of the varicose veins.

The light propagation in turbid biological media is jointly governed by the absorption and scattering properties of the medium. On the other hand, the optical properties (absorption and scattering of light) of the marketed sclerosing substance may be subjects for studies in order to elucidate the mechanisms involved in this particular case in which combined sclerotherapy and light therapy of varicose veins are involved.

The absorption spectra of the medicine, shown in Fig. 7 and Fig. 8, indicate no significant absorption in the UV-VIS spectral range, and very weak peaks in NIR at 900 nm, 1.06 μm, 1.19 μm, 1.69 μm and 1.72 μm; they are the result of the superposed absorption of all the compounds included in the commercially available Aethoxysklerol 2%.

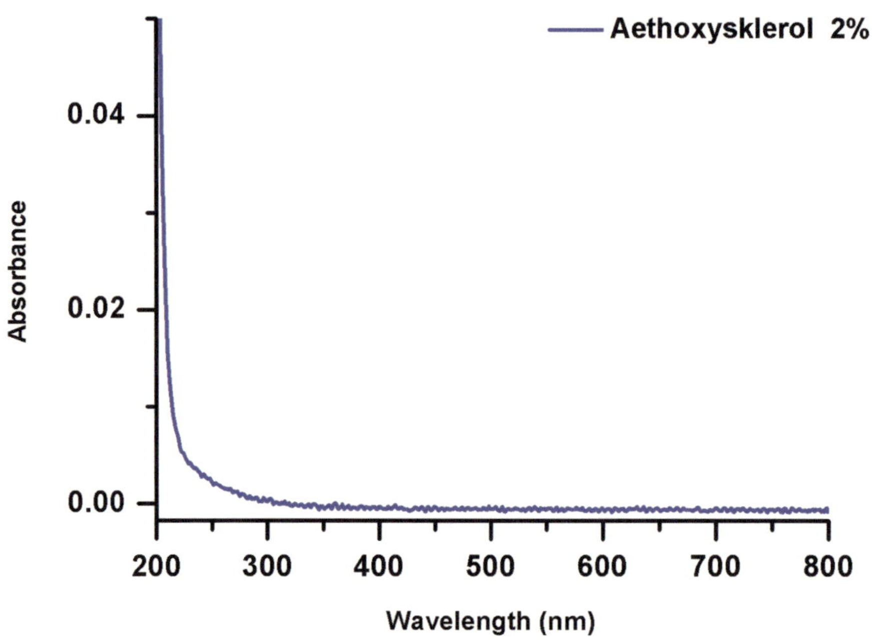

Fig. 7. UV-VIS absorption spectrum of Aethoxysklerol 2%

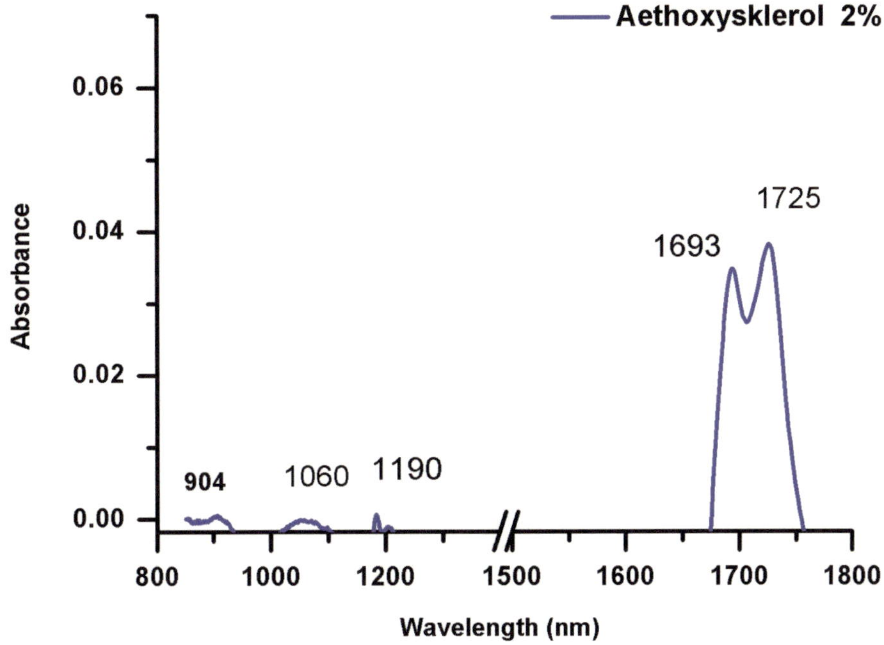

Fig. 8. NIR absorption spectrum of Aethoxysklerol 2%

Also, the FT-IR spectrum of POL shown in Fig. 9 is influenced by the absorption properties of water contained in the commercially presentation of the Aethoxysklerol solution.

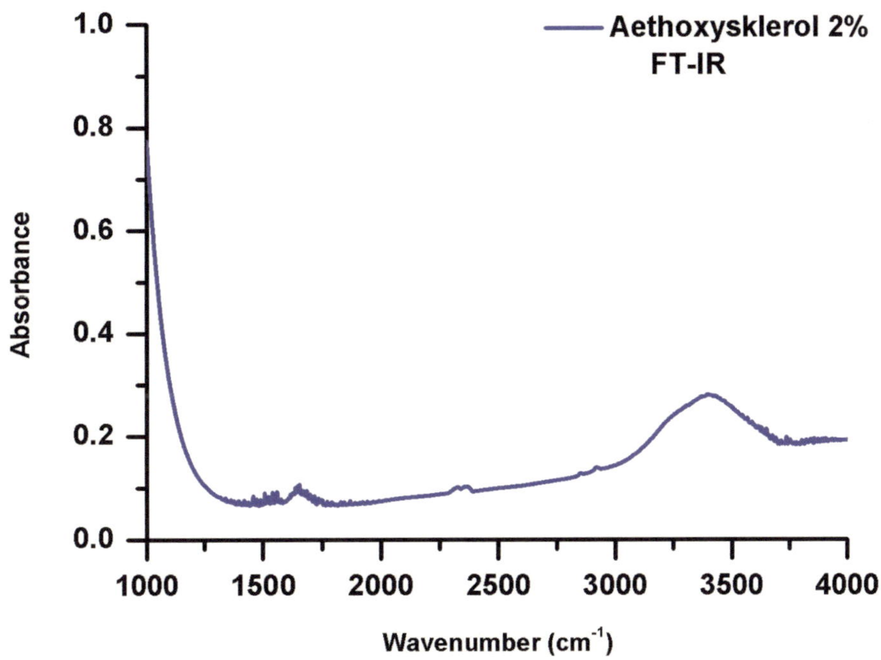

Fig. 9. The FT-IR spectrum of Aethoxysklerol 2%

The ethyl alcohol, for instance, is one of the inactive substances included in the commercially available Aethoxysklerol and has relatively significant absorption peaks at around 900 nm, 1 µm and 1.2 µm as it can be observed in Fig. 10. The absorbance measured in laboratory is in good concordance with literature reports (see Fig.10.a for recent literature reports).

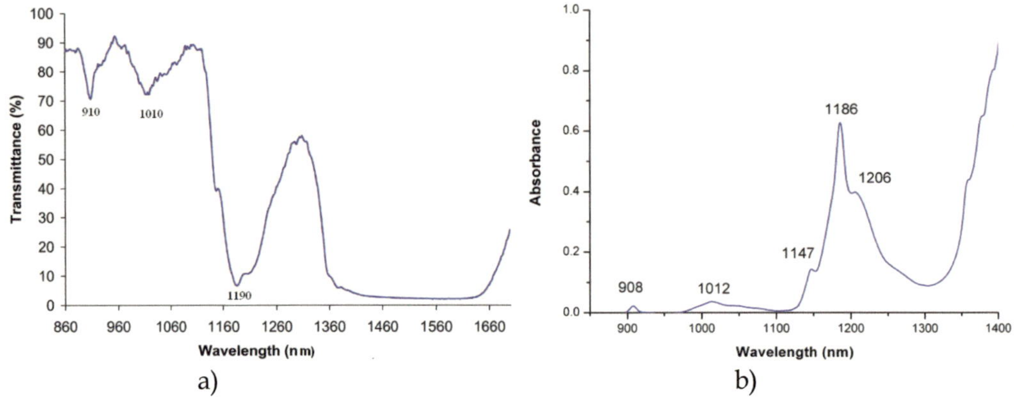

Fig. 10. The spectra of Ethanol in NIR: a) the transmission spectrum (Barun and Ivanov, 2010); b) the absorption spectrum (Smarandache et al., 2010)

Also, water presents broad absorption bands in NIR and middle IR, as it is shown in Fig. 11.

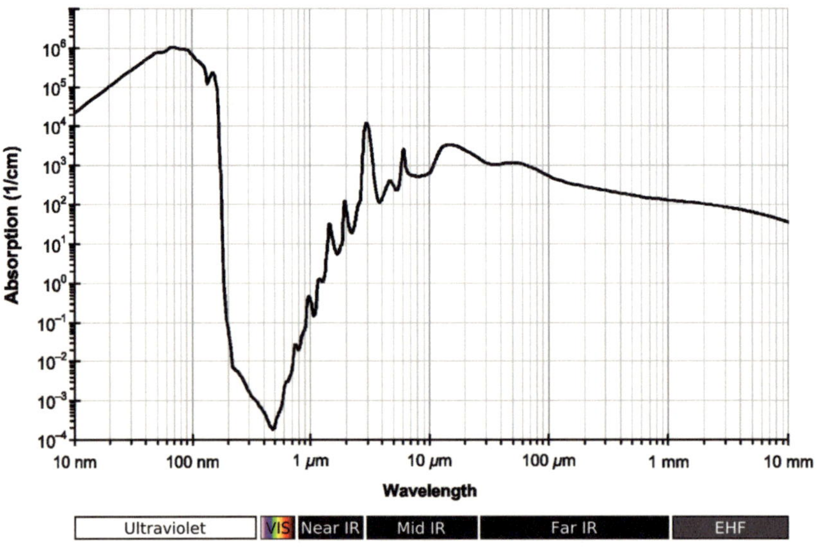

Fig. 11. The absorption spectrum of ultrapure water (Querry and al., 1998)

Some authors have assumed that the molecules of drugs can be photoactivated by assisting the hydrolysis process that results from excitation of vibration levels of the water molecule in the spectral range of 1-2 µm, following the absorption of infrared radiation by the water molecules (Fumarel et al., 2009).

We exposed the commercially available Aethoxysklerol 2% at laser beam emitted by a pulsed Nd:YAG laser at 1.06 µm, between 2 min and 25 min (exposure procedure as shown in Fig. 4). Following the irradiation, the absorption spectra of the medicine was measured (Fig. 12).

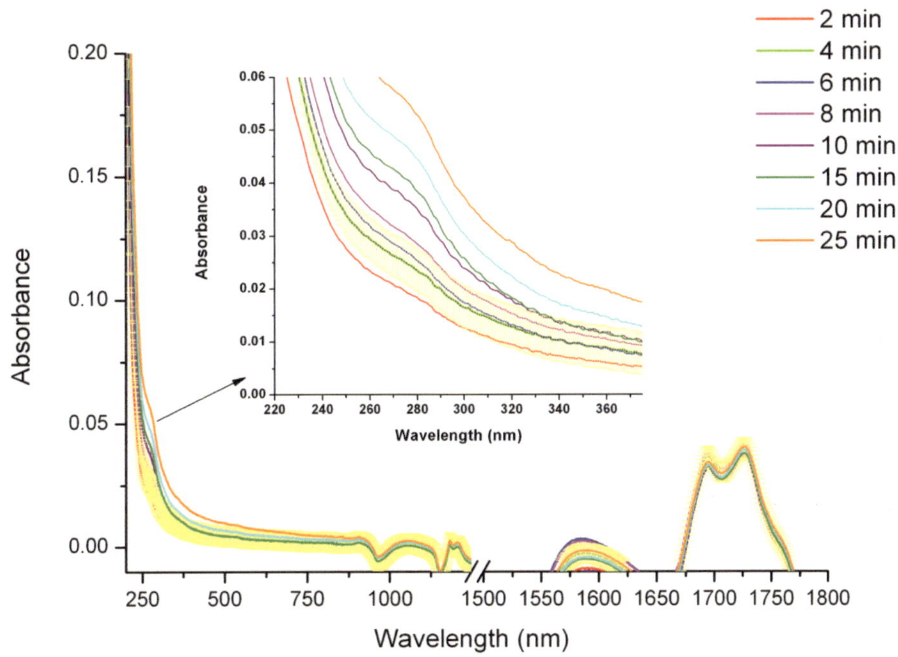

Fig. 12. The absorption spectra of Aethoxysklerol 2% exposed to 1.06 µm Nd:YAG laser radiation.

These spectra indicate that for wavelengths that exceed 500 nm they remain within the measuring error limits (±0.004 %). We assume that in the spectral range (1,000-1,100) nm the absorption is due rather to the main components of the tissue (melanin, water and hemoglobin/ oxihemoglobin); methemoglobin, especially, has the absorption three times greater than oxihemoglobin in this spectral range (Lee et al., 2006), (Fig.13).

Derivatives of blood hemoglobin, molecular oxygen dissolved in all components of the biological tissue, different ferments and other tissue substances that absorb light are considered as primary photoacceptors. Several photoinduced processes are known to take place during irradiation of an organism, such as photodissociation of oxyhemoglobin and the light-oxygen effect (Barun and Ivanov, 2010). Both mechanisms involve the absorption of light and the formation of oxygen in different forms which have a biological effect. The efficiency of these processes depends quantitatively on the way the light propagates and on the absorption coefficients of individual chromophores in the tissue of interest.

Modifications that as a trend are above the error limits are obtained in the spectral range (250-285) nm, as it is shown the detail in Fig. 12. In some cases (exposures at 4 min, 6 min, 8 min) the absorption curves may be considered within the error limits of the measuring system but they evaluate according to the trend.

A possible explanation of this behavior of the curves is that nonlinear absorption effects take place in Aethoxysklerol, such as the absorption of 4 photons at 1.06 μm, which would correspond to a transition at 266 nm. This modification may show that after absorbing four photons at 1.06 μm the Polidocanol molecules change their structure. The interaction mechanisms of laser radiation to the investigated solution are not completely elucidated and it is expected that further studies give a better understanding of them. More, it is expected that the modified Polidocanol is more efficient in destroying the varicose veins than the unirradiated solution (Smarandache et al., 2010).

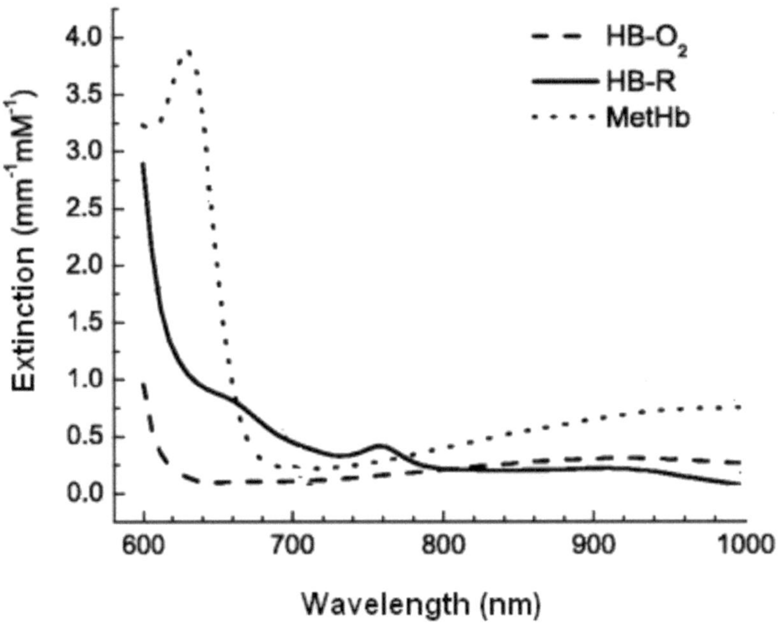

Fig. 13. The absorption spectra of hemoglobin (Hb-R), Oxihemoglobin (Hb-O_2) and Methemoglobin (MetHb) (Lee et al., 2006)

In an attempt to change the chromophore to obtain significant rates of MetHb, while seeking to increase photon absorption, J. Moreno Moraga & all (n.d.) started the combined application of microfoam sclerosants, proven to be efficient for vein sclerosis, with the action of pulsed 1.06 μm Nd:YAG laser beams.

The measurements performed at the National Institute for Laser, Plasma and Radiation Physics at Bucharest (Smarandache et al., 2011) have shown that the use of POL in the foam form increases the optical path of the laser radiation in the foam by light scattering, which leads to an increase of the total absorption, since this is proportional with the product between the absorption coefficient, the optical path length in the sample and the concentration of the absorbent. Laser energy absorption in the foam can be boosted by the multiplication of the impacts of the photons at the collisions with the gas bubbles. Moreover, under these circumstances, the number of changed POL molecules could also increase.

Following the line of increasing the efficacy of the laser beam if the POL is used as foam introduced in the tissue, the foam (produced by the Tessari method and the procedure described in Section 2) and solution samples of POL were investigated by Raman spectroscopy. The obtained Raman signals were more intense for foam than for simple solution samples (Fig. 14).

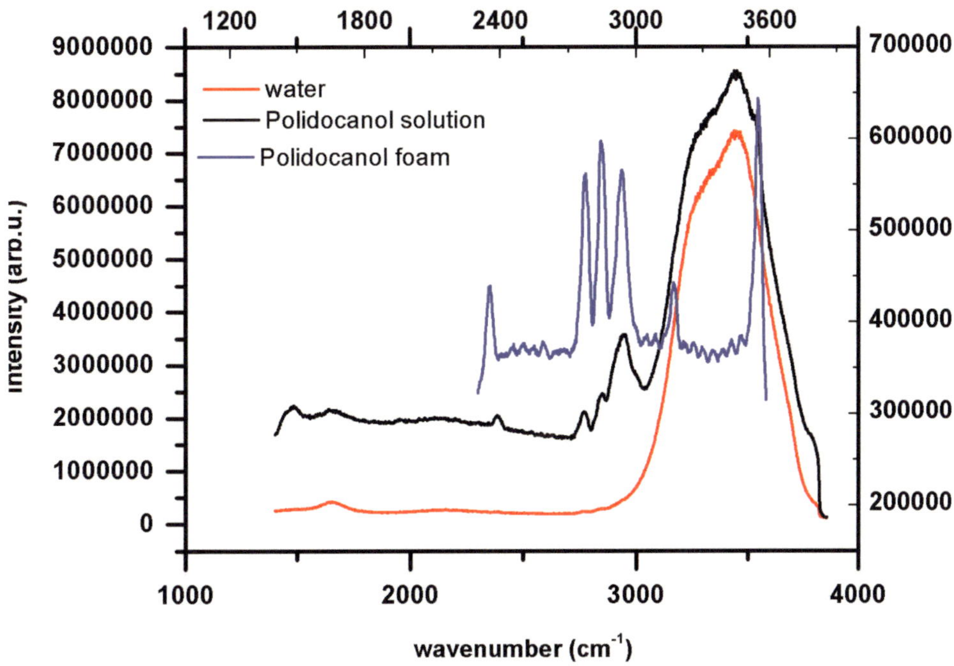

Fig. 14. The compared Raman spectra of the Polidocanol in solution and foam presentation

This laser light scattering enhancement is due to a longer optical path of the laser beam in the foam sample. The Raman vibrational lines corresponding to foam sample are more structured and stronger.

Also, the Raman signals were acquired at different time moments after the preparation of the foam samples. The results indicate that there are some parameters which must be taken into account, such as the bubble dimensions referred to the foam cohesion; these are important with respect to the moment of exposure of the varicose vein injected before with foam POL and exposed in the tissue to laser radiation. As it can be seen in Fig. 15 the best scattering signals were obtained at 2 minutes after foam preparation, in our experimental conditions. After few minutes the foam is destroyed and the intensity of the characteristic vibrational lines decreases drastically.

It is very important that physicians who use laser irradiation of the organism for therapeutic purposes and/or irradiated drugs have a physically well-founded quantitative instrument which would permit an analysis of the significance of the processes involved as a function of the spectral range of the irradiating light, its duration, and power. In addition, the development of a physical basis for the interaction of light with such a special object as biological tissue and the development of recommendations for the optimum interaction parameters are of fundamental scientific interest for medical applications.

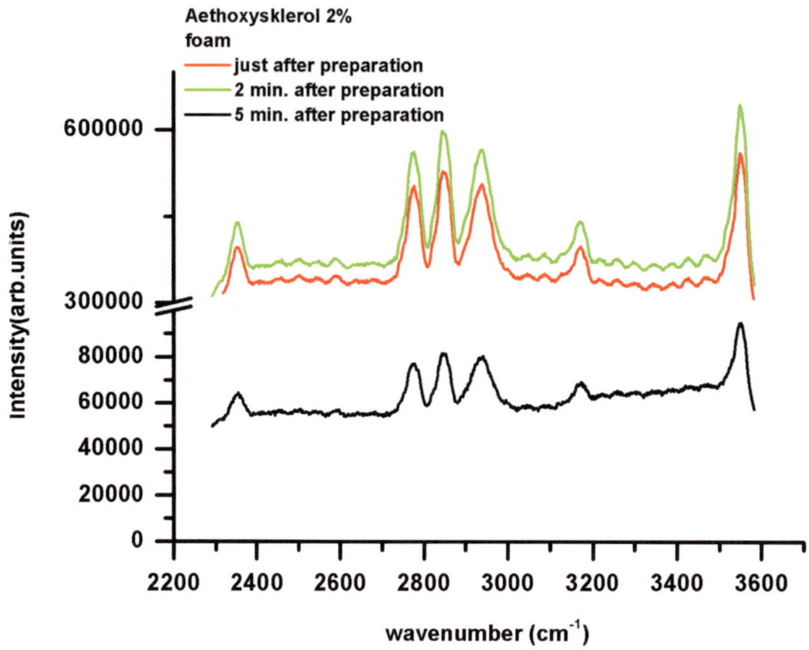

Fig. 15. The influence of foam separation (in two phases: liquid and remaining foam) on the Raman spectra of a POL foam sample

6.2 Clinical results and discussions

Immediately after treatment, darkening of the intravascular pigmentation with perivenous erythema was noticed. These alterations had completely disappeared when patients came back after 6 weeks for the second treatment. In spite of not having used any topical anesthesia, tolerance to treatment was good. Out of the 200 treated patients, 6 patients did not show up for the 2 year assessment, i.e. 3 class I patients, and 3 class III.

Answers to questionnaires regarding results corresponding to the 2 year assessment are presented in Table 4. The results correspond to scores presented by patients (subjectively) and those given by Doctors (objectively). Total satisfaction index (SI) was deduced by adding Very Good and Good scores given by patients. So, SI was 170 out of 194 patients, corresponding to 87,6% that showed up for the 2 years after assessment taking into account that 200 patients received treatment initially.

N° pat.	Vein type according to vessel diameter	PATIENT EVALUATION			DOCTOR EVALUATION		
		Very Good	**Good**	**Fair**	**Very Good**	**Good**	**Fair**
57	Class I	1	46	10	4	48	5
60	Class II	17	41	2	22	37	1
65	Class III	39	26	0	43	22	0

Table 4. Assessment of patients at 2 years

At 18 weeks, hyper-pigmentation and matting were noticed in 8 patients, all of them presenting class I veins. However these complications have disappeared at the 2 year assessment. The greatest success was achieved in class II and III veins. The results obtained in the evaluation at week 18 remained stable even at the 2 year assessment (Fig. 16 - Fig. 18).

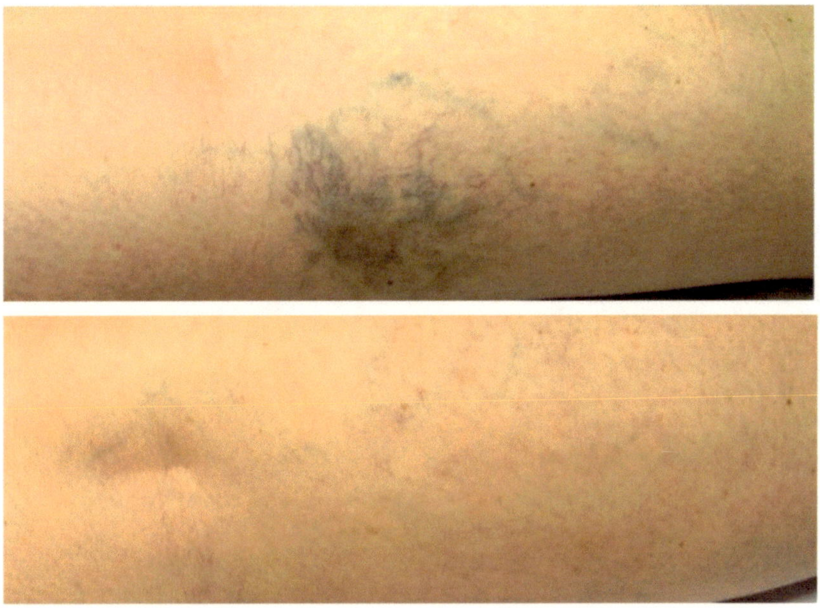

Fig. 16. Vein Class I. Aspect before and at 2 year assessment

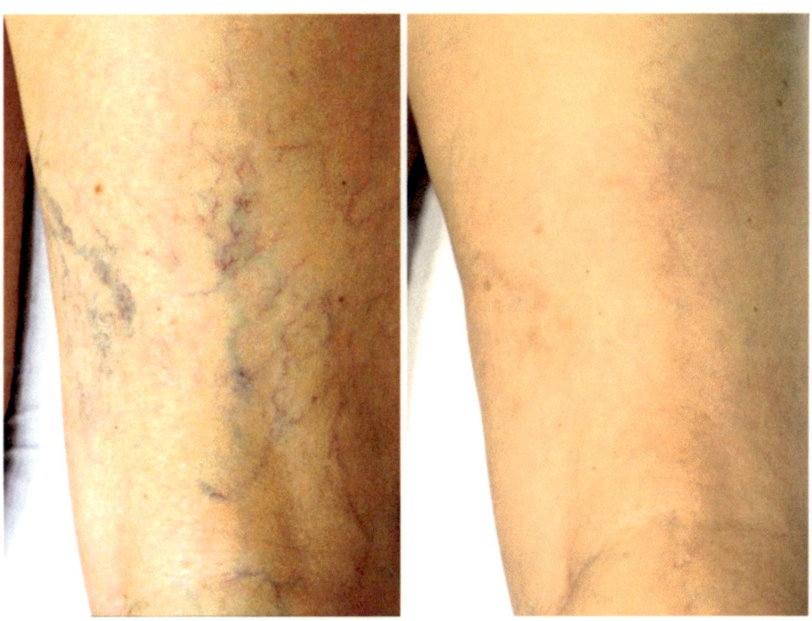

Fig. 17. Vein class II. Before and results 2 years after POL+ laser treatment

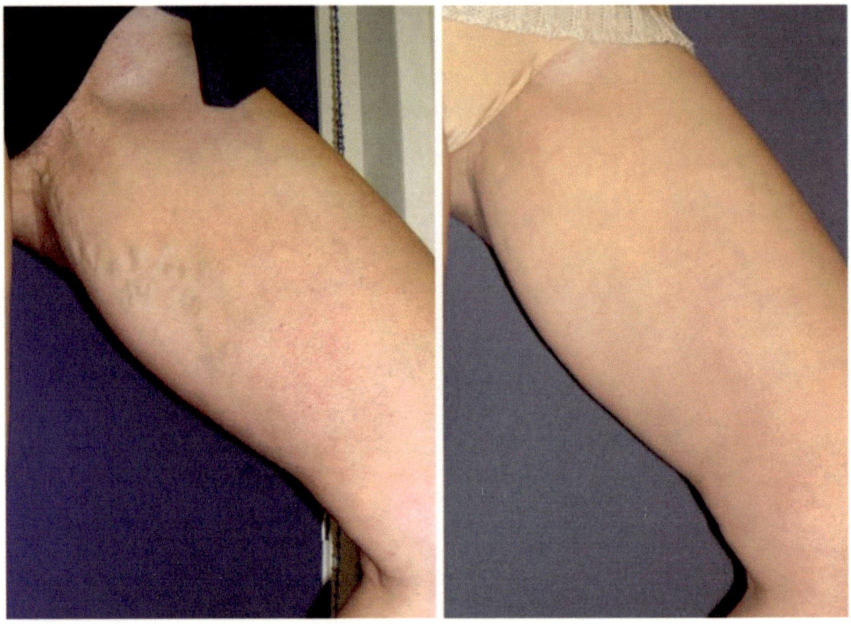

Fig. 18. Vein class III. Patient before and 2 years after treatment with described technique

The reported clinical results show that the use of POL foam combined with the exposure of the patient's tissue which contains it at pulsed laser beams of 1.06 µm wavelength allow obtaining better effects in the treatment of the varicose veins.

7. Conclusions

Clinical results prove that foam sclerotherapy combined with photothermolysis based on laser therapy increases the efficiency of the varicose veins treatment (Trelles et al., 2011).

The foamed form of the detergent sclerosing drugs has resulted in improvements in the efficacy of sclerotherapy. Foam sclerotherapy reduces the dose and concentration of injected drug and assures a better intimate contact of the active substance with the target tissues.

Concerning the foaming procedure, it is recommended to achieve a compromise between the bubbles dimension and the foam stability in time, so that the value of the surface tension is big enough to produce sclerosis of the target tissue and, nevertheless, the foam does not selfdestroys too quickly and dilutes in the blood stream before acting on the vein.

On the other hand, pulsed Nd:YAG lasers emitting at 1,064 nm offer important advantages for treating varicose veins of the lower limbs. They penetrate more and are less absorbed by melanin, which means that they can treat dark phototypes. A laser pulse at 1,064 nm converts Hb into the more spherically shaped MetHb, which has a 4 times higher absorption coefficient. After initial irradiation, further delivering of energy is more effective at heating blood and the surrounding vessel (Kunishige et al., 2007). In order to produce penetration, high fluences are required and hence, treatment is painful and requires cooling and even topical anaesthesia, which are considered as disadvantages of this treatment method.

Comparing the spectroscopic measurements data with the clinical experimental results, we might conclude that the improvement of the action of Polidocanol commercially available as Aethoxysklerol 2% on the varicose vein by exposure of the impregnated tissues with 1.06 µm laser beam is possible due to the following mechanisms:

- for wavelengths around 260 nm, the laser radiation may be absorbed by the Polidocanol proper. The mechanisms of interaction between the veins tissues and the medicine under the influence of laser radiation are not elucidated yet, but it is possible that nonlinear absorption effects take places in the tissue such as absorption of 4 photons at 1.06 µm (which would correspond to a transition at 266 nm), which may be responsible for further effects on the tissue;
- at λ=1.06 µm the absorption may be produced by the Ethyl alcohol and the main chromophores such as hemoglobin, especially methemoglobin and melanin; this may contribute to the sclerosis of the veins in the exposed area but it remains to clarify the possible mechanisms which lead to this effect. The modifications of the molecular structure of Polidocanol may be produced by the Nd:YAG laser radiation once this drug is introduced in the tissue.

The effect of the laser light may be enhanced if the Polidocanol is introduced as foam since than, the light scattering in the tissue becomes more important and the absorption of the laser beam becomes larger. So, we might have in this case a larger number of modified Polidocanol molecules. The Raman spectroscopy measurements prove that the Raman signals were more intense for foam than for simple solution samples. This laser light scattering enhancement is due to a longer optical path of the laser beam in the foam sample. The vibrational lines corresponding to foam sample are more structured and stronger.

As for the proper/most recommended moment to expose the varicose vein injected before with foam POL to laser radiation, the results of Raman spectroscopy study, performed at different moments of foam lifetime, indicate that there are some parameters which must be taken into account, such as bubble dimensions referred to the foam cohesion.

As for the direct clinical application of the Polidocanol, assisted by Nd:YAG laser radiation, the combined treatment of leg veins with Polidocanol (micro)foam and 1,064 nm Nd:YAG laser pulses shows promising possibilities for coagulation of varicles of the lower limbs with efficacy, without the damage of the skin surface. This kind of treatment, using low fluency, permits carrying out treatment sessions on larger areas with patient comfort. Efficacy may be substantially due to Polidocanol injection given to the patient before his/her exposure to laser pulses. The Polidocanol micro foam sclerosant would change the optical absorption of the Nd:YAG emitted at 1,064 nm, increasing the efficacy of the treatment and the vein closure. The obtained promising results are backed by evidence-based clinical outcome, and efficacy has been assessed up to 3 years (Trelles et al., 2011).

Further studies are needed to elucidate the role of Polidocanol microfoam interaction with the Nd:YAG laser beam and both a possible molecular excitation and light scattering phenomenon that will enhance the observed results.

8. References

Alos, J.; Carreno, P.; Lopez, J.A.; Estadella, B.; Serra-Prat, M. & Marinello, J. (2006). Efficacy and safety of sclerotherapy using polidocanol foam: a controlled clinical trial. *Eur J Vasc Endovasc Surg*, Vol.31, No.1, (January 2006), pp. 101–107, ISSN 1078-5884

Altshuler, G.B.; Anderson, R.R.; Manstein, D.; Zenzie, H.H. & Smirnov, M.Z. Extended Theory of Selective Photothermolysis. *Lasers Surg Med,* Vol.29, No.5, (December 2001), pp. 416-432, ISSN 1096-9101

Anderson, R.R. & Parrish, J.A. (1983). Selective photothermolysis: precise microsurgery by selective absorption of pulsed radiation. *Science,* Vol.220, No.4596, (April 1983), pp. 524- 527, ISSN 0036-8075 (print), 1095-9203 (online)

Artemi P. (2007). Pharmacology of Phlebology, *ACP Australasian meeting,* Sydney, Australia, September 18-21, 2007

Barun, V.V. & Ivanov, A.P. (2010). Depth distributions of light action spectra for skin chromophores. *Journal of Applied Spectroscopy,* Vol.77, No.7, (March 2010), pp. 73-79, ISSN 0021-9037 (Print), 1573-8647 (Online)

Black, F.B. & Barton, J.K. Chemical and structural changes in blood undergoing laser photocoagulation. *Photochem Photobiol,* Vol.80, No1, (July 2004), pp. 89-97, ISSN 1751-1097

BTG International Ltd. (2011). Our Pipeline, In: *Development,* September 2011, Available from: http://www.btgplc.com/development/our-pipeline

Cabrera, G.J., Cabrera G.O.J. & Garcia-Olmedo M.A. (1997a). Elargissement des limites de la schlerotherapie: nouveaux produits sclerosants. *Phlebologie,* Vol.50, No.2, (1997), pp. 181–187, ISSN 0031-8280

Cabrera, G.J. & Cabrera, G.O.J. Jr. (1997b). BTG International Limited inventors; assignee Injectable microfoam containing a sclerosing agent. *US patent 5676962.* (October 1997) available at www.patents.com/us-5676962.html

Cabrera, J.; Cabrera, J. Jr. & Garcia-Olmedo, M.A. (2000). Treatment of varicose long saphenous veins with sclerosant in microfoam form: long-term outcomes. *Phlebology,* Vol.15, No.1, (2000), pp.19–23, ISSN 0268-3555

Cabrera, J.; Cabrera, J. Jr; Garcia-Olmedo, M.A. & Redondo, P. (2003). Treatment of venous malformations with sclerosant in microfoam form. *Arch Dermatol,* Vol.139, No.11 (November 2003), pp. 1409–1416, ISSN 1538-3652 (on-line)

Cabrera, J.; Redondo, P., Beccerra, A.; Garrido, C.; Cabrera, J.Jr.; Garcia-Olmedo, M.A.; Sierra, A.; Lloret, P. & Martinez-Gonzalez, M.A. (2004).. Ultrasound-guided injection of polidocanol microfoam in the management of venous leg ulcers. *Arch Dermatol* , Vol.140, (June 2004), pp. 667–673, ISSN 1538-3652 (on-line)

Cavezzi, A. & Frullini, A. (1999). The role of sclerosing foam in ultrasound guided sclerotherapy of the saphenous veins and of recurrent varicose veins: our personal experience. *Australian & New Zealand journal of phlebology,* Vol. 13, (1999), pp. 49–50, ISSN 1441-7766

Cavezzi, A.; Frullini, A.; Ricci, S. & Tessari, L. (2002). Treatment of varicose veins by foam sclerotherapy: two clinical series. *Phlebology,* Vol.17, (2002), pp. 13-18, ISSN 0268-3555

Cavezzi, A. & Tessari, L. (2009). Foam sclerotherapy techniques: different gases and methods of preparation, catheter versus direct injection. *Phlebology,* Vol.24, No.6, (December 2009), pp. 247-251, ISSN 0268-3555

Duffy, D. M. (2010). Sclerosants: A Comparative Review, *Dermatol. Surg.,* Vol. 36, Suppl.2, (June 2010), pp. 1010–1025, ISSN 1524-4725

Eckmann, D.M.; Kobayashi, S. & Li, M. (2005). Microvascular embolization following polidocanol microfoam sclerosant administration. *Dermatol Surg,* Vol.31, No.6, (June 2005), pp. 636–643, ISSN 1524-4725

Eckmann, D.M. (2009). Polidocanol for Endovenous Microfoam Sclerosant Therapy. *Expert Opin Investig Drugs,* Vol.18, No.12, (December 2009), pp. 1919–1927, ISSN 1354-3784

Frullini, A. (2000). New technique in producing sclerosing foam in a disposable syringe. *Dermatologic Surgery,* Vol.26, No.7, (July 2000), pp. 705–706, ISSN 1524-4725 (on-line), 1076-0512 (print)

Frullini, A. (2011). An Investigation into the Influence of Various Gases and Concentrations of Sclerosants on Foam Stability. *Dermatologic Surgery, Vol.37, No.1, (January 2011), pp. 17-18,* ISSN 1524-4725 (on-line), 1076-0512 (print)

Fumarel, R.; Murgoi, G.; Albert, P.; Hurduc, A. & Pascu, M.L. (2009). Increase of Cisplatinum therapeutic index through optical irradiation, *AIP Conf. Proc. of LASER FLORENCE 2008,* Vol.1142, pp.1-7, ISSN 0094-243X, ISBN 978-0-7354-0679-7, Florence, Italy, October 31-November 1, 2008

Gachet, G. (2001). Une nouvelle methode simple et economique pour confectionner de la mousse pour la sclerose echoguidee. *Phlebologie,* Vol.54, No. 1, (2001), pp. 63–65, ISSN 0031-8280

Garcia-Mingo, J. (1999). Esclerosis venosa con espuma: Foam Medical System. *Revista Espanola de Medicina y Cirugia Cosmetica,* Vol.7, (1999); pp.29–31, available at www.cavezzi.it/garciaen.html

Grossweiner, L.I. (2005), *The science of Phototherapy: An Introduction,* Springer, ISBN-13: 978-1402028830

Hamel-Desnos, C.; Desnos, P.; Wollmann, J.C.; Ouvry, P.; Mako, S. & Allaert, F.A. (2003). Evaluation of the efficacy of polidocanol in form of foam compared to liquid form in sclerotherapy of the long saphenous vein. *Dermatol Surg,* Vol.29, No.12, (December 2003), pp. 1170-1175, ISSN 1076-0512 (print), 1524-4725 (electronic)

Hamel-Desnos, C.; Ouvry; Benigni, J.P.; Boitelle, G.; Schadeck, M.; Desnos P. & Allaert, F.A. (2007). Comparison of 1% and 3% Polidocanol Foam in Ultrasound Guided Sclerotherapy of the Great Saphenous Vein: A Randomised, Double-Blind Trial with 2 Year-Follow-up. "The 3/1 Study", *European Journal of Vascular and Endovascular Surgery,* Vol. 34, No.6, (December 2007), pp. 723-729, ISSN 1078-5885

Henriet, J.P. (1997). Un an de pratique quotidienne de la sclerotherapie (veines reticulaires et teleangiectasies) par mousse de polidocanol: faisebilite, resultats, complications. *Phlebologie,* Vol.50, No.2, (1997); pp. 355–360, ISSN 0031-8280

Hsu, T.S. & Weiss, R.A. (2003). Foam sclerotherapy: a new era. *Arch Dermatol,* Vol. 139, No.11, (November 2003), pp. 1494-1496, ISSN 1538-3652 (on-line)

Kunishige, J.; Goldberg, L. & Friedman, P. (2007), Laser therapy for leg veins. *Clinics in Dermatology,* Vol.25, No.5, (September-October 2007), pp. 454–461, ISSN 0738-081X

Lee, J.; El-Abaddi, N.; Duke, A.; Cerussi, A.E.; Brenner, M. & Tromberg, B.J. (2006). *J. Appl. Physiol.* Vol.100, No.2, (February 2006), pp. 615-622, ISSN 8750-7587

Monfreux, A. (1997). Traitement sclerosant des troncs saphenies et leurs collaterales de gros calibre par le methode MUS. *Phlebologie,* Vol.50, No.2, (1997), pp. 351–353, ISSN 0031-8280

Mordon, S.; Brisot, D. & Fournier, N. (2003). Using a "non uniform pulse sequence" can improve selective coagulation with a Nd:YAG laser (1.06 micron) thanks to Met-hemoglobin absorption: A clinical study on blue leg veins. *Lasers Surg Med* , Vol.32, No.2, (2003), pp. 160-170, ISSN 1096-9101 (electronic)

Moreno Moraga, J.; Isarria Marcos, M.J.; Royo de la Torre, J. & Gonzalez Urena, A. (n.d.). Photodynamic therapy in the treatment of varicose veins, Available from http://www.institutomedicolaser.com/archivos/areacientifica/varices_070306.pdf

Nijsten, T.; van den Bos, R.R.; Goldman, M.P.; Kockaert, M.A.; Proebstle, T.M.; Rabe, E.; Sadick, N.S.; Weiss, R. & Neumann, H.A.M. (2009). Minimally invasive techniques in the treatment of truncal varicose veins. *J Am Acad Dermatol,* Vol.60, No.1, (January 2009), pp. 110-119, ISSN 0190-9622

Orbach E.J. (1950). Contribution to the therapy of the varicose complex. *J Int Coll Surg,* Vol.13, No.6, (June 1950), pp. 765–771, ISSN 0020-8868

Ouvry, P.; Allaert, F.A.; Desnos, P. & Hamel-Desnos, C. (2008). Efficacy of Polidocanol Foam versus Liquid in Sclerotherapy of the Great Saphenous Vein: A Multicentre Randomised Controlled Trial with a 2-year Follow-up. *Eur J Vasc Endovasc Surg,* Vol. 36, No.3, (September 2008), pp.366-370, ISSN 1078-5884

Parsi, K.; Exner, T.; Connor, D.E.; Ma, D.D.F. & Joseph, J.E. (2007). In vitro Effects of Detergent Sclerosants on Coagulation, Platelets and Microparticles. *Eur J Vasc Endovasc Surg.*, Vol.34, No.6, (December 2007), pp. 731-740, ISSN 1078-5884

Parsi, K.; Exner, T.; Connor, D.E.; Herbert, A.; Ma, D.D. & Joseph, J.E. (2008). The lytic effects of detergent sclerosants on erythrocytes, platelets, endothelial cells and microparticles are attenuated by albumin and other plasma components in vitro. *Eur J Vasc Endovasc Surg* , Vol. 36, No.2, (August 2008), pp. 216–223, ISSN 1078-5884

Pascu, M.L.; Nastasa, V.; Smarandache, A.; Militaru, A.; Martins, A.; Viveiros, M.; Boni, M.; Andrei, I.R.; Pascu, A.; Staicu, A; Molnar, J. ; Fanning, S. & Amaral, L. (2011). Direct Modification of Bioactive Phenothiazines by Exposure to Laser Radiation. *Recent Patents on Anti-Infective Drug Discovery,* Vol.6, No.2, (May 2011), pp. 147-157, ISSN1574-891X

PubChem. (2011). Polidocanol, In: Compund, August, 2011, Available from: http://pubchem.ncbi.nlm.nih.gov/

Querry, M. R.; Wieliczka, D. M. & Segelstein, D. J. (1998). Water: H_2O, In: *Handbook of Optical Constants of Solids II*, E. D.Palik, (Ed.), 1059–1077, Academic Press 1998, ISBN 0-12-5444422-2, San Diego, CA, USA.

Rabe, E.; Otto, J.; Schliephake, D. & Pannier, F. (2008). Efficacy and Safety of Great Saphenous Vein Sclerotherapy Using Standardised Polidocanol Foam (ESAF): A Randomised Controlled Multicentre Clinical Trial, *Eur J Vasc Endovasc Surg,* Vol.35, No.2, (February 2008), pp. 238-245, ISSN 1078-5885

Rabe, E. & Pannier, F. (2010). Sclerotherapy of Varicose Veins with Polidocanol Based on the Guidelines of the German Society of Phlebology. *Dermatol Surg,* Vol.36, Suppl.2, (June 2010), pp. 968–975, ISSN 1076-0512 (print), 1524-4725 (electronic)

Railan, D.; Parlette, E.; Uebelhoer, N. & Rohrer, T. (2006). Laser treatment of vascular lesions. *Clinics in Dermatology*, Vol.24, No.1, (January-February 2006), pp. 8-15, ISSN 0738-081X

Rao, J. & Goldman, M.P. (2005). Stability of foam in sclerotherapy: differences between sodium tetradecyl sulfate and polidocanol and the type of connector used in the double-syringe system technique. *Dermatol Surg*, Vol.31, No.1, (January 2005), pp. 19–22, ISSN 1076-0512 (print), 1524-4725 (electronic)

Redondo, P. & Cabrera, J. (2005). Microfoam Sclerotherapy. *Seminars in Cutaneous Medicine and Surgery*, Vol.24, No.4, (December 2005), pp. 175-183, ISSN 1085-5629

Rogachefsky, A.S.; Silapunt, S. & Goldberg, D.J. (2002). Nd:YAG laser (1064nm) irradiation for lower extremity telangiectases and small reticular veins: efficacy as measured by vessel color size. *Dermatol Surg*, Vol.28, No.3, (March 2002), pp. 220 -223, ISSN 1524-4725 (electronic)

Ross, E.V. & Domankevitz, Y. (2005). Laser treatment of leg veins: physical mechanisms and theoretical considerations. *Lasers Surg Med*, Vol. 36, No.2, (February 2005), pp. 105-116, ISSN 1096-9101 (electronic)

Sadick, N.S.; Prieto, V.G.; Shea, C.R.; Nicholson, J. & McCaffrey, T. Clinical and Pathophysiologic Correlates of 1064-nm Nd:YAG Laser Treatment of Reticular Veins and Venulectasias. *Arch Dermatol*, Vol.137, No.5, (May 2001), pp. 613-617, ISSN 1538-3652

Sadoun, S. & Benigni, J.P. (1998). The treatment of varicosities and telangiectases with TDS and Lauromacrogol foam. *XIII World Congress of Phlebology 1998, abstract book*, pp. 327, Sydney, Australia, September 6-11, 1998.

Santos, P.; Watkinson, A.C.; Hadgraft, J. & Lane, M.E. (2008). Application of Microemulsions in Dermal and Transdermal Drug Delivery, *Skin Pharmacol Physiol*, Vol.21, No.5, (2008), pp. 246-259

Smarandache, A.; Trelles, M. & Pascu, M.L. (2010). Measurement of the modifications of Polidocanol absorption spectra after exposure to NIR laser radiation. *J Optoelectronics Advanced Materials*, Vol.12, No.9, pp. 1942 - 1945, (2010) ISSN 1454-4164

Smarandache, A.; Nastasa, V.; Militaru, A.; Staicu, A.; Trelles, M.; Moreno-Moraga, J. & Pascu, M.L. (June 2011). Comparison of the experimental techniques used to obtain foams out of medicines solutions, *ISWLA 2011*, May 31 - June 4, 2011, Bran, Romania, Available from: http://iswla.inflpr.ro/CompleteProgram.pdf

Smarandache, A.; Trelles, M.; Staicu, A.; Moreno-Moraga, J. & Pascu, M.L. Laser beams interaction with Polidocanol foam: physical bases, *SELMQ 2011- XIX Congreso Sociedad Espanola de Laser Medico Quirurgico*, 8 - 10 July 2011, Jerez de la Frontera, Spain, 2011, Available from: http://congresos.net/frame.php?id= 1015&web =http : //www.selmq.net/

Suthamjariya, K.; Farinelli, W.A.; Koh, W. & Anderson, R,R. (2004). Mechanisms of microvascular response to laser pulses. *J Invest Dermatol*, Vol.122, No.2, (February 2004), pp. 518-525, ISSN 0022-202x

Tessari, L. (2000). Nouvelle technique d'obtention de la sclero-mousse. *Phlebologie*, Vol.53, No.1, (2000), pp. 129-133, ISSN 0031-8280

Tessari, L. (2001). Extemporary sclerosing foam according to personal method: experimental clinical data and catheter usage. *Int Angiol Suppl*, Vol.20, Suppl.1, (2001), pp. 54.

Tessari, L., Cavezzi, A. & Frullini, A. (2001). Preliminary experience with a new sclerosisng foam in the treatment of varicose veins. *Dermatol Surg*, Vol.27, No.1, (January 2001), pp. 58-60, ISSN 1524-4725 (electronic)

Tibbs, D.J.; Sabiston, D.; Davis, M. & Mortimer, P. (1997). *Varicose Veins, venous disorders and lymphatic problems in the lower limb*, Oxford University Press, ISBN 978-0192627629, USA

Trelles, M.; Allones, I.; Alvareza, X,; Veleza, M.; Buila, C.; Luna, R. & Trelles, O. (2005). Long-pulsed Nd:YAG 1064 nm in the treatment of leg veins: Check up of results at 6 months in 100 patients. *Medical Laser Application,* Vol.20, No.4, (December 2005), pp. 255–266, ISSN 1615-1615

Trelles, M.A.; Weiss, R.; Moreno-Moraga, J.; Romero, C. Velez, M. & Perez, X. (2010). Treatment of legs veins with combined pulsed dye and Nd:YAD lasers: 60 patients assessed at 6 months. *Lasers Surg Med,* Vol.42, No.9, (November 2010), pp.609-614, ISSN 1096-9101 (electronic)

Trelles, M.; Moreno-Moraga, J.; Alcolea, J.; Smarandache, A. & Pascu M.L.. (2011) Laser in leg veins: our personal approach of treatment, In: *Synopsis of Aesthetic Dermatology & Cosmetic Surgery,* M.L. Elsaie, (Ed), Nova Science Publishers Inc, NY, USA, in press

U.S. Department of Health & Human Services, FDA. (2010). FDA Approves Asclera to Treat Small Varicose Veins, In: *Press Announcements,* (September 2011), Available from: http://www.fda.gov/NewsEvents/Newsroom/PressAnnouncements/

Wainwright, M. (2008). Photodynamic therapy: the development of new photosensitisers. *Anticancer Agents Med Chem,* Vol.8, No.3, (April 2008), pp. 280-291, ISSN 1871-5206

Wollmann, J.C. (2004). The history of sclerosing foams. *Dermatol Surg,* Vol.30, No.5, (May 2004), pp. 694-703, ISSN 1524-4725 (electronic)

13

Pulsed Nd:YAG Laser Applied in Microwelding

Vicente Afonso Ventrella
UNESP - São Paulo State University, Mechanical Engineering Department
Brazil

1. Introduction

The first aim of this study was to value the possibility to join, for pulsed Nd:YAG laser welding, thin foils lap joints for sealing components in corrosive environment. Typical problems in lap joint welding of thin foils include excessive distortion, absence of intimate contact between couples, melt drop-through and high levels of residual stress. The second aim of this study was to value the possibility to join thin foils and thick sheets lap joints. Typical problems in lap joint welding of dissimilar thickness include the presence of large void. Experimental investigations were carried out using a pulsed neodymium: yttrium aluminum garnet laser weld to examine the influence of the pulse energy in the characteristics of the weld fillet. The pulse energy was varied from 1.0 to 3.0 J at increments of 0.25 J with a 4 ms pulse duration. The base material used for this study was AISI 316L stainless steel foil with 100μm thickness. The welds were analyzed by optical microscopy, tensile shear tests and micro hardness. The results indicate that pulse energy control is of considerable importance to thin foil weld quality because it can generate good mechanical properties and reduce discontinuities in weld joints. The ultimate tensile strength of the welded joints increased at first and then decreased as the pulse energy increased. The process appeared to be very sensitive to the gap between couples.

The focused laser beam is one of the highest power density sources available to industry today. The welding of metals was one of the first industrial applications of lasers. Laser beam welding is used widely as an important manufacturing process. It can be performed using either pulsed or continuous lasers. A pulsed laser can be used to create weld seams in thin foils by means of overlapping pulses. Pulsed laser processing is expected to be the method of choice because it allows more precise heat control compared with continuous laser processing.

Commercial Nd:YAG lasers for welding applications are available from many supplies. They may be operated in three modes:

1. pulsed mode
2. continuous mode
3. Q-switched mode

Pulsed mode offers a range of pulse length from 0.1 ms to continuous wave (CW) operation. Typical pulse durations for welding applications are 1-20 ms. At the low end of this range, pulse repetition frequencies can approach 1 kHz. Q-switching of the laser output is less useful in welding applications because the pulse duration is much shorter (< 1μs), although pulse repetition frequencies can be high (up to 100 kHz). The higher peak power in these pulses facilitates plasma formation and gas breakdown.

One of the prime advantages of the Nd:YAG laser over the CO_2 laser is the ability to deliver laser radiation through optical fibers. Fortuitously, the 1.06 μm output wave length of the Nd:YAG laser falls within the wavelength range in which glass fibers have low attenuation, so propagation of Nd:YAG laser radiation over distances of as much as several hundred meters is possible with minimal loss.(Duley, 1999)

In general, one important problem in the experimental measurements performed at elevated temperatures or in a corrosive environment is selecting a material that is resistant to chemical attack and is easily formed into the desired shape. For this purpose, industrial product parts and components are covered with stainless steel thin foils or other corrosion-resistant material such as tantalum and Ni alloys. The significance of microtechnology has increased dramatically over the last years, and this has created a growing need for microwelding of thin foils. Furthermore, Nd:YAG pulsed laser welding is expected to be the method of choice because it allows more precise heat control compared with others processes and it reduces the heat affected zone (HAZ), residual stress and the presence of discontinuities.

Materials play an important role in manufactured goods. Materials must possess both acceptable properties for their intended applications and manufacturability. These criteria hold true for micromanufacturing, in which parts have overall dimensions of less than 1 mm. The wide range of materials that can be processed by lasers includes materials for micro-electronics, hard materials such as tungsten carbide for tool technology and very weak and soft materials, such as polymers for medical products. Even ceramics, glass and diamonds can be processed with laser technology to an accuracy better than 10 μm. In comparison with classical technologies, laser processes are generally used for small and medium lot sizes but with strongly increased material and geometric variability (Gillner et al., 2005).

Industrial product parts and components are being made smaller to reduce energy consumption and save space, which creates a growing need for microwelding of thin foil less than 100 μm thick. For this purpose, laser processing is expected to be the method of choice because it allows more precise heat control compared with arc and plasma processing (Abe et al., 2005).

There is a trend toward increased steel microwelding applications in the medical device manufacturing industry; these require spot sizes down to 25 μm and even smaller. Applications include sensors with very thin membranes (where no thermal deformation is allowed), microbonded wires with diameters of about 15 μm and welding of markers on to stents. As medical devices become smaller in size, new challenges will appear that laser welding will have to address (Tolinski, 2008).

Welding with a pulsed Nd:YAG laser system is characterized by periodic heating of the weld pool by an incident high peak power density pulsed laser beam incident that allow

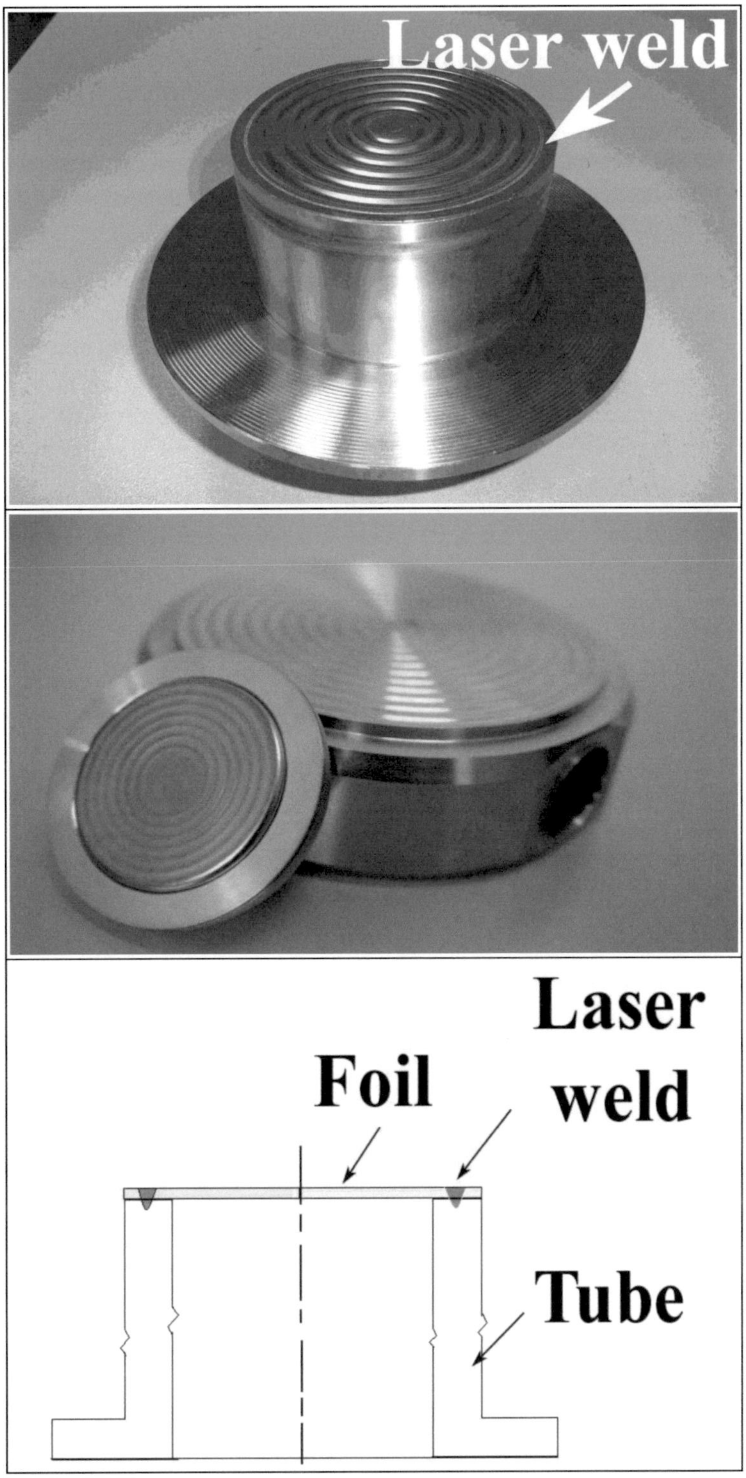

Fig. 1. Chemical seal welded by pulsed Nd:YAG laser.

melting and solidification to take place consecutively. The welding speed is defined by the overlap, the pulse repetition rate and the focus diameter. However, due to the very high peak power density involved in pulsed laser welding, the solidification time is shorter than that using a continuous laser or conventional welds. A combination of process parameters such as pulse energy [E_p], pulse duration [t_p], repetition rate [R_r], beam spot size [Φ_b] and welding speed [v] determines the welding mode, that is, conduction or keyhole (Ion, 2005; Duley, 1999 and Steen, 2005).

Research examining the Nd:YAG laser for continuous welding, pulsed welding, dissimilar sheet welding and coated sheet welding has been published. Kim et al. (2001) reported successful welding of Inconel 600 tubular components of nuclear power plant using a pulsed Nd:YAG laser. Berretta et al. (2007) using a homemade Nd:YAG Pulsed Laser System studied dissimilar welding of austenitic AISI 304 and martensitic AISI 420 stainless steel. Ping and Molian (2008) utilized a nanosecond pulsed Nd:YAG laser system to weld 60 μm of thin AISI 304 stainless steel foil.

This study investigates the use of an Nd:YAG laser operating in pulsed mode for welding a 100 μm thick AISI 316L stainless steel thin foil. The effect of pulse energy on weld joint characteristics is studied, and a discontinuity-free welding structure with good mechanical properties is proposed. Figure 1 shows a chemical seal where 316L thin foil is joined with a thickened body by pulsed Nd:YAG laser welding.

2. Experimental study

This study used a pulsed Nd:YAG Laser System. The experimental set up of the laser system is shown in Fig. 2 and Fig. 3. Figure 2 shows a thin foils lap joint and an example of penetration shape by pulsed Nd:YAG laser welding in AISI 316L. Figure 3 shows a thin foil / thick sheet lap joint and an example of penetration shape by pulsed Nd:YAG laser welding in AISI 316L.

AISI 316L was selected as the base metal for welding experiment with the following composition (wt.%): C-0.03, Cr-17.28, Ni-13.0, Mn-0.80, Si-0.75, P-0.045, S-0.003, Mo-2.3. The base material used for this study was thin foil with a thickness of 100 μm and thick sheet with 3.0mm. It was cut to a size of 20mm x 44.5mm. The experimental results were analyzed on the basis of the relationships between pulse energy and weld bead geometry, the presence of discontinuities and mechanical properties. The specimens were prepared and cleaned to ensure that all samples presented the same surface conditions with a homogeneous finish.

To evaluate the influence of the pulse energy, welding was performed using specimens positioned as lap joints. They were welded with a beam spot size (Φ_b) and beam angle (A_b) of 0.2 mm and 90 degrees, respectively. The pulse energy (E_p) varied from 1.0 to 2.25 J at increments of 0.25 J with a 4 ms pulse duration (t_p). Figure 4 shows a top view of AISI 316L Pulsed Nd:YAG laser welded joint. The specimens were held firmly using a jig, as shown in Fig. 5, to fixture and prevent absence of contact and excessive distortion. Fixturing is extremely important for thin-section laser welding. Tolerances were held closely to maintain joint fitups without allowing either mismatch or gaps.

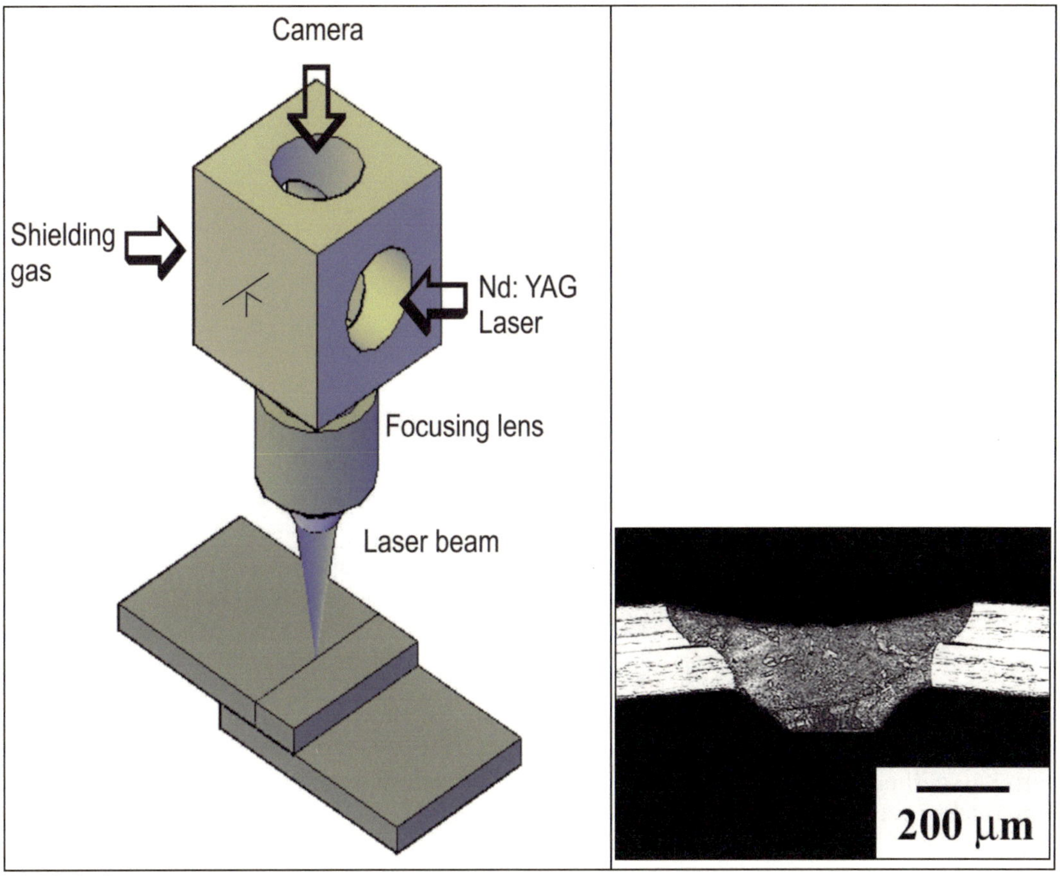

Fig. 2. Experimental set up of the pulsed Nd:YAG laser system and a macrograph of thin/thin foil welded joint.

The specimens were laser-welded in an argon atmosphere at a flow rate (F_r) of 10 l/min. Back shielding of the joint was not necessary because AISI 316L is not an oxidizable metal like Al and Ti. None of the specimens were subjected to any subsequent form of heat treatment or machining. After welding, the specimens were cut for the tensile-shear tests, as shown in Fig. 6. Finally, part of the cut surfaces was prepared for metallographic inspection by polishing and etching to display a bead shape and microstructure. Metallographic samples were prepared by electrolytic etching (2.2 V, 20 s) with a solution of 50% nitric acid. The bead shape measurements were made using an optical microscope with an image analysis system. Figure 7 shows a schematic illustration of the transverse joint section with the analyzed geometric parameters.

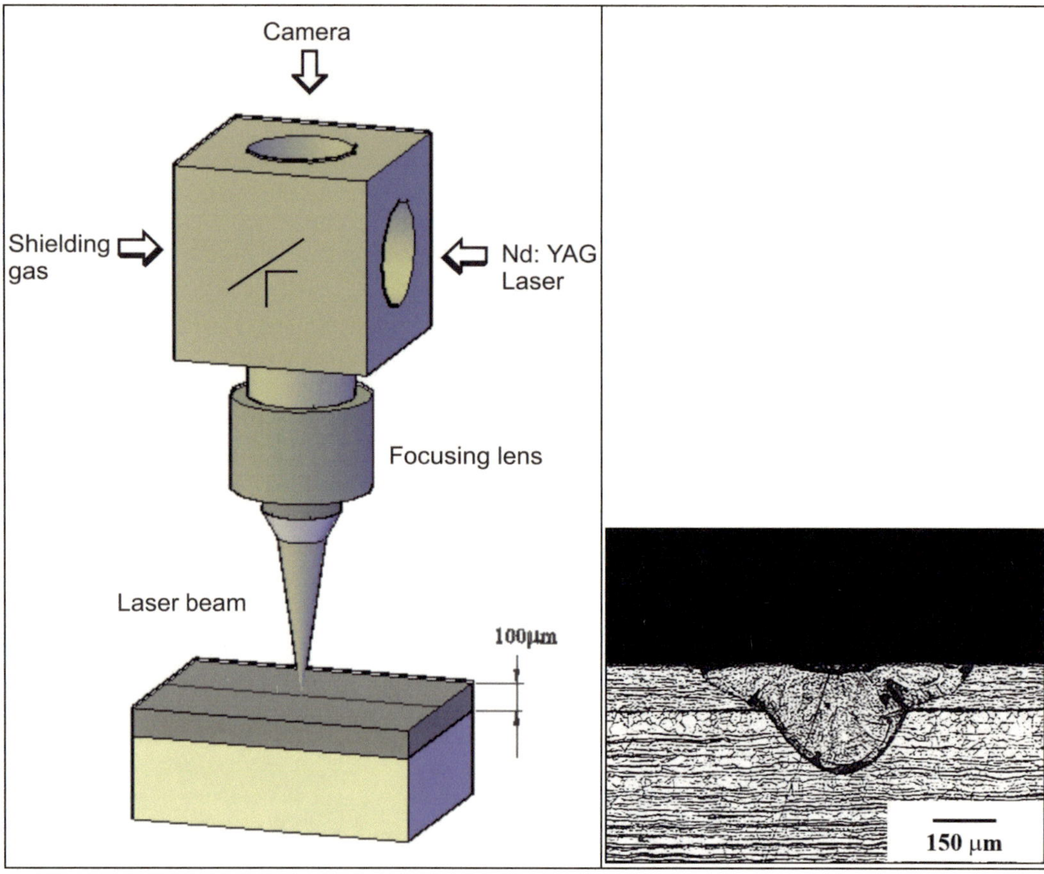

Fig. 3. Experimental set up of the pulsed Nd:YAG laser system and a macrograph of thin foil/thick sheet welded joint.

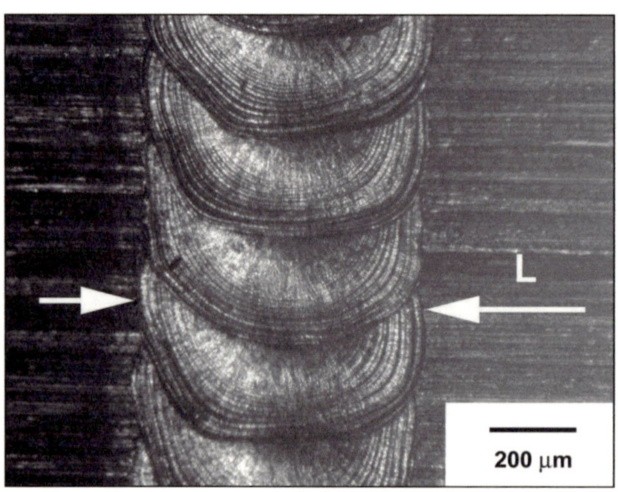

Fig. 4. Top view of Pulsed Nd:YAG laser welded joint. (L-bead width).

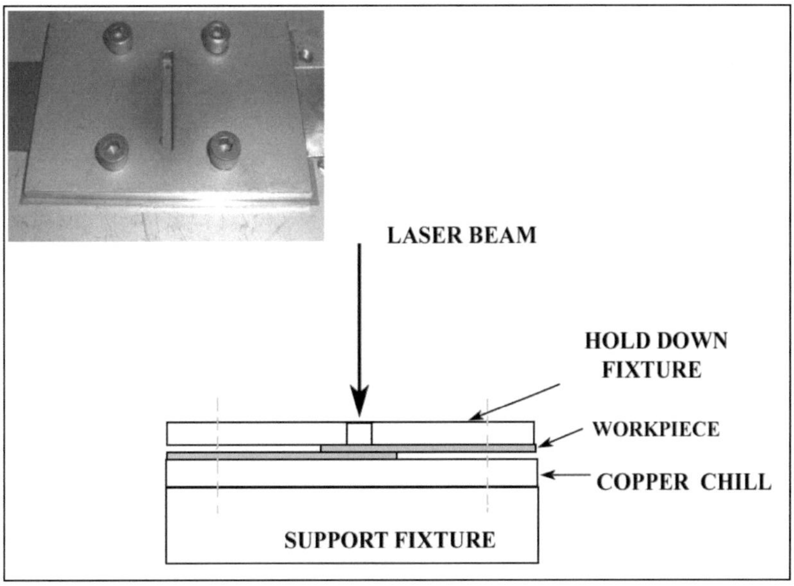

Fig. 5. Schematic of the hold-down fixture developed to hold the thin foils.

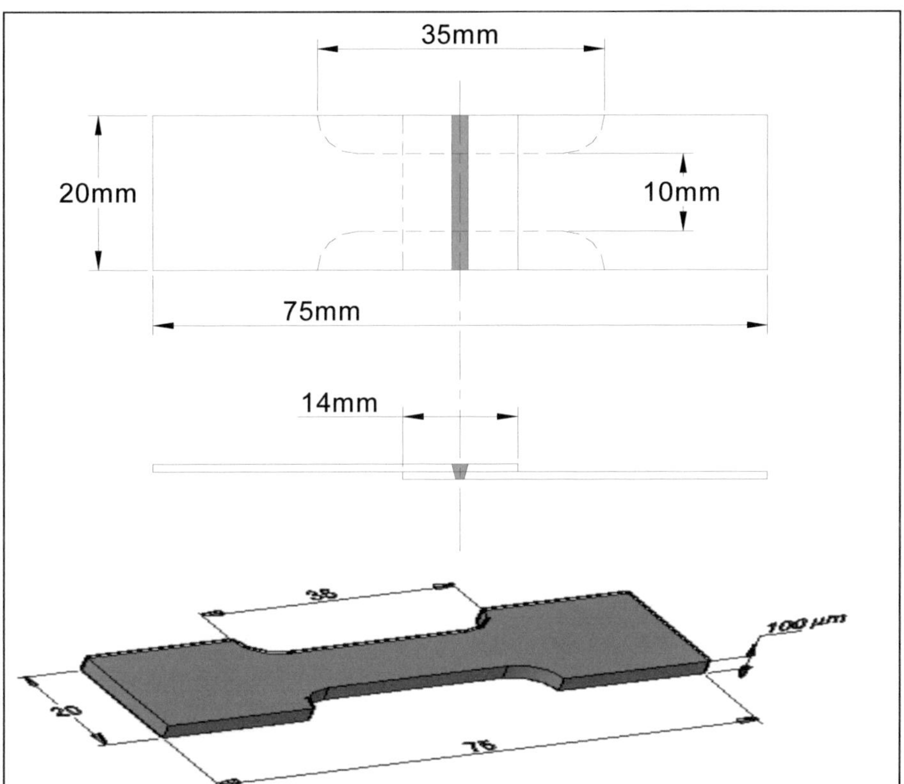

Fig. 6. Lap joint configuration to extract the tensile-shear test specimen.

The strength of the welds was evaluated using Vickers microhardness and tensile shear strength tests. Microhardness tests were performed on a transverse section of the weld bead, parallel to the surface of the thin foils, in the region next to the connection line of the top foil. Microhardness tests identify possible effects of microstructural heterogeneities in the fusion zone and in the base metal. The reported data were the average of five individual results. For the tensile shear test, specimens were extracted from welded samples, and the width of the samples was reduced to 10 mm to lower the load required to fracture them.

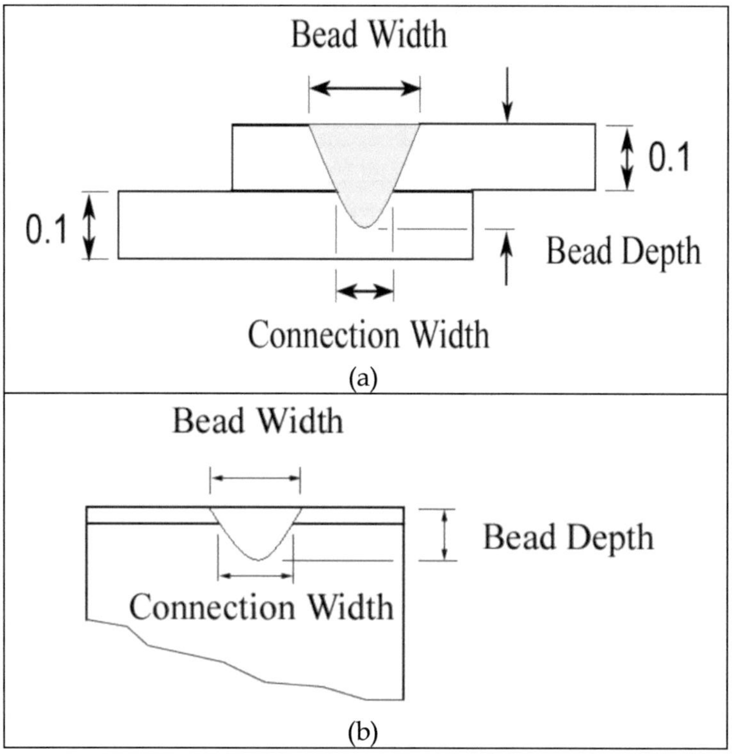

Fig. 7. Schematic of the joint transversal section. Thin/thin foils lap joint(a) and thin foil/thick sheet lap joint (b).

3. Results and discussion

The optimum performance of the seam welded joint for chemical seals employed in corrosive environment applications was determined using AISI 316L stainless steel thin foil (100 µm) and an Nd:YAG pulsed laser with different pulse energies (E_p). Figure 8 shows a top view of a welded specimen (E_p = 1.5 J) and it prepared for tensile shear test. It is clearly noticeable in the detail region (Fig.8) the no detectable defect existed on the surface of the weld bead or adjacencies. This good surface appearance was observed in specimens welded with pulse energy from 1.0 to 1.75J. The weld beads showed characteristic of pulsed laser welding. No welding cracks were found in any of the welds; this may be partly due to the good crack resistance of base the metal and the correct welding parameters. No discontinuities were observed in the fusion metal of the beads, which demonstrates the efficiency of the shielding gas in preventing oxidation, large porosities and gas inclusions, which cause poor weld quality. All specimens were welded in the conduction mode.

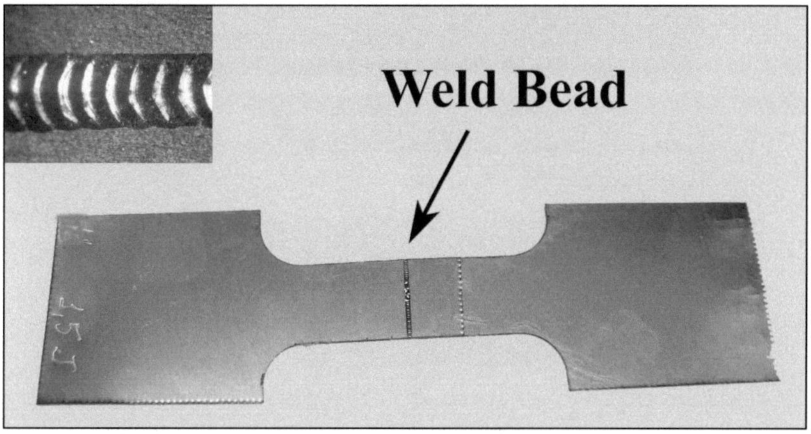

Fig. 8. Welded specimen prepared for the tensile shear test.

The cross section macrostructures of lap laser welds as a function of pulse energy (E_p) are summarized in Fig. 9. In Figs. 9a and 9b (specimens with 1.0 and 1.25J pulse energy, respectively) no penetration at the bottom sheet and no depression at the top of the bead was observed, probably due to insufficient energy to bridge the couple. It is clearly noticeable in Figs. 9a and 9b the presence of a small gap between couple. The macrostructures showed that melting started at the surface irradiate by the laser beam and the molten pool grew continuously to axial and radial axis. Due to the thin thickness, low pulse energy of the laser beam and the presence of a small gap, the molten pool just grew in the radial direction of the top foil resulting in a no bonded joint with the weld morphology observed in Figs. 9a and 9b. Depth-to-width ratio of the fusion zone of these specimens was about 0.2. Gaps between foils and gaps in the connection line increase stress and thus are detrimental to weld quality in terms of mechanical properties. As reported in the literature (Kawarito et al., 2007) the presence of wider gaps in welded joints result in more deeply concave underfills. Specimens welded in the present work with low pulse energy, 1.0 and 1.25 J, present no underfill because the molten material did not have enough time to fill the gap. When the pulse energy was increased on the other specimens, a connection region between the foils was observed, as shown in Figs. 9c and 9d (welded with 1.5 and 1.75 J, respectively). Both specimens present no underfill but an excess of molten material at the root. The penetration depth increased from 133 to 200 μm (full penetration) as the pulse energy increased from 1.5 to 1.75 J. Both joints present an intimate contact between the couples (absence of gap). These specimens present excellent conditions for laser seam welding. In Figs. 9e and 9f (specimens with 2.0 and 2.25 J pulse energy, respectively), an increase occurred with a depression at the top and a penetration bead. The concavity increased proportionally to the pulse energy (E_p). Moreover, it was evident that specimens welded with 2.0 and 2.25 J pulse energy undergo deformation during joint welding, which causes a large bending moment. Areas near the heat source of the upper foil are heated to higher temperatures and thus expand more than areas away from the heat source or regions of the lower foil. After the foil cools to the initial temperature, the final deformation remains. Like the material heated by the laser beam, the irradiance did not cause the material reach its boiling point; no significant amount of surface material was removed.

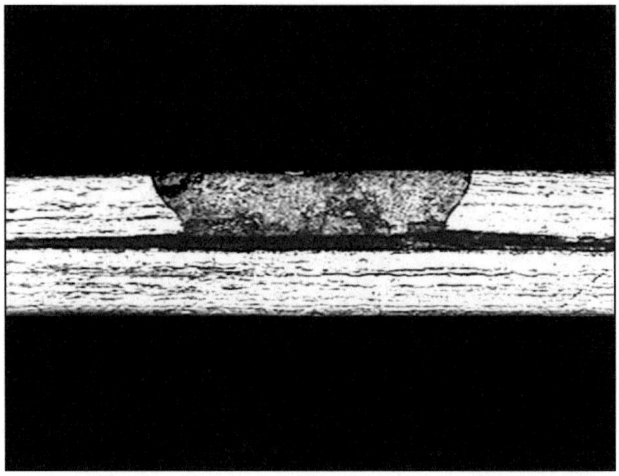

(a)

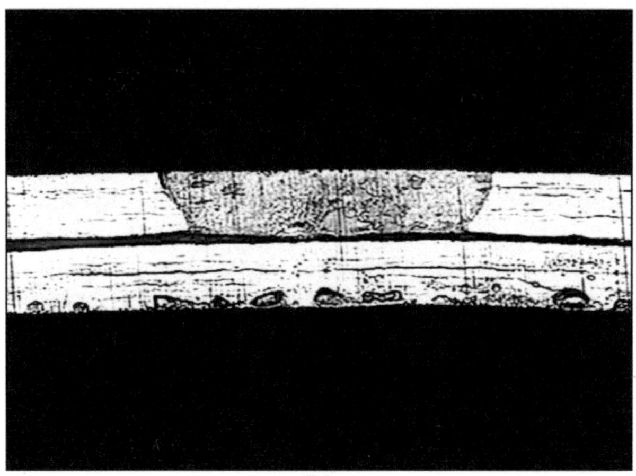

(b)

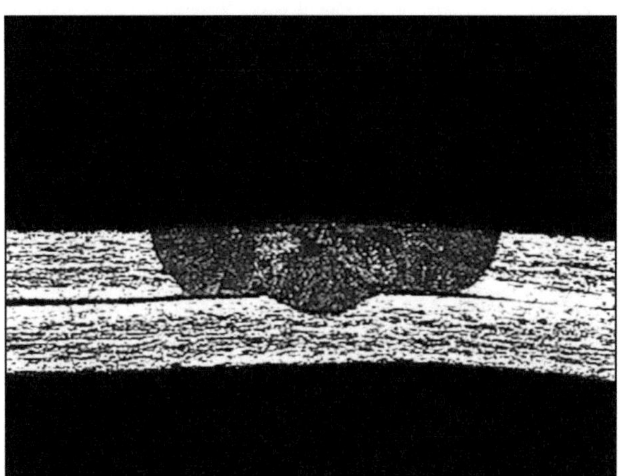

(c)

(d)

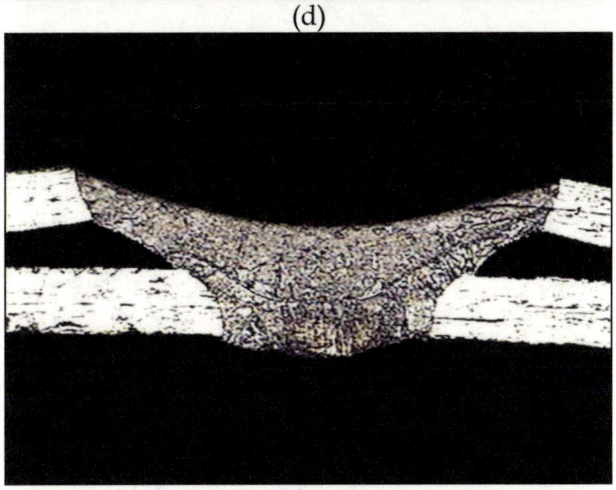

(e)

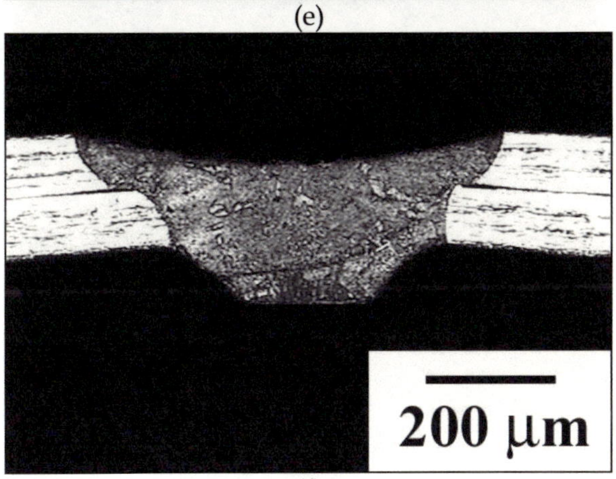

(f)

Fig. 9. Cross sections of thin/thin foil lap joints made with pulsed Nd:YAG laser welding with different pulse energies (E_p): a) 1.0 J, b) 1.25 J, c) 1.50 J, d) 1.75 J, e) 2.0 J and f) 2.25 J. All figures have the same magnification as shown in f. (Ventrella, 2010)

The cross section macrostructures of AISI 316L thin foil/thick sheet lap laser welds as a function of pulse energy (E_p) are summarized in Fig. 10.

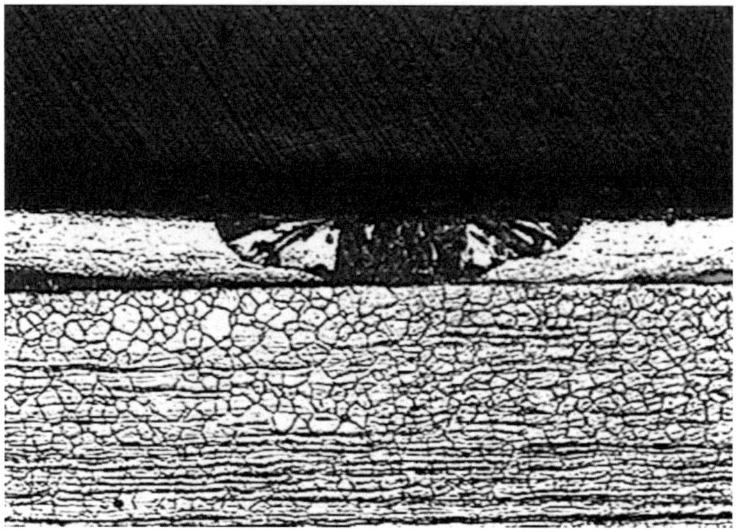

(a)

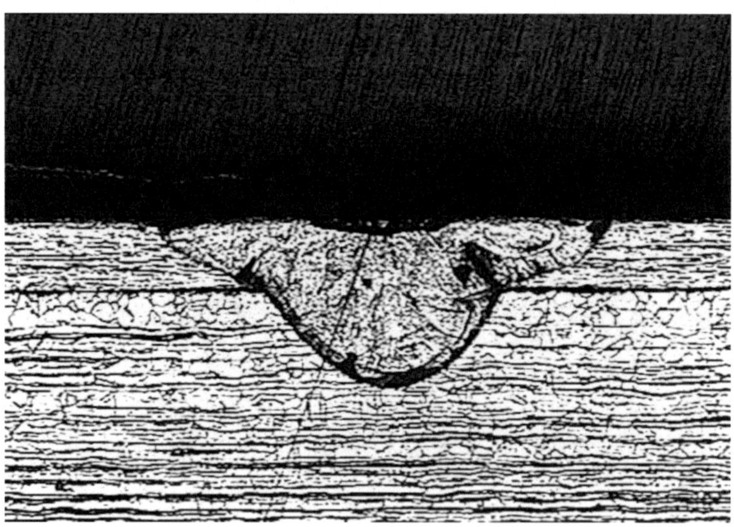

(b)

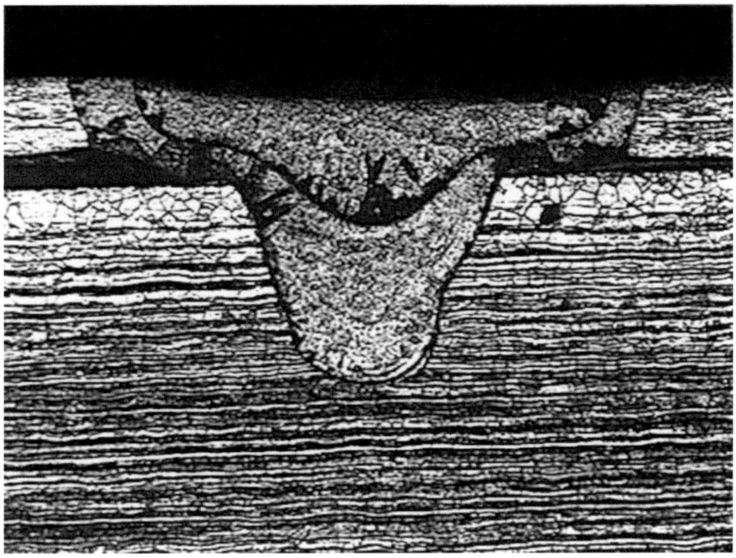

(c)

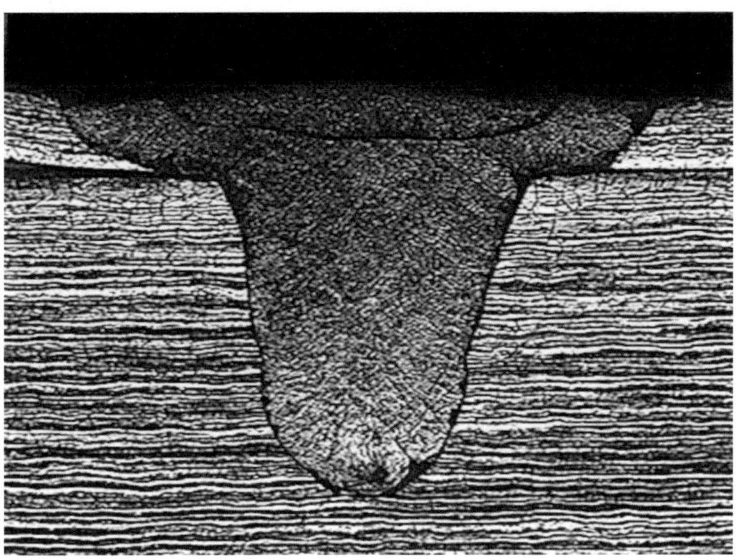

(d)

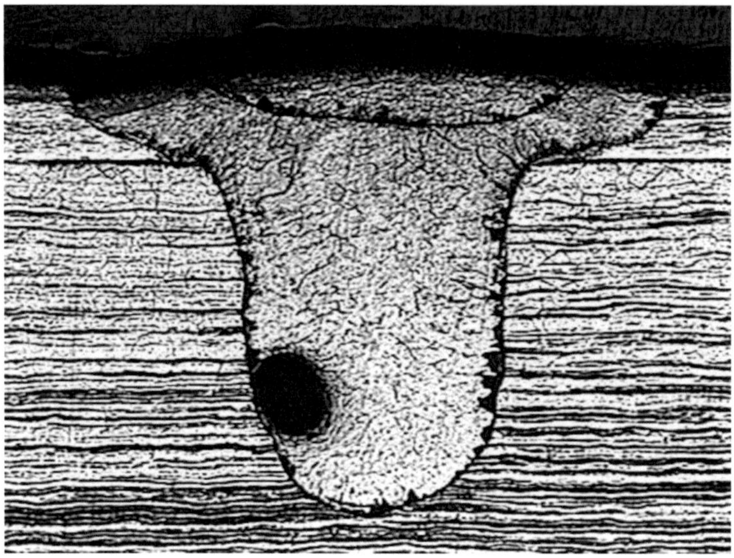

(e)

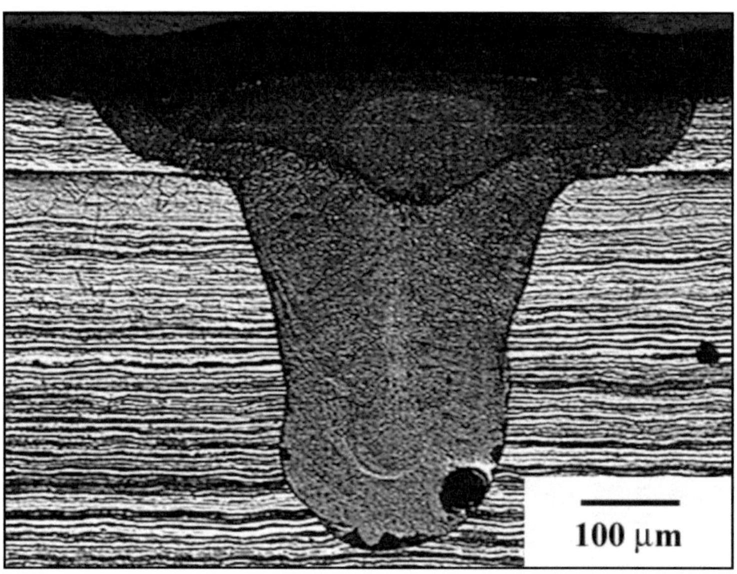

(f)

Fig. 10. Cross sections of thin foil/thick sheet lap joints made with pulsed Nd:YAG laser welding with different pulse energies (E_p): a) 1.50 J, b) 2.0 J, c) 2.25 J, d) 2.5 J, e) 2.75 J and f) 3.0 J. All figures have the same magnification as shown in f.

The relationship between pulse energy and weld metal geometry of AISI 316L thin foil/thick sheet is summarized in Fig. 11.

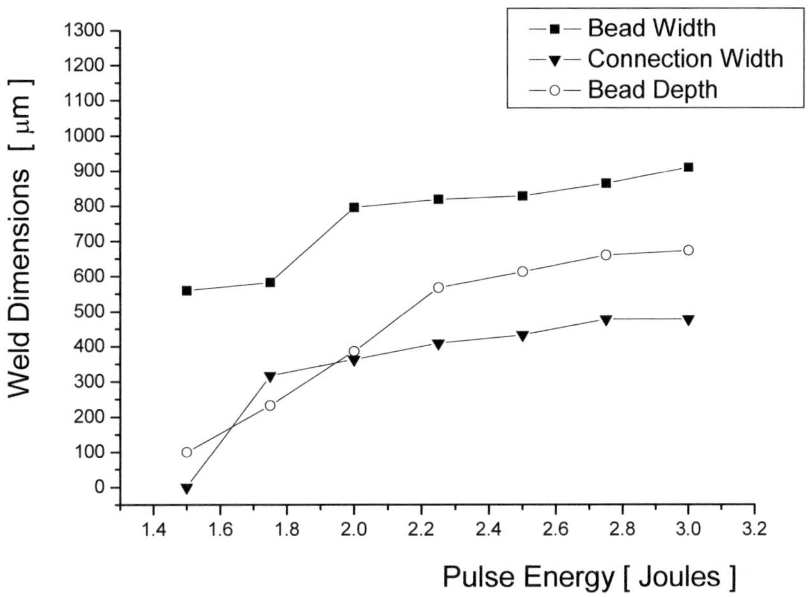

Fig. 11. Weld dimension versus pulse energy [E_p] of thin foil/thick sheet lap joints made with pulsed Nd:YAG laser welding with different pulse energies [E_p].

Figures 10e and 10f show large voids. They consist in empty zones delimited by the molten zone boundary as if the molten metal had solidified before collapsing into the keyhole entirely. These voids seem to be the result of a lack of metal filling up the keyhole at the end of the laser pulse, associated with very fast solidification.

The macroscopic examination of the cross sections of all specimens also indicated that the weld pool morphology is essentially symmetrical about the axis of the laser beam. This symmetry at the top and bottom was observed in all joints independent of the pulse energy, which suggests a steady fluid flow in the weld pool; however, as the pulse energy increased, a high depression formed at the top.

Liao and Yu (2007) and Manonmani et al. (2007) reported that, in pulsed laser welding with the same incident angle of AISI 304 stainless steel the characteristic lengths of the weld increased as the laser energy increased. The same tendency was observed in the present work as the pulse energy increased to 2.0 J. With pulse energy greater than 2.0 J, excessive burnthrough occurred and an excess of melt material at the root region was observed. This reduced the radial conductive heat transfer at the top foil, and the bead width decreased slightly, as shown in Fig. 9f.

The relationship between pulse energy and weld metal geometry of AISI 316L thin/thin foil is summarized in Fig. 12. The bead width increased from 480 to 750 μm as pulse energy varied from 1.0 to 2.0 J. This indicated that when the laser beam interacts with the specimen, it creates a liquid melt pool by absorbing the incident radiation. This bead width variation is a result of the higher pulse energy; a high amount of material is molten and then propagates through the base material. In the presence of high pulse energy, part of the melt material passed through the joint, which increased the concavity at the top of the weld, the excess of weld metal at the root and the heat-affected zone extension. On the other hand, at a pulse energy of 2.25 J, the molten metal volume decreased, and deep concave underfills occurred. Figure 9f shows a macrostructure between a fusion zone and base metal in the case of 2.25 J of pulse energy.

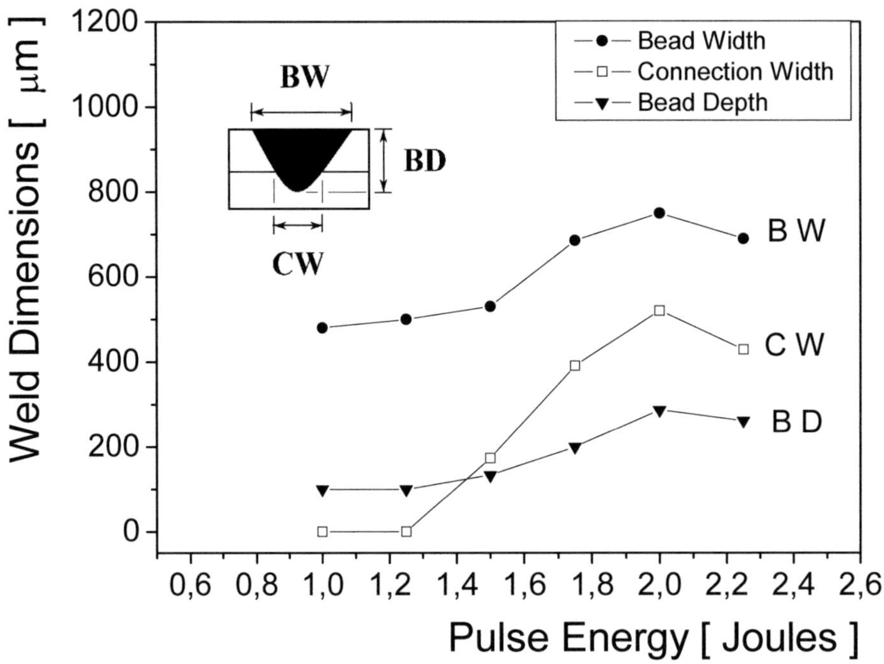

Fig. 12. Weld dimension versus pulse energy [E_p] of thin/thin foil lap joints made with pulsed Nd:YAG laser welding with different pulse energies [E_p].

These macrostructural results indicate that weld metal characteristics are sensitive to pulse energy variation. To obtain an acceptable weld profile, intimate contact between couples is necessary. The presence of an air gap in the weld joint restringes heat transfer between the workpieces. This results in a lack of fusion of the bottom element or the formation of a hole on the superior element of the joint. As the pulse energy increases, the concavity at the top of the weld and the excess of material at the weld root increase; consequently, the weld joint is weakened. Moreover, higher pulse energy extended the heat-affect zone.

Lack of fusion between couples was observed at pulse energies bellow 1.5 J. Moreover, lack of fusion between couples was expected to be suppressed by using high pulse energy until the molten pool bridged the gap. The presence of holes was observed at a pulse energy above 2.25 J. No welding cracks were observed in any of the welds, which suggests that welding cracks are not sensitive to variations in the pulse energy. Instead, welding cracks are a function of the Cr_{eq}/Ni_{eq} ratio. The presence of spatters, porosities and undercut was not found in the welded joints investigated with optical microscopy.

The HAZ displays the effect of temperature cycling to a peak temperature, T_{max}, which is less than the melting point but may be sufficient to initiate other transformations. The cooling rate of steel, in the HAZ, may approach 1000°C/s but varies with location, as does T_{max}. The grains in the heat-affected zone were coarsened as the pulse energy [E_p] increased. This phenomenon can be explained by the cooling rate change. An increase in the pulse energy decreased the cooling rate. A slower cooling rate during solidification allowed more time for grain coarsening. In laser welding the heat-affected zones are much narrower than those generated by conventional welding process and are not likely to be sensitized to corrosion.

Analyses of the Heat-Affect Zone (HAZ) showed that it increased as the pulse energy increased. For all tested specimens the top foil HAZ is larger than the bottom foil HAZ, as shown in Figs. 13a and 13b. This is strongly dependent with the beam energy, conductive heat transfer and with a perfect contact between couple (presence or not of air gap). It is different processing material with a determined thickness than two foils of the same material with half of the thickness in a lap joint configuration. Air gap acts as a barrier to heat transfer. In this situation most of the beam energy can be consumed on the side exposed to the laser beam, which results in a large heat-affected zone or bead perforation.

Figure 14 shows a microscopic examination of the heat-affect zone extension of the top and bottom foils of the specimen welded with 2.25 J. The grains in HAZ are coarsened, and the extension of the heat-affected zone is approximately 50 and 25 μm at the top and bottom foils, respectively. This observation in thin foils seems to contradict the expected narrow heat-affected zone for the laser welding process. The top and bottom HAZ width of the weld becomes obviously larger when the pulse energy (E_p) increases. The HAZ width difference between the top and the bottom becomes smaller as the pulse energy increases, so the cooling rate decreases with increasing pulse energy. Therefore, when the pulse energy is higher, a higher volume of metal is melted and the welding heat has more time to be conducted into the bottom from the top.

Light micrographs of the specimens showed a fine-grained microstructure that is essentially cellular-dendritic in the fusion zone. This type of microstructure is a result of high cooling rates, which are typical of the laser welding process. The formation of a given solidification structure morphology is determined by the G/R ratio (G=temperature gradient, R=growth rate) during solidification. Cellular growth structures form rather than dendritic structure if G/R ratio is high (Molian, 2007).

Figure 14 illustrates typical microstructures of AISI 316L austenitic stainless steel weld joint. Figure 14a shows the fusion line solidification structure at the top of the weld where the un-melted base metal grains act as substrates for nucleation of the fusion zone columnar grains (epitaxial growth), which are perpendicular to the fusion boundary. Figure 14b shows the heat affected zone at the bottom of the joint where the effects of the large thermal gradient in this region are evident. Comparing thin and thick foil welding, it can be concluded that the grains in the solid state coarsen with decreasing parent metal thickness. This shows that the volume of the parent metal plays an important role during the welding thermal cycle. As the material volume decreases, the time to cooling increases and the heat-affected zone appearance coarsens. This indicates that in thin foil welding, heat-affected zone control is of considerable importance for welded joint quality.

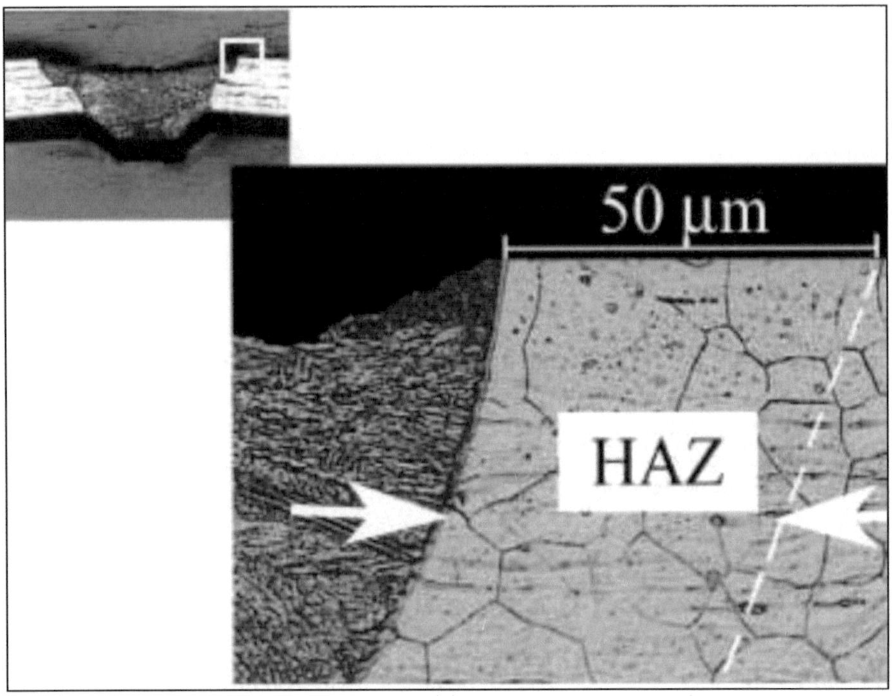

(a)

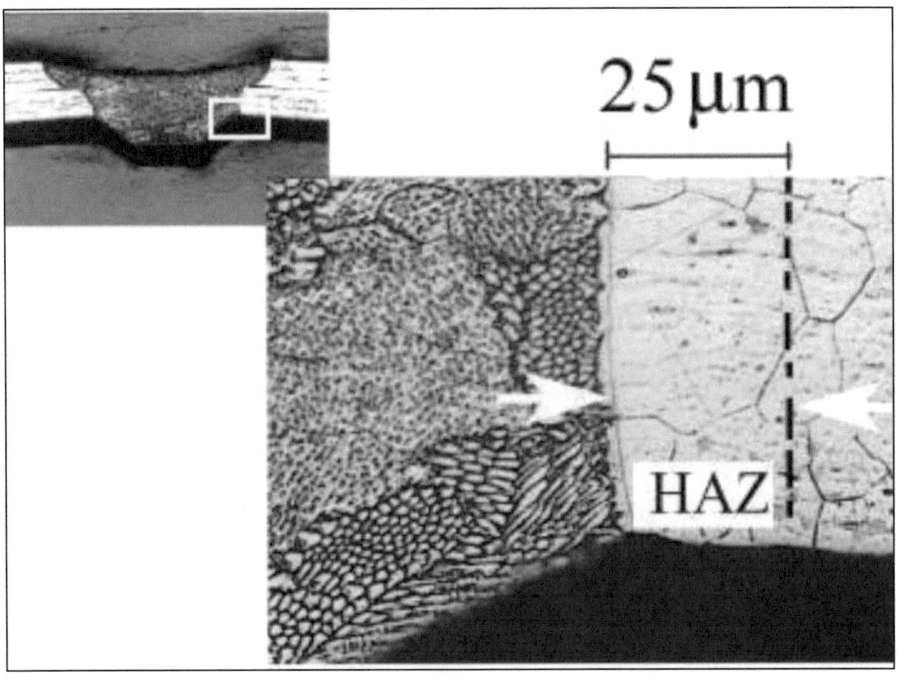

(b)

Fig. 13. Heat-Affect Zone of the top foil (a) and bottom foil (b) of the specimen welded with 2.25 J. (Ventrella, 2010)

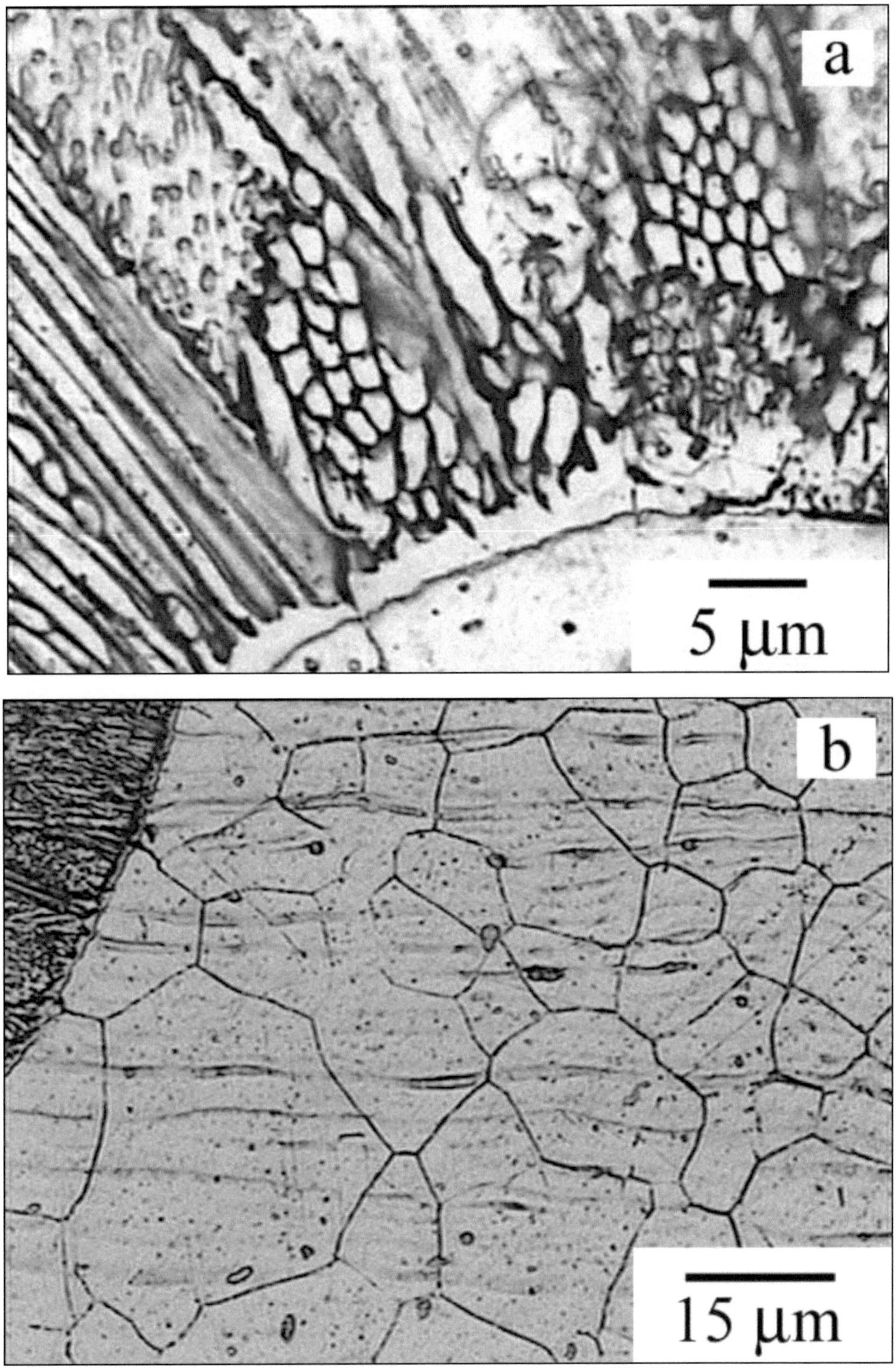

Fig. 14. Typical optical microstructures of the fusion zone (a) and HAZ (b) of an AISI 316L welded joint (Ventrella, 2010).

First, the failure of all specimens occurs in the region of the parent metal, next to the fusion line of the top foil. This is expected because hardness and tensile strength values are known to be related. The ultimate tensile strength (UTS) tends to increase at first and then decrease as the pulse energy (E_p) increases. The relationship between pulse energy and tensile shear strength of welded joints is summarized in Fig. 15. Specimens welded with a pulse energy lower than 1.5 J were not bonded because the pulse energy was too low, and the molten pool did not have enough time to propagate to the bottom foil; incomplete penetration occurred. Otherwise, when the specimens were welded with pulse energy greater than 1.75 J, excessive underfilling and burnthrough was observed. Perforations in the weld bead were observed with pulse energy higher than 2.25 J. The maximum value of UTS, obtained with a pulse energy of 1.75 J, was up to 95% of the parent metal.

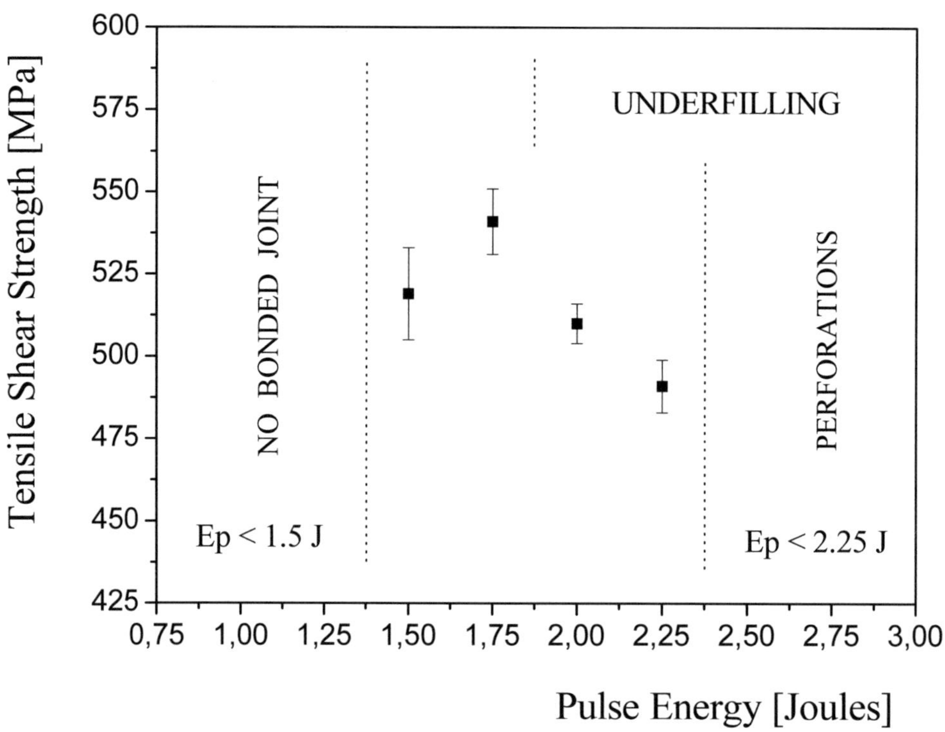

Fig. 15. Relationship between pulse energy and tensile shear strength.

Scanning electron microscopy observations of the welded joint showed that the weld surface region next to the rupture line displayed a sequence of sags due to differential deformations that occur in the weld bead as a result of different microconstituent orientations.

The tensile properties of the welded joint affected by pulse energy (E_p) can be explained by macro and microstructural analyses. As the pulse energy increases, the grains in the weld metal and in the HAZ become coarser. The heat-affected zone extension increases too. Discontinuities become more severe. Some precipitates can be present intergranularly and even continuously along the grain boundary. These microstructural changes contribute to a weakness of the weld joint, which reduces the tensile properties as reported in the literature (Quan et al. 2008).

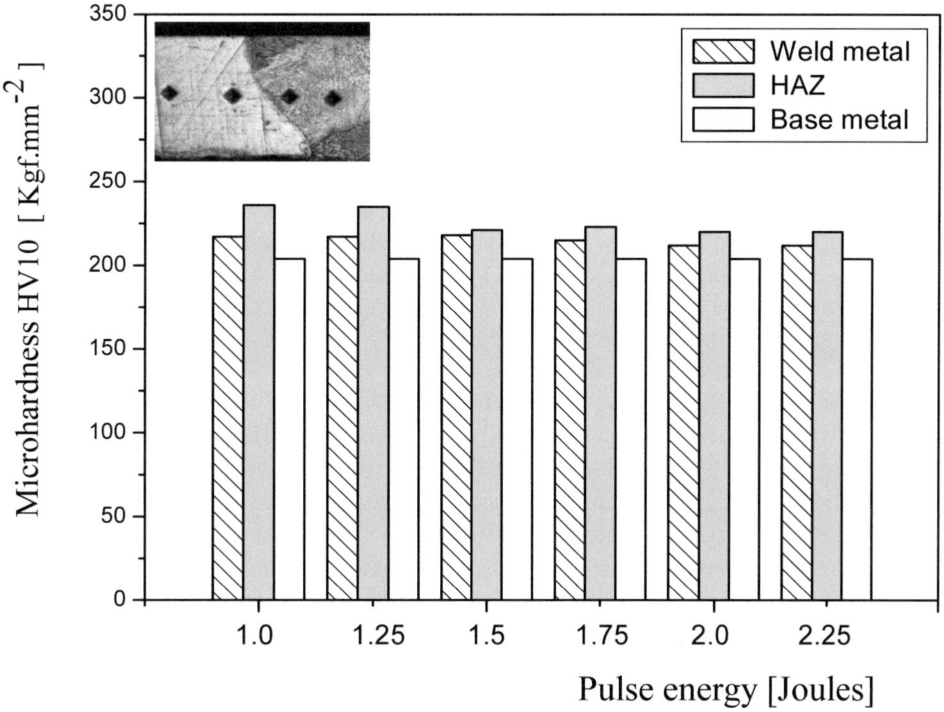

Fig. 16. Hardness profile of base metal, HAZ and weld metal of AISI 316L stainless steel thin foils as a function of pulse energy.

Hardness profiles of the base metal, heat-affected zone and weld metal of AISI 316L stainless steel thin foils as a function of pulse energy are shown in Fig. 16. No significant difference between hardness of weld metal and the heat-affected zone was obtained; the hardness of the heat-affected zone was slightly higher than that of the weld metal regardless of the pulse energy. Base metal hardness was always lower than that of HAZ and weld metal. These results are valid for all joints. This is expected because the mechanical properties of steel, in general, are based on its microstructures (Abdel, 1997).

As can be seen in Fig. 16, the microhardness values decreased as the joint energy increased (1.0 J to 2.25J) to maximum energy (2.25 Joules). During the solidification of the fusion zone, the material generally loses original strength that is induced by strain hardening. Microhardness profiles in welded joints obtained with lower energy show increasing of the hardness in the fusion zone and a finer microstructure that is induced by rapid cooling.

In summary, the most acceptable weld bead was obtained at a pulse energy of 1.75 J(thin/thin joint) and 2.0 J (thin/thick joint), where the molten pool bridged the couple and the weld bead profile showed minimum underfill and maximum depth (full penetration). The tensile shear test exhibited 541 Mpa. No undercut and minimum porosity were observed. No evidence of hot cracking was observed in the weld metal and this is attributed to the rapid solidification conditions typical of the pulsed Nd:YAG laser welding process. Large voids delimited by the molten zone boundary were observed during thin foil/thick sheet welding with high pulse energy. This is related to the molten metal that had solidified before collapsing into the keyhole entirely. The metal surface locally melts and vaporizes. Then, a deep and narrow keyhole appears and traps the laser beam inside the material itself. The keyhole grows vertically and it is surrounded by the molten metal that is ejected from the bottom to the top of the keyhole, where it gathers in a crown shape. Finally, the surrounding molten metal collapses into the keyhole and solidifies.

4. Conclusions

The pulsed Nd:YAG laser welding process has been employed to join AISI 316L thin/thin foil and thin foil/thick sheet. In general, the results obtained from this study demonstrate that is possible to weld 100 µm thickness AISI 316L thin foils, in terms of microstructural and mechanical reliability, by precisely controlling the laser pulse energy. The better performance was due to the high quality joint; a joint marked by full penetration, no underfill and free from microcracks and porosity. This was obtained at an energy pulse of 1.75 J (thin/thin joint) and 2.0 J (thin/thick joint), a repetition rate [R_r] of 39 Hz and a 4 ms pulse duration. This reflects one of the most notable features of pulsed laser welding compared with other processes; welding with low heat input. The work also shows that the process is very sensitive to the gap between couples which prevents good heat transfer between the foils. The shape and dimensions of the thin foil weld bead observed in the present work depended not only on the pulse energy, but also on the presence of gaps between foils. Bead width, connection width and bead depth increased as the pulse energy increased, and then decreased at the end because of burntrougth. The ultimate tensile strength (UTS) of the welded joints initially increased and then decreased as the pulse energy increased. The specimen welded with 1.75 J attained the maximum tensile shear strength. In all the specimens, fracture occurred in the top foil heat-affected zone next to the fusion line. The microhardness was almost uniform across the parent metal, HAZ and weld metal. A slight increase in the fusion zone and heat-affected zone compared to those measured in the base metal was observed. This is related to the microstructural refinement in the fusion zone, induced by rapid cooling. Large voids delimited by the molten zone boundary were observed in thin foil/thick sheet joints welded with high pulse energy.

5. Acknowledgments

The authors gratefully acknowledge the financial support of CNPq and Mechanical Engineering Department of Sao Paulo State University - UNESP.

6. References

Abdel, M.B., 1997. Effect of laser parameters on fusion zone shape and solidification structure of austenitic stainless steels. Materials Letters. 32, 155-163.

Abe, N., Funada, Y., Imanada, T., Tsukamoto, M., 2005. Microwelding of thin stainless steel foil with a direct diode laser. Transaction of JWRI. 34, 19-23,

Berretta, J.R.; Rossi, W.; Neves, M.D.M.; Almeida, I.A. and Junior, N.D.V., 2007. Pulsed Nd:YAG laser welding of AISI 304 to AISI 420 stainless steels. Optics and Lasers in Engineering. 45, 960-966.

Duley, W.W., 1999. Laser Welding, John Wiley & Sons, New York, pp. 67-94.

Gillner, A., Holtkamp,J., Hartmann,C., Olowinsky,A., Gedicke,J., Klages,K., Bosse,L., Bayer,A., 2005. Laser applications in microtechnology. Journal of Materials Processing Technology. 167, 494-498.

Girard, K.;Jouvard, J.M.; Naudy, Ph., 2000. Study of voluminal defects observed in laser spot welding of tantalum. J. Phys. D: Appl. Phys. 33 , 2815-2824.

Ion, J.C., 2005. Laser Processing of Engineering Materials, Elsevier, Oxford, pp. 327-329, 395-399.

Kawarito, Y., Kito, M. and Katayama, S., 2007. In-process monitoring and adaptive control for gap in micro butt welding with pulsed YAG laser. Journal of Physics D: Applied Physics. 40, 183-190.

Kim, D.J.; Kim, C.J. and Chung, C.M., 2001. Repair welding of etched tubular components of nuclear power plant by Nd:YAG laser. Journal of Materials Processing Technology. 14, 51-56.

Liao, Y. and Yu, M., 2007. Effects of laser beam energy and incident angle on the pulse laser welding of stainless steel thin sheet. Journal of Materials Processing Technology. 190, 102-108.

Manonmani, K., Murugan,N. and Buvanasekaran,G., 2007. Effects of process parameters on the bead geometry of laser beam butt welded stainless steel sheets. International Journal of Advanced Manufacturing Technology. 32, 1125-1133.

Molian, P.A., 1985. Solidification behaviour of laser welded stainless steel. Journal of Materials Science Letters. 4, 281-283.

Ping, D. and Molian, P., 2008. Q-switch Nd:YAG laser welding of AISI stainless steel foils. Materials Science & Engineering A. 486, 680-685.

Quan,Y.J., Chen,Z.H., Gong,X.S. and Yu,Z.H., 2008. Effects of heat input on microstructure and tensile properties of laser welded magnesium alloy AZ31. Materials Characterization. 59, 1491-1497.

Steen, W.M., 2005. Laser Material Processing, third ed. Springer, London, pp. 160-165.

Tolinski, M., 2008. Lasers seal the deal in medical. Manufacturing Engineering. 140, 14-20.

Ventrella,V.A.; Rossi,W.; Berretta,J.R., 2010. Pulsed Nd:YAG laser welding of AISI stainless steel thin foils. Journal of Materials Processing Technology, 210, 1838-1843.

14

Single and Double Laser Pulse Interaction with Solid State – Application to Plasma Spectroscopy

Richard Viskup
Johannes Kepler University Linz, Institute of Applied Physics, Linz Austria

1. Introduction

When a nanosecond laser pulse with sufficiently high laser energy is focused into a solid-state surface, pulsed laser ablation of the material and subsequently laser induced plasma formation occurs. An emitting light from a plasma is a rich source of information and its optical emission is something like "fingerprints" of emitting atomic, ionic and molecular species, composed in the irradiated sample. This emission can be used for the basic spectrochemical analysis of the studied material's composition. Laser induced plasma spectroscopy is an emerging technique for fast and accurate compositional analysis of many different materials (Viskup, 2010), and is nowadays, very often used in diverse applications. However, the issue of sensitivity and detection limits are still open question. A main goal of this chapter is to shown the possibility to enhance an optical emission from plasma and improve a limit of detection of trace elements detected by optical emission from a measured sample. For this purpose, the advantage of double laser pulse irradiation is very convenient for plasma re-heating or re-excitation (Viskup et al., 2010).

The double-pulse irradiation was proposed for more efficient production of analyte species in excited states. It uses two laser pulses for excitation with a defined delay time between pulses typically in the order of ns or µs. One of the first articles about the double-pulse configuration was published by (Cremers et al., 1984). Collinear double pulse studies of solid state in a gas atmosphere were first reported in the mid nineties, in (Sattmann et al., 1995). The technique has been re-investigated especially in the last decade for a broad range of materials and for many different measurement parameters, including the use of ultrashort femtosecond or picosecond laser pulses in (Noll et al., 2004; Semerok et al., 2004; Babushok et al., 2006). The technique splits into two main streams: First the collinear double-pulse geometry (Gautier et al., 2005: Peter et al., 2007), where two laser beams are delivered on the same target surface along the same optical path, and the second the orthogonal geometry (Cristoforetti et al., 2006; Choi et al., 2009), where only one pulse ablates the sample surface. The double-pulse irradiation approach has received some attention in the scientific community due to the considerable signal enhancement observed for some materials.

This chapter is dedicated to single and double laser pulse interaction with main emphases into application to spectroscopy. Research has been dedicated to systematic studies of collinear nanosecond double-pulse measurements, and to double pulse plasma plume characterization. A comparison between single pulse and a double pulse laser irradiation by means of time resolved plasma plume photography and plasma spectroscopy are presented. The optimal parameters for double pulse exposure, especially for pulse to pulse separation time is considered. Results of this research should contribute to a better understanding of single and double nanosecond laser pulse interaction with solid-state material. By setting up of optimal parameters for double pulse laser interaction, it is possible to improve plasma emission from studied sample for industrial or scientific applications.

2. Characterization of nanosecond laser induced plasma

2.1 Plasma formation

When a nanosecond laser pulse is focused into solid-state surface with sufficient energy of the radiation to induce material vaporization, a dense vapour plume can be formed. This vapour consists of clusters, molecules, atoms, ions, electrons and scatter radiation. Species, which are ejecting from the surface, take along some kinetic and internal energy. Clusters are formed due to collective effects and droplets formation are related to hydrodynamic perturbations, big material fragments are ejected due to stress relaxation. This phenomenon can affect also the average energy per atom required for material removal, which can be then significantly smaller. Thermalization of species which are leaving from the surface is performed via collisions within few mean free paths from the sample surface, which is known as Knudsen layer. The plume generated at the surface has a strongly forwarded direction due to high temperature and pressure gradients in the axial direction, compared to gradients in the lateral directions. The simplest model (Anisimov et al., 1993) considers the expansion of the vapour beyond the Knudsen layer, and is described by an adiabatically expanding gas. Model predicts the temperature gradients within the plume with increasing distance from the source. In general the species leaving from the substrate surface generate recoil pressure on the substrate. In the presence of a molten surface layer, generated by the focused laser beam irradiation the recoil pressure expels the liquid from the surface. This is schematically shown in Figure 1. The ablated material may also generate a shock wave and the vapour plume absorbs the scatter incident laser radiation.

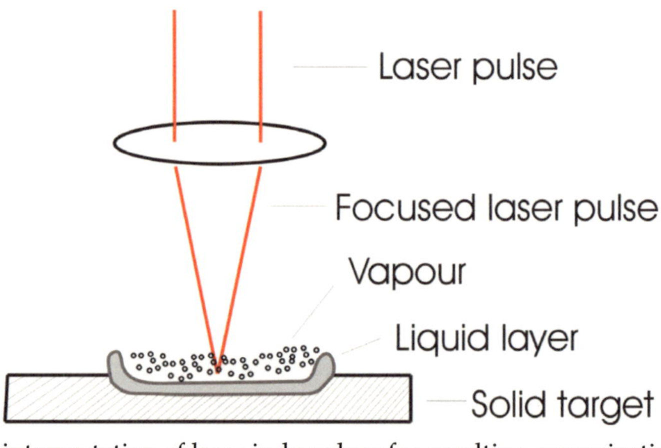

Fig. 1. Schematic interpretation of laser induced surface melting, vaporization and liquid phase expulsion.

With increasing laser light intensity an increasing fraction of atoms and molecules become ionized. After the substantial ionization, it can be described as plasma. With formation of the plasma, it strongly absorbs the incoming laser radiation and hence, shields the substrate. With increasing laser pulse intensity the plasma plume expands in volume and its forward direction becomes more dominant. With further increase, the plasma decouples from the substrate and propagates in direction towards the incident laser beam. This is often termed as laser supported absorption wave (Bäuerle, 2011). If the propagation velocity is subsonic we denoted it as laser supported combustion wave. In case that the velocity is supersonic, we talk about the laser supported detonation waves.

We can distinguish these main mechanisms of laser induced ns plasma. First, as the laser pulse is impinging to the sample surface, it produces heat on the surface and optical breakdown. Subsequently ionized plasma is formed. Ablated vapour and plasma expands into the ambient atmosphere, producing shock waves, followed by optical emission and particle ejection (Viskup et al., 2009). Later the cooling effect and visible crater produced as a side effect on a surface. Photography of ns laser induced plasma from Nd:YAG laser in air atmosphere is shown in Figure 2.

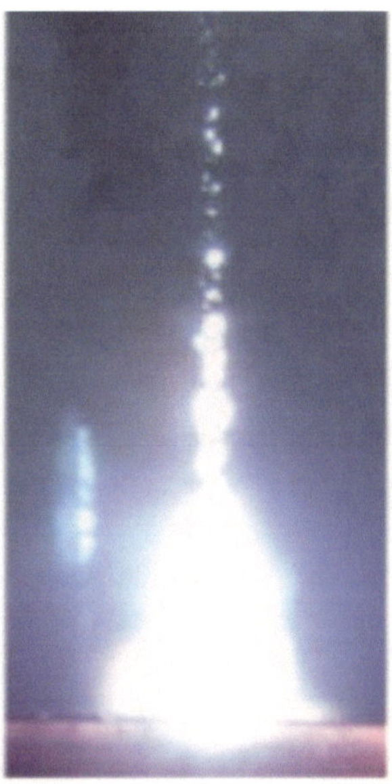

Fig. 2. Photography of ns laser ablation from Nd:YAG laser in air atmosphere. Ablation of material leads to ejection of cluster particles that may be irradiated by subsequent laser pulses.

2.2 Ionization

Ionization of atoms, molecules and dissociation of molecules within a gas at temperature T occurs at values of $k_B.T$ much lower than the ionization potential. Within a gas, collision between the species - electrons, atoms, molecules results in a certain degree of ionization, ξ. By considering dynamic equilibrium, where the rate of generation is equal to rate of recombination, degree of ionization can be expressed by the Saha equation

$$\frac{\xi^2}{1-\xi} = \frac{2g_i}{g_a N_g}\left(\frac{m_e k_B T}{2\pi\hbar^2}\right)^{\frac{3}{2}} \exp\left(-\frac{E_i}{k_B T}\right) \tag{1}$$

where $N_g = N_i + N_a$ with N_i and N_a the number densities of ions and atoms. The degree of ionization is given by $\xi = N_i / N_g$, and g_i and g_a denote the degeneracy of states for ions and atoms, m_e is the electron mass, k_B is the Boltzmann constant, $\hbar$ is the Planck constant and E_i is the ionization energy. However within the laser matter interaction the Saha equation has only limited validity. Laser light may directly ionize species through sequential or coherent multiphoton excitation or via collisions with electrons accelerated within the laser field - called impact ionization. The diffusion processes of electrons out of the plasma plume and strong non-equilibrium condition have to be also considered.

2.3 Optical breakdown

In principle a fast electron results in two slow electrons, which are again accelerated by the laser radiation, and this cycle is repeating again. In the initial stage of optical breakdown, the concentration of electrons increases exponentially. These processes increase with the number of energetic electrons, which can again ionize atoms or molecules. This avalanche like ionization of the plasma due to laser irradiation is called optical breakdown. In order to estimate the threshold laser intensity for optical breakdown it is necessarily to consider the electron energy losses due to collisions with atoms and electrons. It has been proved that laser-induced breakdown occurs in the range of 10^9 -10^{13} W.cm^{-2}, in air gas at atmospheric pressure and with ns laser pulses (Capitelli, 2004). These values are about three orders of magnitude higher than the values observed in front of solid or liquid targets. This is because the ionized hot material vapours which are present on a target surface. With increasing temperature near the target, the light intensity change due to reflections and the electric field at the fringe of the target surface is enhanced. With high pressures, the electron energy losses are mainly due to collisions with atoms. An estimation of intensity for optical breakdown in ambient gas I_p^{opt} can be made by using following equation:

$$I_p^{opt} \approx \frac{m_e^2 c \omega_c^2 E_i}{2\pi m_g e^2} \propto p^2 \tag{2}$$

where m_e is the electron mass, m_g is the mass of the gas particles, p is the gas pressure and ω_c is the collision frequency.

2.4 Plasma plume expansion

Plasma expansion into vacuum

In a vacuum the plasma plume undergoes free expansion and reaches some final constant velocity. This process can be thought in good approximation to be adiabatic. The fastest expansion occurs along the shortest axis of the plume where the highest pressure gradients are.

Plasma expansion into an ambient atmosphere

The expansion dynamics of the plasma plume change significantly, under ambient gas. This occurs generally since gas pressure exceeds 0.1mbar. The ejected species from the sample substrate behave like a piston. They compress and heat the ambient gas medium. This phenomenon causes an external shock wave pressure, which push the ambient gas away from the contact surface. This situation is schematically shown in Figure 3.

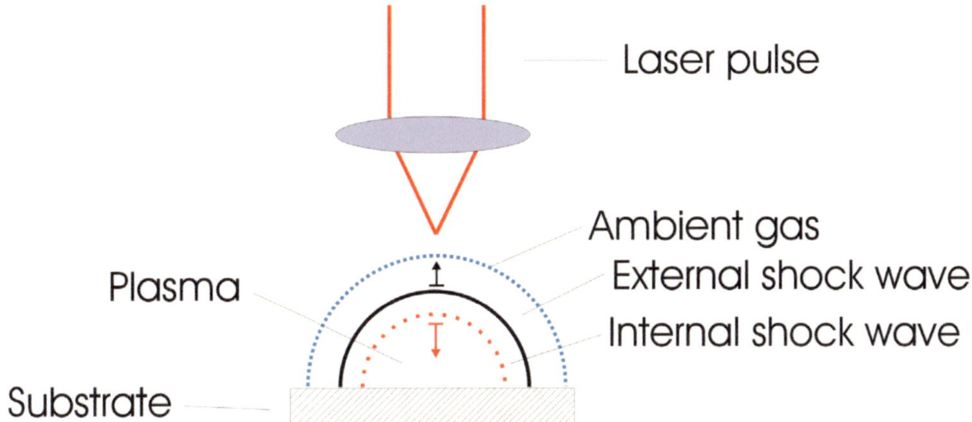

Fig. 3. Schematics showing the expansion of a spherical plasma plume in an ambient medium.

2.5 Expansion models

We can consider different models for the plasma plume expansions dynamics. In analytical model of plasma expansion is assumed that the equations of the gas dynamics can be applied. In this model the Knudsen-layer is not involved and a mixing of the plume and the ambient gas is not considered. The condensation, recombination and excitation are not included. The plasma expansion is of spherical symmetry. Input parameters are initial energy and mass of the plume. Losses due to sample heating, evaporation and plasma shielding are infinitesimal. The overall kinetic energy is conserved in the initial free and adiabatic expansion. The velocity of free expansion scales as

$$v_f \approx \left(\frac{E}{M}\right)^{1/2} \tag{3}$$

where E is the initial energy and M is the mass. This velocity of free expansion is in the range of $10^6 - 10^7 cm.s^{-1}$. Because the plume acts like a piston, an internal shockwave is formed at the contact surface and moves in opposite direction than the plume. Later it propagates in outward direction. This oscillation of the internal shock wave front homogenizes an entire plasma plume. After the plume slow down, its energy becomes mostly thermal and obeys the point blast wave expansion. This occurs when the rest mass in the external shock wave becomes comparable to the mass within the plume.

The radius of the shock wave expansion at medium pressures is given by (Arnold, 1999):

$$R_{SW} = \xi_0 \left(\frac{E_0}{\rho_0} \right)^{\frac{1}{5}} t^{\frac{2}{5}} \tag{4}$$

where ξ_0 is a constant dependent on specific heat capacity. This model describes the explosive energy release E_0 through a background gas density ρ, while it neglects viscosity and predict complete propagation. For calculation of plasma plume expansion radius is also often used the Drag model,

$$R_{DF} = R_0 \left(1 - \exp(-\beta t)\right) \tag{5}$$

where R_0 is the stopping distance of the plasma plume and β is slowing coefficient $(R_0 \beta = v_0)$ which depend on velocity v_0 of ablated species. This model in the contrast with the previous, predicts that plasma will eventually come to rest due to resistance from collisions with the background gas.

2.6 Validation of the model

To experimentally validate this model an iron oxide sample has been irradiate by Nd:YAG laser at fundamental wavelength $\lambda = 1064$ nm, with energy $E_L = 100$ mJ in air ambient gas. The variation of plasma emission intensity perpendicular to the target surface measured at different delay times are shown in Figure 4(a). The intensity profiles were taken in the plume centre and the intensities were normalized to the highest intensity measured for each delay time. The profiles showed a maximum intensity I_{max} followed by a strong intensity gradient which marked the front of the bright plume. The position for maximum intensity and the plume front shifted to larger distances at longer delay times. An effective plume length R was derived from plume images taken at different time t employing an intensity

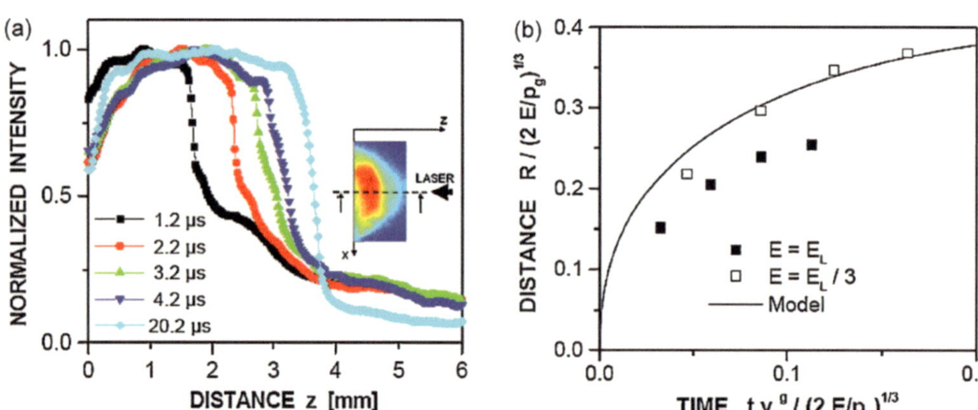

Fig. 4. Intensity profile of plasma emission of iron oxide sample after laser ablation ($\lambda = 1064$ nm, $E_L = 100$ mJ) in air, Fig. 4(a). The dashed line indicates the position of line cut through intensity maps (inset). Comparison of the temporal evolution of iron oxide plasma plume length with model curve shown in Fig. 4(b). Solid and open symbols are obtained assuming an initial plume energy of $E = E_L$ and $E = E_L/3$, respectively (Viskup et al., 2010).

criterion $I = 0.5I_{max}$. The variation of plume length with time is shown in Figure 4(b) in normalized coordinates, with the initial plume energy E, the gas background pressure $p_g = 96.900Pa$, and the sound velocity in air $v_0^g = 343 \mathrm{m/s}$. The solid symbols were calculated from the measured plume length assuming an initial plume energy that was equal to the laser pulse energy ($E = E_L = 100$ mJ). The open symbols were calculated for a lower plume energy ($E = E_L/3$) assuming that processes such as melting, vaporization and ejection of materials consumed a large part of the laser pulse energy. The solid line is calculated by a model describing spherical plume expansion into ambient gas. For solid bulk materials such as stainless steel, metal oxides and polymers a good agreement was found between the plume length measured at low background pressure and the model (Huber et al., 2000). At higher pressure additional effects such as hydrodynamic instabilities at the expansion front become relevant (Heitz et al., 2003).

2.7 Ablation mechanisms

Pulsed laser ablation by nanosecond laser pulses has been analyzed on the basis of various models. Depending on initial activation we recognize phototermal ablation, photochemical ablation or combination of both processes - photophysical ablation. These models usually include only single dominant mechanism, therefore they permits to analyze experimental results only partially, depending on material in narrow range of parameters.

The processes involved in phototermal ablation are initiated with single or multi-photon excitation of the irradiated material surface. If the excitation energy is rapidly transformed into heat, it causes the temperature increase at the surface. This affects the overall optical properties of the irradiated material surface and consequence the absorbed laser power. The temperature rise can result in thermal material ablation or vaporization with or without the surface melting. Another possible way is that temperature rise induces stress which can produce explosive type of ablation. The stresses can also change the optical properties and thereby influence the temperature rise.

For photochemical ablation, when the photon energy is sufficiently high, laser excitation can lead to direct bond breaking. Accordingly from the irradiated surface the atoms, molecules clusters or fragments of materials are desorbed into ambient atmosphere. The indirect way to ablation through light induced defects which can photochemically dissociated bonds; this can build up a stress resulting in mechanical ablation. Both these ways can apply in principle without any change at temperature in surface.

In case of photophysical ablation, process both thermal and non-thermal mechanisms contribute to ablation rate. It can be a system in which the lifetime of electronically excited species or broken bonds is sufficiently long. During this time the species can desorbs by the temperature rise from the surface, well before the total excitation energy is entirely dissipated into a heat. The overall processes of photophysical ablation are influenced by thermal or non-thermal defects, stress and volume changes.

Mention possible mechanisms are not complete, due to additional development of plasma emission, mainly the electrons and ions. These can cause additional electric fields, which may change the activation energy for thermal desorption or direct bond breaking.

3. Single laser pulse irradiation

3.1 Basic parameters

Laser induced plasma is a pulsed source, and therefore the optical spectrum evolves rapidly in time. The emission from laser induced plasma generated by single nanosecond laser pulse in air atmosphere is characterized by different processes. An important time periods after laser plasma formation during which emissions from different species predominate are: strong continuum emission, formation of ionized species, emission of neutrals and finally emission of molecules. Depending on measurement is possible to carried out time-resolved detection of optical plasma emission. Corresponding spectra can be monitored by using a gateable detector. To enhanced laser ablation processes, commonly measurements are conducted by using repetitive single pulse irradiation, which mean a series of individual laser pulses, which are formed on the surface of the investigated sample, with laser pulse repetition rate a few hertz.

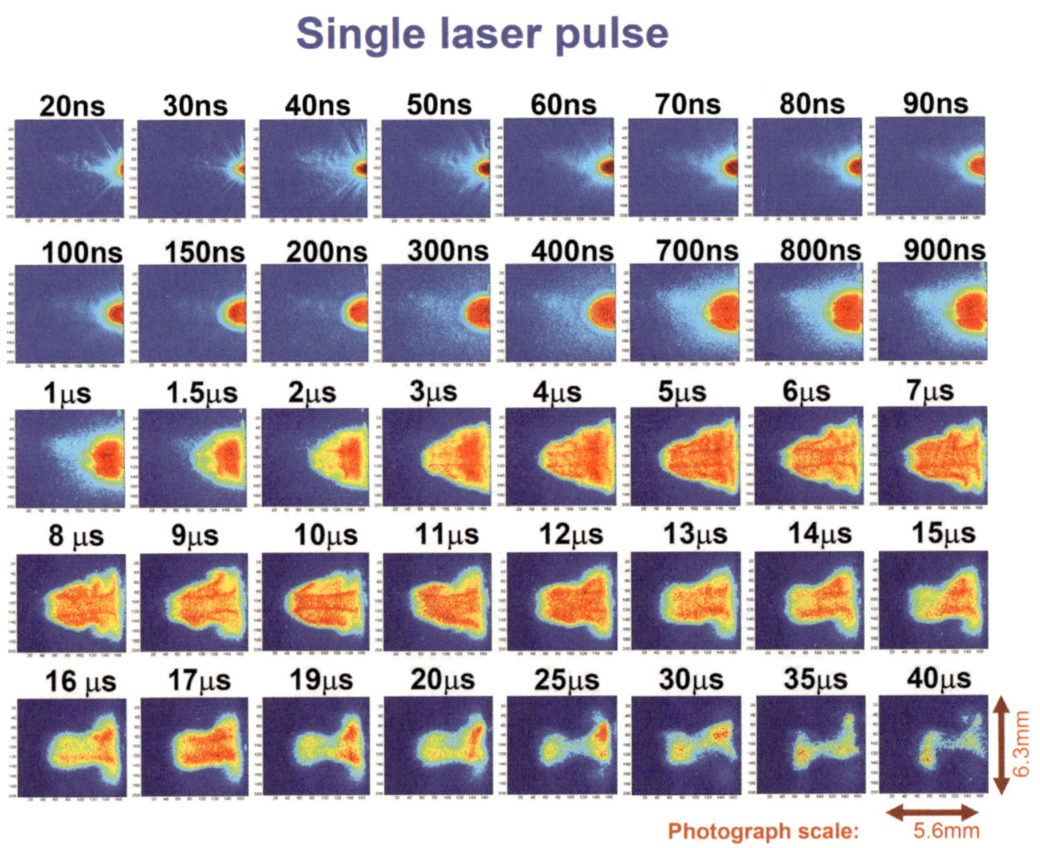

Fig. 5. Time resolved photography of plasma plume expansion from PA (Ultramid®, black) into the air ambient atmosphere. Laser pulse energy E ≈170 mJ, λ_L=532nm, integration time 10ns, photograph scale 5.6mm × 6.3mm. The images are shown in false colour technique encoding the emission intensity normalized to the maximum intensity (dark red) of each image.

3.2 Time resolved plasma photography of single pulse interaction

Time resolved photography of plasma plume expansion from PA (Ultramid®, black) into the air ambient atmosphere is shown in Figure 5. The images are taken at different delay times after the laser pulse ranging from $\Delta\tau_D$=20ns to $\Delta\tau_D$=40µs as indicated above the images. The integration time of the ICCD was set to 10ns. Each image was made with a separate laser pulse. The scale bar 5.6mm × 6.3mm is the same for all images. The images are shown in false colour technique encoding the emission intensity normalized to the maximum intensity (dark red) of each image.

Time resolved photography shows relatively homogeneous evolving plasma plume with approximately hemispherical shape and clearly defined rim, given by the shock front for delay times up to about 1µs after the laser pulse. After this time delay, the rim of the plasma becomes irregular and the plasma starts to detach from the sample surface forming a "mushroom cloud", which is mostly pronounced for a delay time of $\Delta\tau_D$=10µs after the laser pulse. For even later delays, the plasma develops into a perturbed and heterogeneous plume, which is not detectable for the camera any more after about $\Delta\tau_D$=40 µs. An estimation of the absolute intensities of the images is possible with help of the data in Figure 6, which were extracted from the same time resolved photography images. In this figure, the temporal development of not spectrally resolved intensities of the laser-induced plasmas is shown for single-pulse excitation with different time delay. The mean intensities of the recorded images were measured with the fast time-resolved ICCD camera. The spectral range was limited to wavelengths above about 350 nm, which is given by the cut-off of the glass camera objective employed. From this graph we can see that, the plasma intensities decay fast in the first $\Delta\tau_D$=1µs or $\Delta\tau_D$=2µs and then is stabilize on a lower level for considerably longer times, which is only slightly decaying. This is in consistent with the time-resolved plasma plume images, shown in Figure 5.

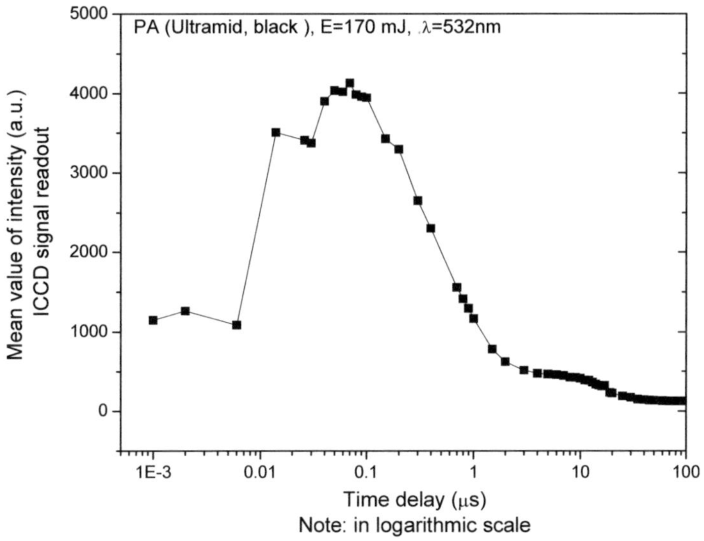

Fig. 6. The mean intensity value derived from time resolved ICCD photography of plasma (PA Ultramid®, black) plume expansion for single laser pulse. The integration time of the ICCD is set to 10 ns for this evaluation.

4. Double or multiple laser pulse irradiation

4.1 Methods of double pulse irradiation

In laser induced plasma spectroscopy technique, very important factors are precision and sensitivity of obtained spectra. Recently, new studies have lead to investigations of multiple-pulse irradiation which can greatly enhanced plasma emission. It has been found that dual-pulse irradiation technique can significantly give enhancement to plasma emission and hence improve the emission from the weak elements. The main idea of double pulse method is based on the formation of pre-ablation plasma on the surface of the studied material, while the second pulse, which is incoming several nanoseconds or microseconds behind the first laser pulse is further enhancing the plasma emission. The important parameter for double pulse laser interaction is the temporal difference between the arrival of the second laser pulses at the target Δt_L, also called interpulse delay or pulse to pulse separation time.

There are two main methods in double pulse irradiation. In first method the first laser pulse generates laser induced plasma at the sample surface, and second laser pulse re-heats a plasma (increased plasma volume, temperature, ion density, expansion and ablation rates, longer decay times) or re-excite material ablated by first laser pulse. In the second method 1st laser pulse generates low pressure region above the sample surface and 2nd pulse generates plasma that expands into low pressure region. This can be generally realised by four commonly used beam geometries for double pulse irradiation measurements, which are schematically shown in Figure 7.

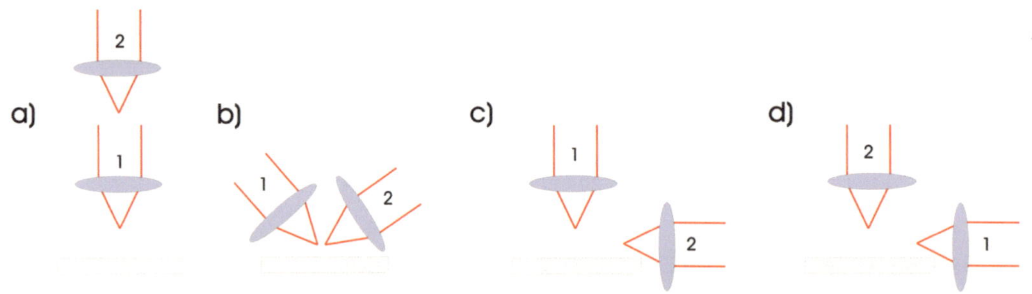

Fig. 7. Commonly used beam geometries for double pulse excitation: a) collinear dual pulse b) crossed beam dual pulse c) orthogonal reheating d) orthogonal pre-ablation dual pulse.

- Collinear dual pulse arrangement, where both pulses are focused upon the same point on the sample.
- Crossed beam dual pulse configuration, where separate two laser pulses under different angles are focused into the same point on the sample surface.
- Orthogonal pre-ablation spark dual pulse configuration in which air plasma is formed above the sample surface prior to ablation.
- Orthogonal reheating configuration, where an air plasma is formed after ablation

4.2 Time resolved plasma plume photography of double pulse interaction

Time-resolved photographic images of collinear double-pulse (DP) plasma plume expansion dynamics from polyamide (Ultramid®, black) into air ambient atmosphere are shown in

Figure 8. Delay between two laser pulses is Δt_L = 20ns. The images are taken at different delay times after the first laser pulse, up to $\Delta\tau_D = 90\mu s$ as indicated above the images. The integration time of the ICCD was set to 10ns. Pictures starts from single laser pulse exposure ($\Delta\tau_D = 10ns$ after the first pulse) and in $\Delta\tau_D = 20$ ns after the first pulse, the second laser pulse is entering. At this time delay, all the energy from first laser pulse is already deposited in polyamide target and plume is expanding. From time resolved photography pictures we can assume, that energy transfer between both laser pulses performed very smoothly, without subsequent disturbances of plasma plume. Plasma plume continuously absorbs energy and continues to growth forming relatively homogeneous evolving plasma plume with approximately hemispherical shape profile up to $\Delta\tau_D = 4\mu s$, similar, as it was in the case of single laser pulse exposure (Figure 5). Later in $\Delta\tau_D = 7\mu s$ the "mushroom" like structure phase of expanding plasma plume is in not so clear and evident as it was in the single pulse mode. For delays of $\Delta\tau_D = 10\mu s$ and longer, the perturbations occur resulting in

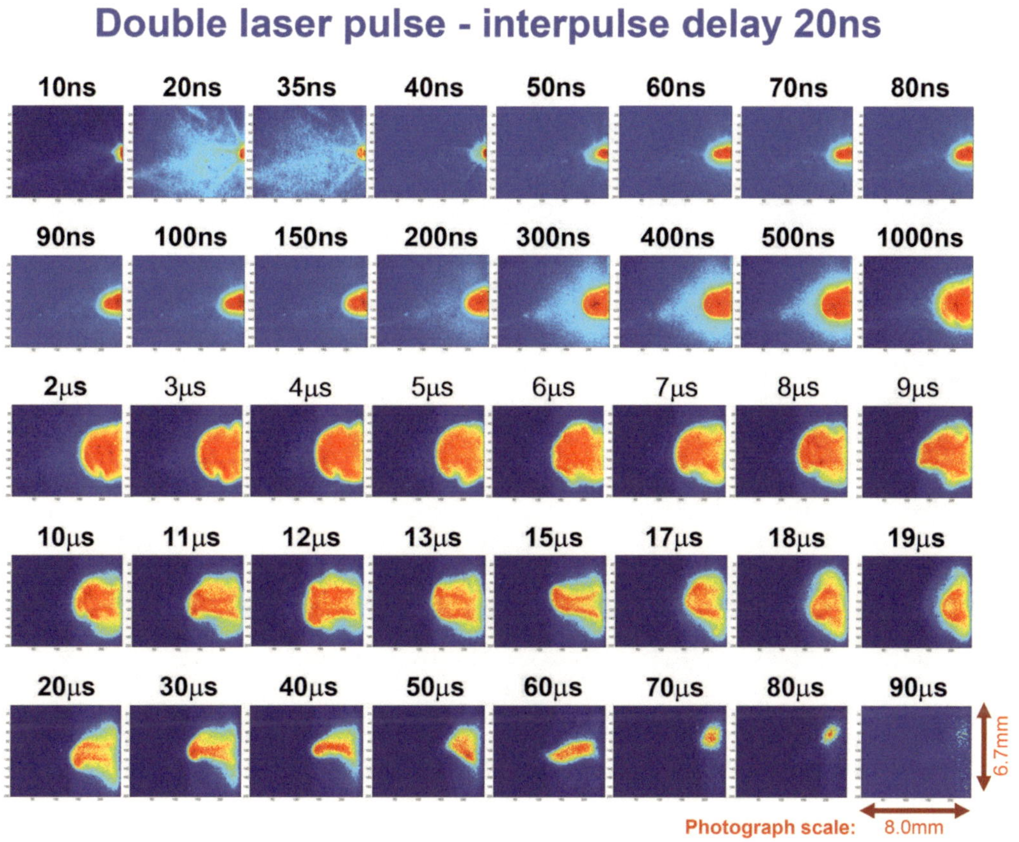

Fig. 8. Time-resolved photographic images of double-pulse (DP) plasma plume expansion from polyamide (PA - Ultramid®, black) into air ambient atmosphere. Delay between two laser pulses is 20 ns, laser pulse energies of $E_1 \approx E_2 \approx 170$ mJ, λ_L=532nm, integration time 10ns, photograph scale 8.0mm × 6.7mm. Pictures start from single pulse exposure ($\Delta\tau_D = 10ns$) and in $\Delta\tau_D = 20$ ns the second laser pulse is entering.

a heterogeneous plasma plume, which finally declines after about $\Delta\tau_D = 90\mu s$. Generally, the time of the different expansion phases of the plasma plume for double pulse excitation are longer than in the single-pulse case.

The first three images in Figure 9 show single-pulse plasma expansion. Then at a delay time of 1 μs after the first laser pulse, the sample is irradiated with the second laser pulse forming a new plasma channel in the single-pulse plasma. Pictures reveals plasma plume conjunction ($\Delta\tau_D$ = 1030ns ÷1060ns), after impact of the second laser pulse to already formed plasma, generated by first laser pulse. The plasma channel expands first vertically to the sample surface ($\Delta\tau_D = 1050$ ns) and then expands mainly in the transverse direction ($\Delta\tau_D = 1100$ ns to $\Delta\tau_D = 5\mu s$) resulting in an appearance homogeneous disk-shaped plasma

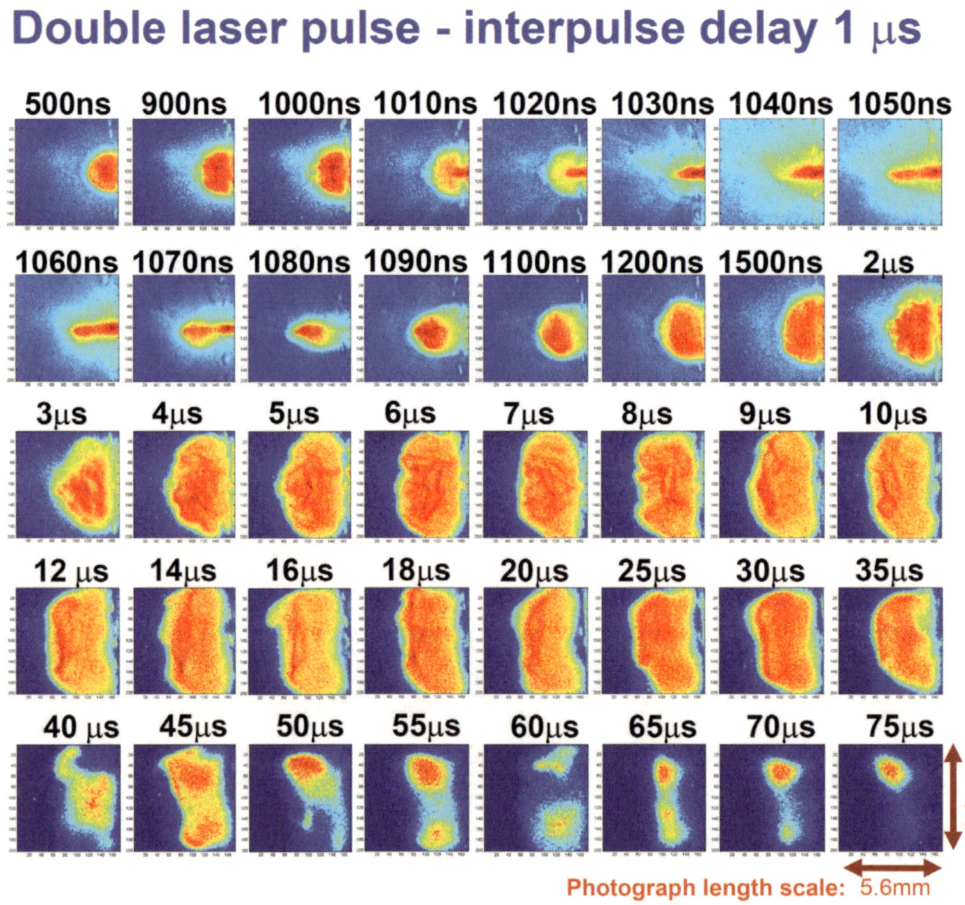

Fig. 9. Time-resolved photographic images of double-pulse (DP) plasma plume expansion from PA (Ultramid®, black) into air ambient atmosphere. Delay between two laser pulses is 1μs, laser pulse energies of $E_1 \approx E_2 \approx 170$ mJ, λ_L=532nm, integration time 10ns, photograph scale 5.6mm × 6.3mm. Pictures start from single laser pulse exposure (500ns, 900ns), followed by second laser pulse in $\Delta\tau_D = 1\mu s$.

with a height of about 3 mm and a diameter of about 7 mm. The size of the plasma remains relatively stable up to a delay time $\Delta\tau_D = 30$ μs. Afterward, it starts to decay from around $\Delta\tau_D = 40$ μs up to approximately $\Delta\tau_D = 80$ μs.

Comparing plasma plume length from Figure 8 and Figure 9 in vertical direction, both cases, plume's length are approximately identical. Whereas in transverse direction, the plume size aggrandise more than double, in the case of DP with Δt_L=1 μs interpulse delay.

Figure 10 shows the plasma expansion images for double-pulse excitation with interpulse delay Δt_L = 10 μs. The first image at a delay time $\Delta\tau_D = 9.999$ ns is the single-pulse plasma of the first laser pulse, which appears again similar to a "mushroom cloud". The following images show the plasma after double-pulse excitation. The picture ($\Delta\tau_D = 10.010$ns) reveal evolving small second plasma (after irradiation of the target by second laser pulse) in the vicinity of the target, inside the already well formed "mushroom" plume, created by first laser pulse.

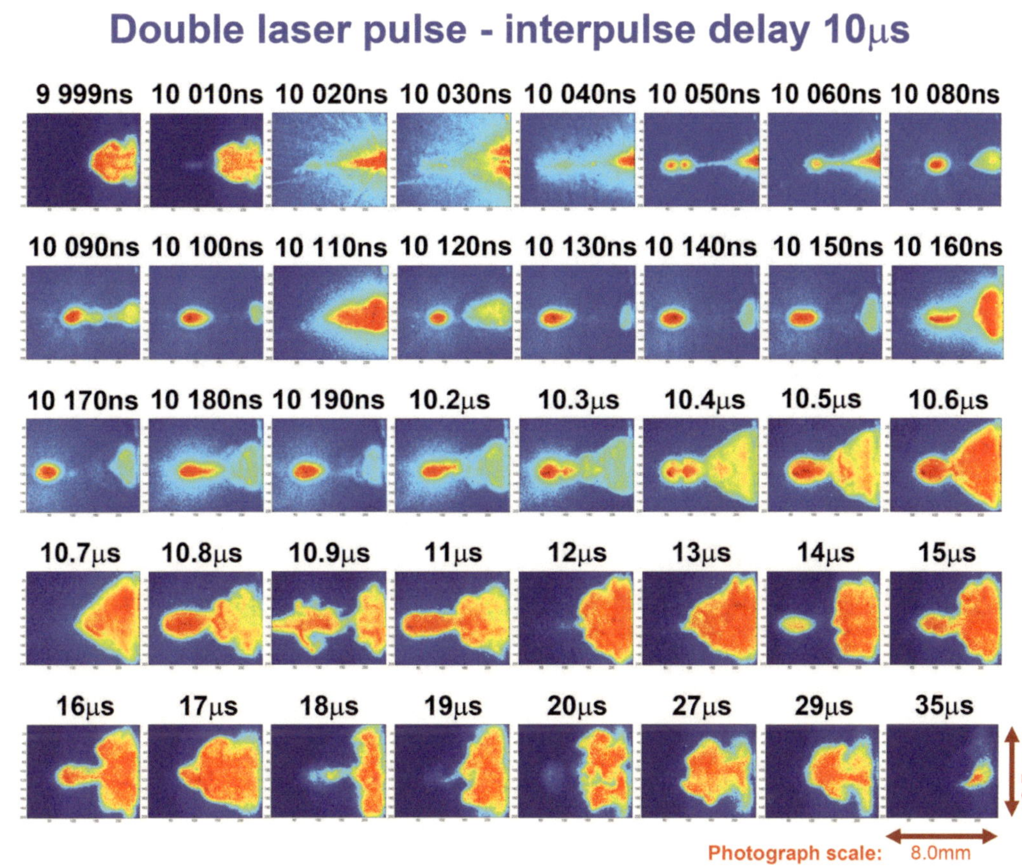

Fig. 10. Time-resolved photographic images of double-pulse (DP) plasma plume expansion from PA (Ultramid®, black) into air ambient atmosphere. Delay between two laser pulses is 10μs, laser pulse energies of $E_1 \approx E_2 \approx 170$ mJ, λ_L=532nm, integration time 10ns, photograph scale 8.0mm × 6.7mm. Pictures start from single laser pulse exposure (9 999ns) and since 10μs the second laser pulse is entering.

Conjunction of plasma plume in $\Delta\tau_D = 10.020$ns takes a place, plasma from second laser pulse further continue to grow. In $\Delta\tau_D = 10.050$ns plume starts to split into islands or droplet, connecting to channels, which last up to approximately $\Delta\tau_D \approx 10.200$ns. During this evolving time delay might happen that these two plasmas are again joining (see $\Delta\tau_D = 10.110$ns) or re-joining. Both plasmas are further expanding in $\Delta\tau_D = 10.400$ns. In later time delay $\Delta\tau_D = 18\mu$s plume starts to decay but persist up to $\Delta\tau_D = 40\mu$s.

We can outline, that for Δt_L =10μs pulse separation we obtained unstable heterogonous plasma expansion behaviour. This distortion in plasma plume (islands or droplet) can be also due to the interaction of the incoming second laser beam with ejected solid or liquid particles resulting from the first irradiation. Therefore plasma generated by second laser pulse divers and created structure can appear with randomized re-joining shape. Because each image is made with a separate laser pulse, the evaluation of the dynamic behaviour based on similarities of consecutive images is not clearly visible. Plume does not last longer in this case, and is approximately comparable with duration of single pulse irradiation.

The temporal development of not spectrally resolved intensities of the laser-induced plasmas is shown in Figure 11 for single-pulse and double-pulse excitation with different interpulse-delay times, Δt_L. The mean intensities of the recorded images were measured with the fast time-resolved ICCD camera.

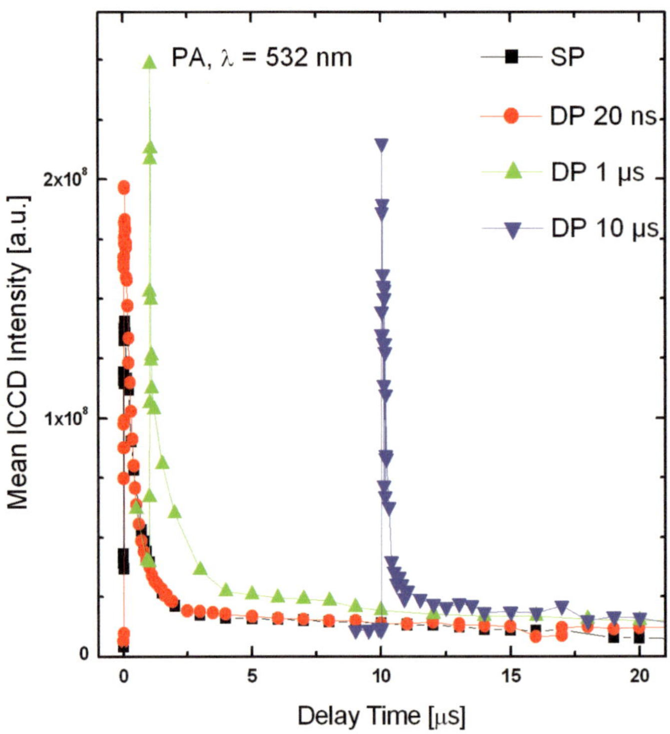

Fig. 11. Single-pulse (SP) and double-pulse (DP) mean intensity derived from time resolved ICCD photography of plasma plume expansion. The integration time of the ICCD is set to 10 ns. Interpulse delay is Δt_L= 20 ns, Δt_L= 1 μs, and Δt_L= 10 μs.

For the double-pulse excitations with Δt_L = 1μs and Δt_L=10μs, only the decay after the second pulse and the time shortly before the second pulse are shown. The images for single-pulse and double-pulse excitation with Δt_L = 1μs and for double-pulse excitation with Δt_L = 20 ns and Δt_L=10μs were recorded in two independent experiments with slightly different gain and magnification. This was corrected by multiplication of a factor in order to obtain consistent overlaps of the different curves. In all four cases, the plasma intensities decay fast in the first 1μs or 2μs and then stabilize on a lower level for considerably longer times, which is only slightly decaying. The intensity of the double-pulse plasma with Δt_L = 1μs starting in the shoulder of the single-pulse excitation is considerably more intense than the single-pulse plasma and also the double-pulse plasmas with Δt_L = 20 ns and Δt_L=10μs.

5. Emission processes in plasma

5.1 Basic mechanisms

In plasma atoms and ions particles undergo transitions between their quantum states through collisional and radiative processes. Among these processes the most important are spontaneous radiative transitions and collisional transitions induced by electron impact collisions.

After absorption of the intense short laser beam, the plasma has sufficient energy to emit radiation via free-free, free-bound and bound-bound mechanisms. Free-free or Bremsstrahlung emission occurs when a free electron interacts with the Coulomb potential of the ions, and radiates a continuum electromagnetic spectrum. The second process also produces a continuum electromagnetic spectrum due to electron transitions from initial free electron states to bound electron states. This mechanism is known as recombination radiation. The last emission mechanism produces a line spectrum as a result of transitions between discrete bound levels of ionized and neutral atoms.

Significant contribution to plasma emission is due to electron-ion recombination processes. The recombination processes gives rise to continuum emission at photon energies greater than the ionization energy of the recombining ion. In plasma where exists simultaneously two or more ionized states, spectral emission is characterized by characteristic jumps, related to different ionization energies, also known as recombination edges.

5.2 Line emissions

Transitions of bound electrons from excited states to lower states of ions and neutrals emit radiation with discrete energy. This could take place in the ground state where strong emission is produced or between excited states where relatively weak radiation is emitted. The emitted intensity of an atomic, ionic, or molecular spectral line is proportional to the population of the upper level, to the transition probability, and to the energy of the quantum. Due to different processes in the plasma, the line shows a spectral profile instead of sharp wavelength emitted over a spectral range. According to Boltzmann statistics the proportionality of the optical emission line intensity I_λ^{ij} corresponding to the transition at certain wavelength between two energy levels E_u and E_l of an atom or ion is given by the formula

$$I_{\lambda}^{ul} = F.A_{ul}.h.\nu_{ul} \frac{n_i}{Z_i} g_{i,u} \exp\left(-\frac{E_{iu}}{kT_{ex}}\right) \qquad (6)$$

where

F is the factor depending upon experimental setup;
A_{ul} is the corresponding transition probability per unit time
h is the Planck's constant;
ν_{ul} is the frequency of the photons emitted due to transition from upper excited level u to the lower level l;
n_i is the concentration of the chemical species i;
Z_i is the partition function of the chemical species i calculated at T_{ex}
$g_{i,u}$ is the statistical weight of the upper excited state of the chemical species i;
E_{iu} is the energy of the upper excited state of chemical species i;

It the case of using this equation it is necessary to assume that plasma is homogeneous, is in local thermal equilibrium and also that plasma is optically thin, which mean that emitted photons are not re-absorbed.

The power radiated per unit volume, solid angle and frequency interval, corresponding to the transition from the upper state to the lower state. Level populations and shape of line emission depend on the physical conditions in the plasma and are determined by the dynamical balance of several collisional and radiative processes inside the plasma. Three physical processes which influence the shape of line emission are: a) natural broadening due to the finite lifetime of the states involved in the transition, b) Doppler broadening due to thermal motion, c) pressure broadening due to the interaction of the radiating systems with the rest of the plasma and d) Stark broadening due to charged perturbers.

5.3 The effect of double pulse laser interaction to plasma emission

Figure 12 shows single-pulse versus double-pulse plasma optical spectra of polyamide with interpulse-delay times, Δt_L, of 20 ns, 1 μs, and 10 μs. In spectra are dominating emission from atomic and ionic C, N, H, O, Ca, or Mg lines originating from the PA, additives in the PA, or the surrounding air atmosphere. Strong lines are visible from carbon main matrix compound (C I at 247.86 nm), trace element calcium (Ca II at 393.37nm), and H(α) Balmer series line at 656.3nm. The lines are seated on a pronounced continuum with a periodical structure, especially for the single-pulse spectrum and the double-pulse spectrum with Δt_L = 20 ns. The periodical structure of the continuum results are from the different diffraction orders of the Echelle spectrometer, which has for each diffraction order a lower sensitivity at the rim compared to the centre of the order. Additionally, molecular emission of moderate and weak intensity is visible, e.g., the CN violet band with band-head at 388 nm and the C_2 Swan band with band-head at 516nm. With longer interpulse-delay times the continuum gets weaker and some atomic lines get sharper. This is especially visible for the width of the H(α) line. However, most of the other atomic lines are pronounced for the double-pulse spectrum with Δt_L = 1 μs. They are considerably weaker for the single-pulse spectrum, which has only half of the total laser energy, but also for the double-pulse spectra with Δt_L = 20 ns and Δt_L=10 μs.

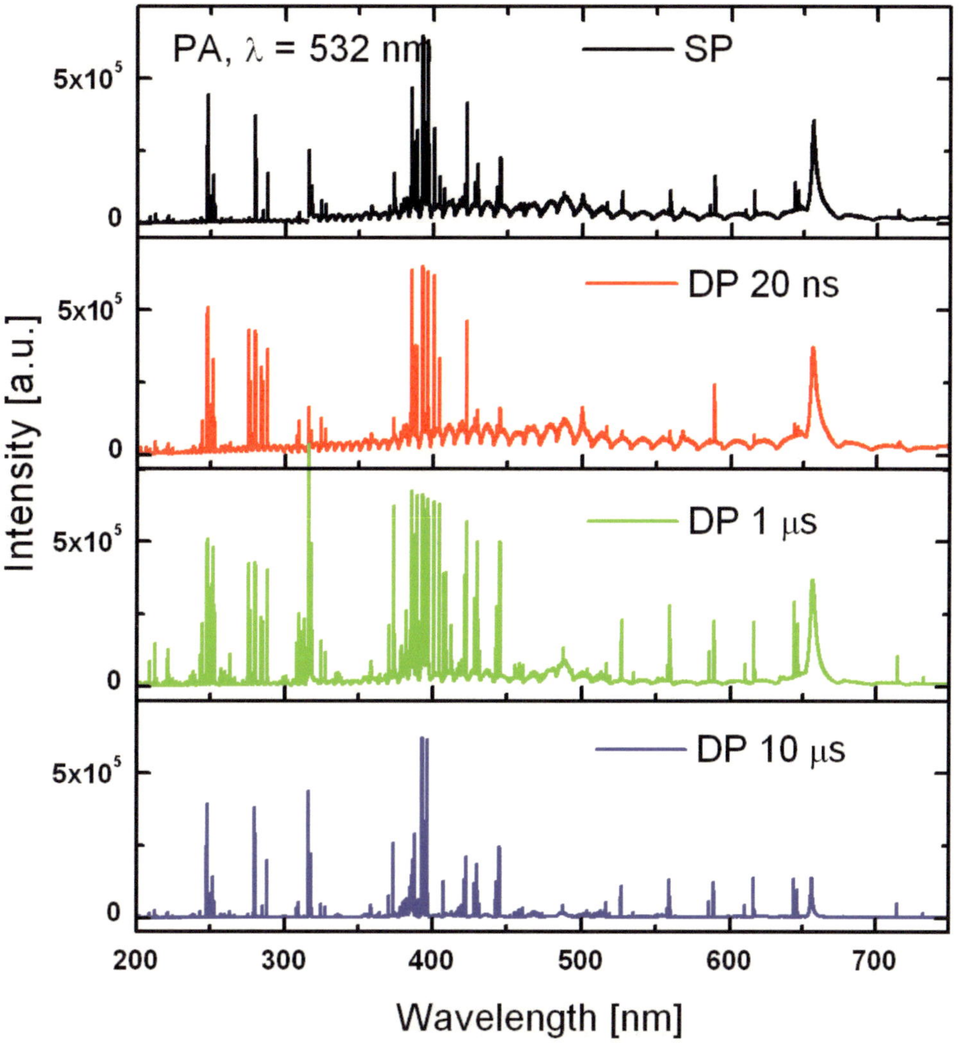

Fig. 12. Single-pulse and double-pulse optical spectra of polyamide plasma. ($E_1 \approx 170$ mJ or $E_1 \approx E_2 \approx 170$ mJ, λ_L=532nm, $\Delta\tau_G$=25μs, $\Delta\tau_D$=500ns). Delay times between the pulses: Δt_L of 20 ns, 1 μs, and 10 μs. All spectra are accumulated over 20 laser pulses.

6. Conclusions

This chapter has presented experimental results on single and double ns laser pulse interaction with solid state with application to plasma spectroscopy. We have applied a collinear nanosecond double pulse exposure of target materials to enhance an optical emission from plasma. Beside of these studies, a plasma from single and double pulse irradiation has been analysed by means of time resolved plasma plume photography and spectroscopy. We compare these results and we find out a very good relation between shape of plasma plume and plasma emission.

From single laser pulse interaction the time-resolved photography shows a relatively homogeneous evolving plasma plume with approximately hemispherical shape and clearly defined rim, given by the shock front for delay times up to ~1μs after the laser pulse. For later time delay, the rim of the plasma plume becomes irregular and the plasma starts to detach from the sample surface forming a "mushroom cloud", which is most pronounced for a delay time about ~10μs after the laser pulse. For even later delays, the plasma develops into a perturbed and heterogeneous plume, which is not detectable for the camera any more after ~40μs. The main atomic and ionic emission signal in studied plasma last approximately ~7μs, while the molecular emission from plasma last approximately ~50μs in air atmosphere.

To enhance a plasma emission from single pulse, we performed collinear nanosecond double-pulse excitation. Special attention has been given to interpulse-delay between 20ns to 10 μs. The similar measurement of time resolved plasma plume photography and time resolved spectroscopy were performed. It has been find out that after double pulse laser impact with short interpulse delay (20ns), plasma seems to expand also relatively homogenously and last significantly longer, compared with to only single pulse exposure. In a case of very long interpulse delay approximately 10μs, formed plasma plume is very heterogonous and plasma emission last similar time as single pulse plasma.

Double-pulse excitation can result in enhanced emission of atomic and ionic spectral lines. We observed a strong influence of pulse-to-pulse delay on the results of 532 nm double-pulse irradiation. We attribute this interdependence to the different plasma expansion dynamics observed by time-resolved photography with different pulse-to-pulse delays. A further enhancement in signal intensity and sensitivity seems to be feasible by optimization of the timing of the interpulse-delay times between the laser pulses. However, the optimal time parameters depend on the specific type of the irradiated material and the individual elemental spectral line under investigation.

7. Acknowledgment

This work has been supported by the Christian Doppler Research Association (Austria). Author would like to thank Prof. Johannes David Pedarnig and Prof. Johannes Heitz for valuable discussions.

8. References

Anisimov S. I., Bäuerle D. & Luk'yanchuk B. S. (1993). Gas dynamics and film profiles in pulsed-laser deposition of materials, *Physical Review B*, Vol.48, pp. 12076–12081.

Arnold N. (1999). Spherical expansion of the vapor plume into ambient gas: an analytical model, *Applied Physics A*, Vol.69, pp. 87.

Babushok V.I., DeLucia F.C., Gottfried J.L., Munson C.A. & Miziolek A.W. (2006). Double pulse laser ablation and plasma: laser induced breakdown spectroscopy signal enhancement, *Spectrochimica Acta Part B: Atomic Spectroscopy*, Vol.61, pp. 999–1014.

Bäuerle D. (2011). *Laser Processing and Chemistry* (4thed ed.), Springer, ISBN 978-3642176128, Berlin.

Capitelli M. (2004). Laser induced plasma expansion: theoretical and experimental aspects, *Spectrochimica Acta Part B: Atomic Spectroscopy*, Vol.59, pp. 271.

Cremers D.A., Radziemski L.J. & Loree T.R. (1984). Spectrochemical analysis of liquids using the laser spark, *Applied Spectroscopy*,Vol.38, pp. 721–729.

Cristoforetti G., Legnaioli S., Pardini L., Palleschi V., Salvetti A. & Tognoni E. (2006). Spectroscopic and shadowgraphic analysis of laser induced plasmas in the orthogonal double pulse pre-ablation configuration, *Spectrochimica Acta Part B: Atomic Spectroscopy*, Vol.61, pp. 340–350.

Gautier C., Fichet P., Menut D., Lacour J.L., Hermite D. L. & Dubessy J. (2005). Main parameters influencing the double-pulse laser-induced breakdown spectroscopy in the collinear beam geometry, *Spectrochimica Acta Part B: Atomic Spectroscopy*, Vol.60, pp. 792–804.

Heitz J., Gruber J., Arnold N., Bäuerle D., Ramaseder N., Meyer W. & Hochörtler J. (2003). In situ analysis of steel under reduced ambient pressure by laser-induced breakdown spectroscopy, *Proceedings of SPIE, XIV International Symposium on Gas Flow, Chemical Lasers, and High-Power Lasers*, Vol.5120, pp. 588, ISBN: 9780819449801.

Huber N., Gruber J., Arnold N., Heitz J. & Bäuerle D. (2000). Time-resolved photography of the plasma-plume and ejected particles in laser ablation of polytetrafluoroethylene, *Europhysics Letters*, Vol. 51, pp. 674.

Choi S., Oh M., Lee Y., Nam S., Ko D. & Lee J. (2009). Dynamic effects of a pre-ablation spark in the orthogonal dual-pulse laser induced breakdown spectroscopy, *Spectrochimica Acta Part B: Atomic Spectroscopy*, Vol.64, pp. 427–435.

Noll R., Sattmann R., Sturm V. & Winkelmarm S. (2004). Space and time-resolved dynamics of plasmas generated by laser double pulses interacting with metallic samples, *Journal* of Analytical Atomic *Spectrometry*, Vol.19, pp. 419–428.

Peter L. & Noll R. (2007). Material ablation and plasma state for single and collinear double pulses interacting with iron samples at ambient gas pressures below 1 bar, *Applied Physics B*, Vol.86, pp. 159–167.

Sattmann R., Sturm V. & Noll R. (1995). Laser-induced breakdown spectroscopy of steel samples using multiple Q-switch Nd:YAG laser pulses, *Journal of Physics D: Applied Physics*, Vol.28, pp. 2181–2187.

Semerok A. & Dutouquet C. (2004). Ultrashort double pulse laser ablation of metals, *Thin Solid Films*, Vol.453, pp. 501–505.

Viskup R., Praher B., Stehrer T., Jasik J., Wolfmeir H., Arenholz E., Pedarnig J. & Heitz J. (2009). Plasma plume photography and spectroscopy of Fe - oxide materials, *Applied Surface Science*, Vol.255, No.10, pp. 5215-5219.

Viskup R. (2010). Laser Assisted Diagnostics of Industrial Materials, PhD. Dissertation, Johannes Kepler University Linz, Austria.

Viskup R., Praher B., Linsmeyer T., Scherndl H., Pedarnig J.D. & Heitz J. (2010). Influence of pulse-to-pulse delay for 532 nm double-pulse laser-induced breakdown spectroscopy of technical polymers, *Spectrochimica Acta Part B: Atomic Spectroscopy*, Vol.65, pp. 935.

15

Laser Drilling Assisted with Jet Electrochemical Machining

Hua Zhang
School of Mechanical Engineering, Nantong University, Nantong, Jiangsu
China

1. Introduction

Laser drilling is a noncontact, precise and reproducible technique that can be used to form small diameter and high-aspect ratio holes in a wide variety of materials. Laser drilling is most extensively used in the aerospace, aircraft, and automotive industries. The most important application of laser drilling in the aerospace industry is the drilling of a large number of closely spaced effusion holes with small diameter and high quality to improve the cooling capacity of turbine engine components. Drilling rates as high as 100 holes/s can be achieved in production environment by coordinating the workpiece motions with pulse period of pulsed laser source. Laser drilling does not pose substantial problems at high angles of incidences. Laser drilling is also well suited for the nonconducting substrates or metallic substrates coated with nonconducting materials where the electric discharge machining is limited. In addition, recently the laser drilling of composite materials such as multilayer carbon fiber composites for aircraft applications is attracting increasing interest due to potential advantages of rapid processing, absence of tool wear, and ability to drill high-aspect ratio holes at shallow angles to the surface.

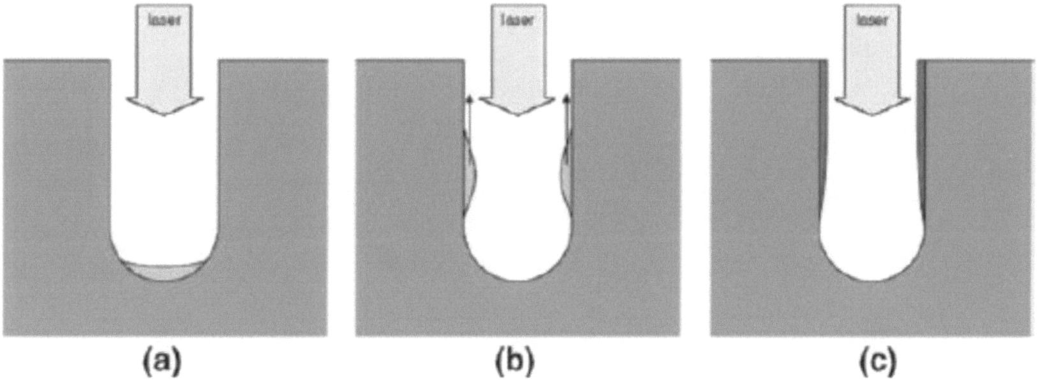

Fig. 1. Physical processes during percussion laser drilling. (a) Melt pool formation (b) splashing out of molten material, and (c) formation of recast layer.

The physical processes occurring during laser drilling are shown in Fig.1. The process can be split up into three stages. Initially a thin region of molten material is formed by absorption of laser energy at target surface. After some time the surface of this melt pool reaches vaporization temperature. The sudden expansion of the vapor evaporating from the surface eventually leads to the splashing stage when the melt pool is pushed radially out by the recoil pressure. Melt expulsion occurs when the pressure gradients on the surface of the molten material from the hole. On its way of escaping out some part of this molten material may re-solidify at the wall. These molten materials re-solidify adhering to the sidewall of the hole as a thin layer referred to as the 'recast layer'. Also, often spatter of molten material on the surface surrounding the entrance and exit of the hole will occur during the laser drilling process. The spatter and recast layer are defects of laser drilling and limit the application. Therefore, the elimination of these defects is one subject of intense research in laser machining.

This chapter contains a new typical hybrid process of laser drilling assisted with jet electrochemical machining (JECM-LD) for the minimization of recast and spatter.

2. Principle

JECM-LD combines two different sources of energy simultaneously: energy of photons (laser drilling) and energy of ions (ECM). The main aim of combining a jet electrolyte with laser beam is to obtain high process quality by reducing the recast layer and spatter produced in laser drilling. The jet electrolyte is aligned coaxially with a focused laser beam and creates a noncontact tool-electrode. The focused laser beam and the jet electrolyte are acting on the same surface of workpiece synchronously.

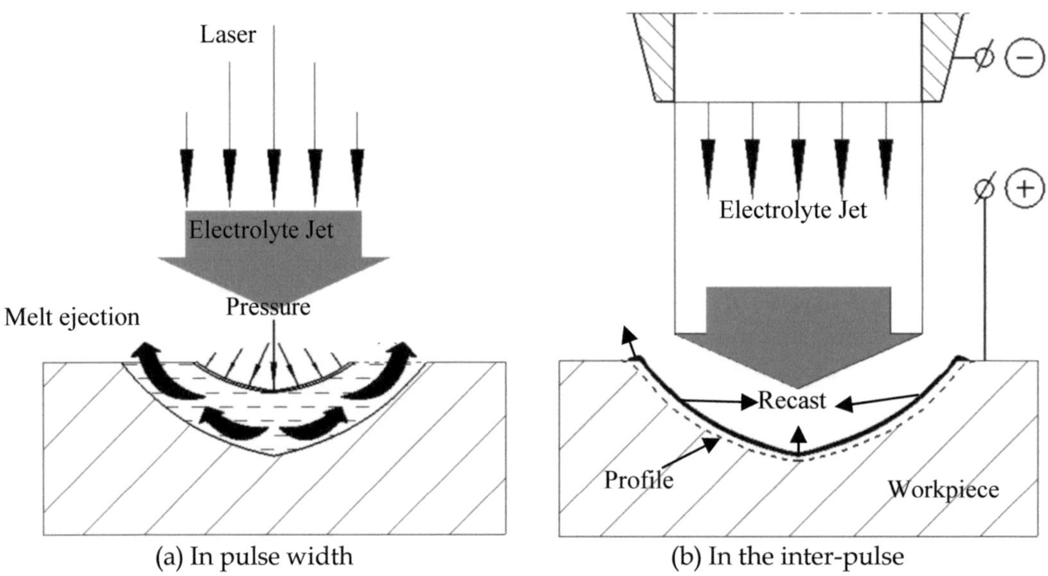

Fig. 2. Principle scheme of JECM-LD

In the course of JECM-LD, material is removed mainly by laser drilling in pulse width. The defects are overcome by the effect of jet electrolyte which consists of effective cooling to workpiece, transporting debris and electrochemical reaction with materials in the inter-pulse of laser. Fig.2 illustrates the principles of JECM-LD.

3. Measurement of laser attenuation in electrolyte

JECM-LD aligns a jet electrolyte coaxially with a laser beam. The laser beam needs to transmit in the jet electrolyte before being focused on the machining area. Electrolyte is a neutral salt solution, which is attenuable to laser energy by absorption and scattering. The property of laser attenuation in electrolyte is the key factor of JECM-LD.

3.1 Theoretical background

Electrolyte is a neutral salt solution with the solute of sodium salt and the solvent of pure water. The solution has a number of particles such as water molecules, metallic ions and suspended matter. These particles have significant effects on the laser attenuation in the electrolyte with absorption and scattering. The total attenuation coefficient $\mu(\lambda)$ of solution is given by

$$I_x(\lambda) = I_0(\lambda)\exp[-\mu(\lambda)x] \tag{1}$$

Where $I_0(\lambda)$ is a source of intensity and $I_x(\lambda)$ is the intensity transmitted a distance x in solution. The total attenuation coefficient is the sum of the absorption and scattering coefficients. In terms of the absorption coefficient $\alpha(\lambda)$ and the total scattering coefficient $\beta(\lambda)$, $\mu(\lambda)$ can been written as

$$\mu(\lambda) = \alpha(\lambda) + \beta(\lambda) \tag{2}$$

The absorption coefficient is correlative with the refraction of solution. The index of refraction $n(\lambda)$ and the extinction coefficient $k(\lambda)$ of solution are, respectively, the real and imaginary parts of its spectral complex refractive index is written as

$$\vec{n}(\lambda) = n(\lambda) - ik(\lambda) \tag{3}$$

The absorption coefficient $\alpha(\lambda)$ is determined by $k(\lambda)$ and the expression is given by

$$\alpha(\lambda) = 4\pi k(\lambda) / \lambda \tag{4}$$

The total scattering coefficient includes two parts of Rayleigh scattering coefficient $\beta_R(\lambda)$ and Mie scattering coefficient $\beta_M(\lambda)$, which can be written as

$$\beta(\lambda) = \beta_R(\lambda) + \beta_M(\lambda) \tag{5}$$

The scattering coefficient is determined by the radius of the particles in solution. A dimensionless parameter q ($q = 2\pi R / \lambda$) is used to be token of particle size. When $q < 0.1$, the scattering is Rayleigh scattering, contrariwise, which is Mie scattering. The electrolyte of this work is an industrial reagent of sodium chloride or sodium nitrate. The electrolyte has different sizes (Radius: 5~10μm) suspended particles result from undissolved matters involved in the industrial reagent. With the higher concentration, the solution has more suspended particles and the effect of scattering is more evident. So the laser attenuation in electrolyte is directly affected by the concentration of solution. The temperature of solution is also a factor to influence the attenuation coefficient. Here two parameters of Ψ_C and Ψ_T are used to represent the influence of the concentration and temperature, then Eq. (1) can be deduced a new expression

$$I_x(\lambda) = I_0(\lambda)\exp\{-[\alpha(\lambda) + \beta_M(\lambda) + \Psi_C(C - C_0) + \Psi_T(T - T_0)]x\} \quad (6)$$

3.2 Measurement and results

A measurement of laser attenuation in electrolyte has been carried out based on the theoretical analysis mentioned above. The laser systems used in the measurement are a CW semiconductor green laser (wavelength: 532 nm) and a pulsed Nd:YAG laser (wavelength: 1064 nm). The other equipments include a sample quartz glass cell, a heater and a photometer. The sample electrolytes are the neutral sodium solution. The solutes are industrial sodium chloride (Standard no.QB2238.2- 2005 of China National Light Industry Council) and industrial sodium nitrate (Standard no.GBT4553-2002 of China). The concentration of the solution is in terms of linear increase with 2%, 8%, 14% and 20%.

During the measurement, the first step is measuring the source intensity of laser transmitted in the empty sample cell and the result is taken as I_0, then measured the intensity I_x transmitted a distance x in sample solution with different concentration and temperature. Finally, it is straightforward to invert Eq. (6) and find the attenuation coefficient $\mu(\lambda)$, the concentration influence coefficient Ψ_C and the temperature influence coefficient Ψ_T.

Fig.3 shows a light path of green laser transmitting in electrolyte. The brightness of the path is higher with the increase of the solution concentration, which is result of scattering of suspended particles in the electrolyte. It's shown that the reason of green laser attenuation in the electrolyte is mainly scattering.

Figs.4 (a) and (b) show, respectively, the variety of green laser and infrared laser attenuation coefficient in electrolyte at 25℃. It is shown that the laser attenuation coefficient is linear with increase of concentration.

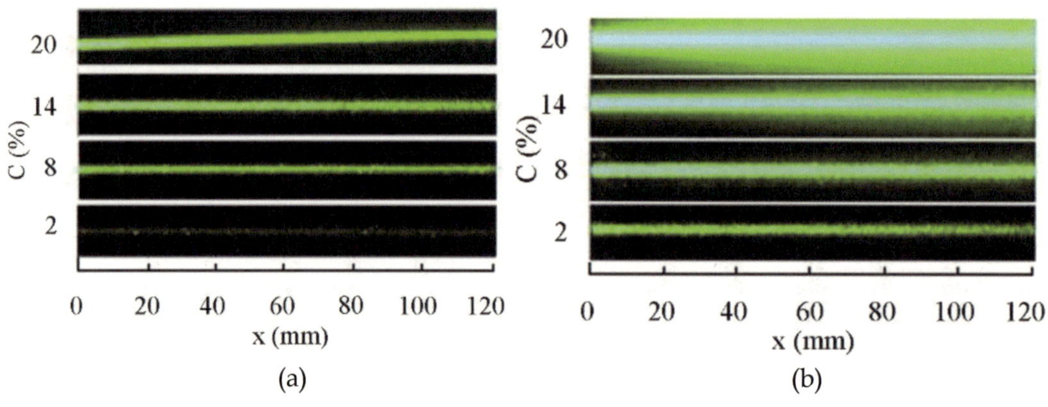

Fig. 3. Light path of green laser propagation in electrolyte (a) in sodium chloride electrolyte; (b) in sodium nitrate electrolyte

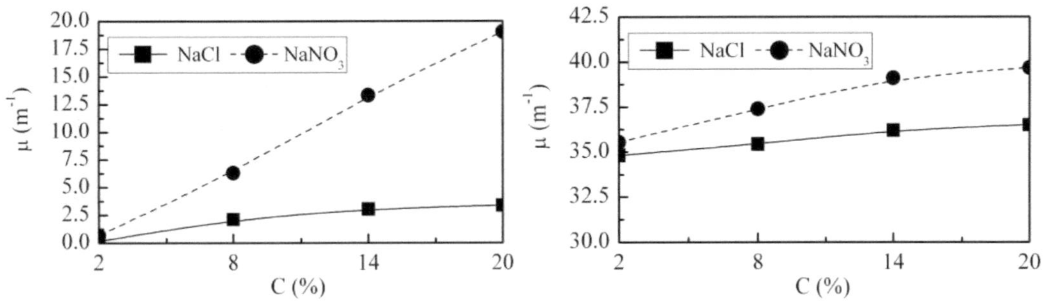

Fig. 4. Attenuation coefficient for electrolyte vs. laser: (a) green laser; (b) infrared laser.

Pure water is most transparent to the visible laser, especially, to green-blue laser of wavelength in the range 470–560 nm. Absorption of infrared laser in pure water is evident. The absorption coefficients of green laser (532 nm) and infrared laser (1064 nm) are 0.0355 and 34.2438m^{-1} respectively. It is shown that the attenuation of green laser in the electrolyte is mainly due to scattering; however, infrared laser is absorbed in electrolyte compared with pure water. The infrared laser attenuation coefficients in various concentrations of electrolyte are higher than 34.5m^{-1}. Thus the preferred laser in JECM-LD is the green laser or ultrashort infrared laser with high power density.

4. Measurement of mechanical effects during pulsed laser and metals interaction in neutral solution

JECM-LD can be added on purpose to gain better results: to avoid redeposition of debris, to cool the material, or to increase plasma pressure. The improvement of plasma pressure is a major physical phenomenon, which is the reason of mechanical effect during JECM-LD.

4.1 Principle of test on mechanical of laser in neutral solution

As shown in Fig.5, a test setup is comprised of optical parts and measuring parts. A pulsed Nd: YAG laser was used operating at two wavelengths of 1064nm and 532nm. The neutral

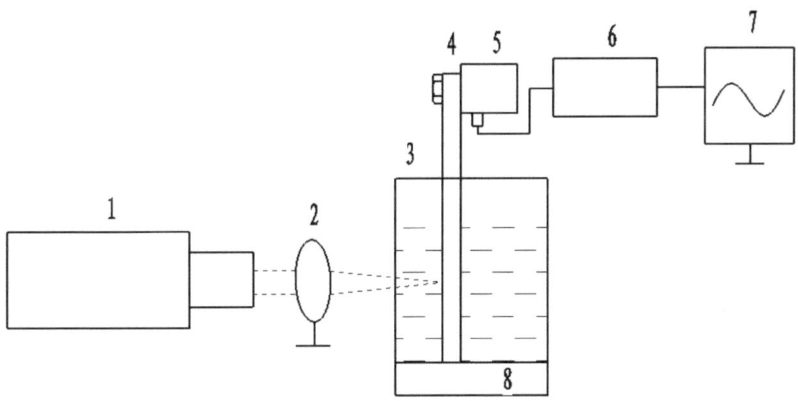

Fig. 5. Schematic diagram of test setup for mechanical effect of laser and metal interaction
1. Laser 2. Lens 3. Quartz glasses sink 4. Metal target 5. Piezoelectric accelerometer 6. Charge amplifier 7. Oscilloscopes 8. Base

solution is sodium nitrate solution and the concentration is 20%. The focal length of lens is 100mm. The thickness of metal target is 0.5mm. The metal target is fixed on the base in quartz glasses tank and forms a cantilever. A piezoelectric accelerometer is fixed on the free end of the cantilever to detect vibration acceleration signals. These signals are output on the oscilloscope through a charge amplifier.

When a pulsed laser is focused on axis position of the cantilever, the laser pulse will produce an instant pressure on the target. Then the cantilever makes a lateral vibration, as shown in Fig.6.

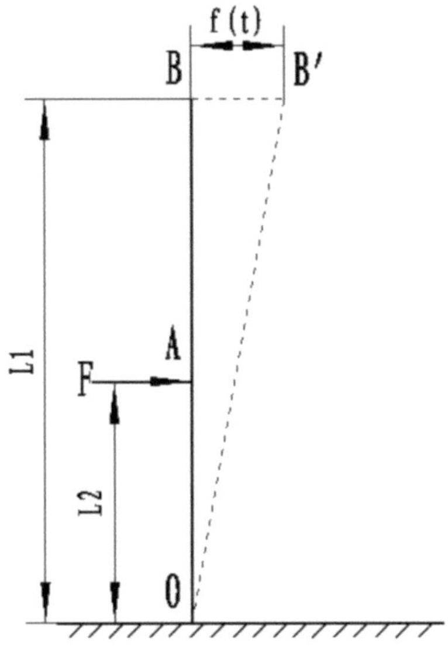

Fig. 6. Schematic diagram of vibrating cantilever

The piezoelectric accelerometer is comprised of piezoelectric element, mass, preload spring, et al. It has the same vibration as the cantilever, and then the piezoelectric element is loaded with an inertia force of the mass. According to Newton's Second Law, the inertia force is written as,

$$F = m \cdot a \tag{7}$$

Where F is inertia force, a is acceleration.

According to vibration theory, the cantilever can be seen as a freedom vibration system with a single degree, the movement of free end can be describe as,

$$f(t) = A_0 \cdot e^{-\beta t} \sin(\omega_n t + \varphi) \tag{8}$$

Where A_0 is initial amplitude, β is damping coefficient, ω_n is natural frequency, φ is initial phase angle. The acceleration of free end can be obtained by the second differential equation (8),

$$a(t) = (\beta^2 A_0 e^{-\beta t} + A_0 e^{-\beta t} {\omega_n}^2) \sin(\omega_n t + \varphi) - 2\beta A_0 e^{-\beta t} \omega_n \cos(\omega_n t + \varphi) \tag{9}$$

When t=0, the acceleration of free end is the maximum,

$$a_{\max} = (\beta^2 + \omega_n{}^2)A_0 \tag{10}$$

Initial amplitude A_0 is the maximum deformation of the cantilever. According to mechanics of materials, A_0 should be written as,

$$A_0 = \frac{FL_2{}^2}{3EI}(3L_1 - L_2) \tag{11}$$

Where E is modulus of elasticity, I is moment of inertia. Substitution of equation (11) into equation (10) yields,

$$a_{\max} = F \cdot (\beta^2 + \omega_n{}^2)\frac{L_2{}^2}{3EI}(3L_1 - L_2) \tag{12}$$

Because β and ω_n are constant, the impact force has certain proportion to the maximum acceleration. Used $K_a = \frac{L_2{}^2}{6EI}(3L_1 - L_2)(\beta^2 + \omega_n{}^2)$ as a constant, equation (12) can be written as,

$$a_{\max} = K_a \cdot F \tag{13}$$

According to the piezoelectric effect, the charge generated by the piezoelectric element has a certain proportion to the acceleration of the free end,

$$q_{\max} = S_q \cdot a_{\max} \tag{14}$$

Where, S_q is charger sensitivity. According to theory of circuit, the output of the charge amplifier is calculated as function of charge q with the following equation,

$$U_0 = -\frac{q}{C_f} \tag{15}$$

Where, C_f is feedback capacitance. Substitution of equation (13) and (14) into equation (15) yields,

$$U_0 = -\frac{S_a K_a}{C_f}F \tag{16}$$

Therefore, the value of signal on oscilloscope can be used to calculate the impact force by the laser and metal interaction.

4.2 Measurement and results

Fig.7 and Fig.8 show, respectively, the mechanical signals of pulsed green laser and infrared laser interactived with metal. The environments included being in air, being under $NaNO_3$ solution with depth 2mm and 20mm. Used laser energy per pulse is 150mJ. It is shown that

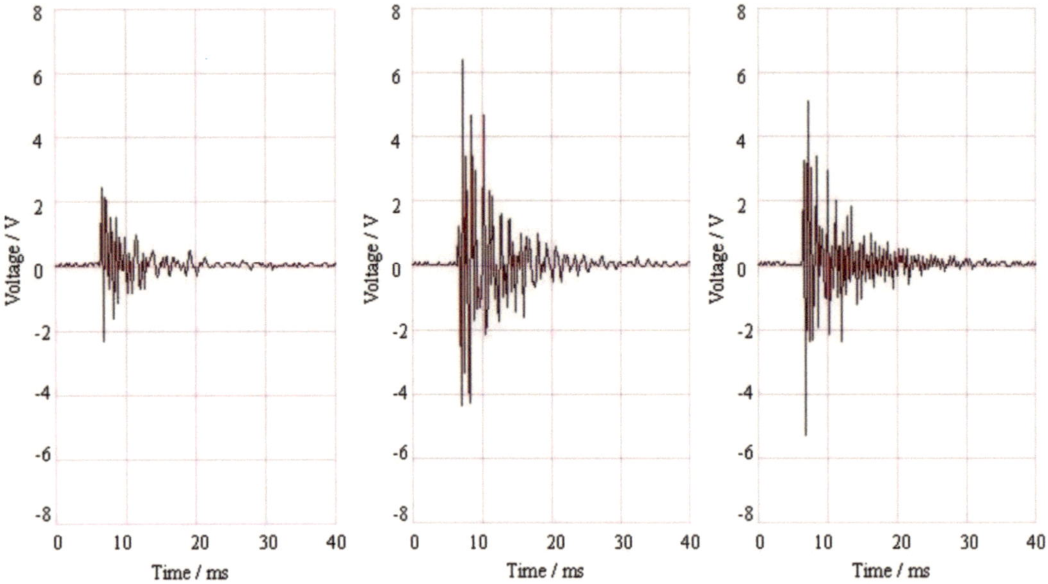

(a) in air (b) under $NaNO_3$ solution(depth=2mm) (c) under $NaNO_3$ solution(depth=20mm)

Fig. 7. Mechanical signals of pulsed green laser and metal interaction (150mJ/per pulse)

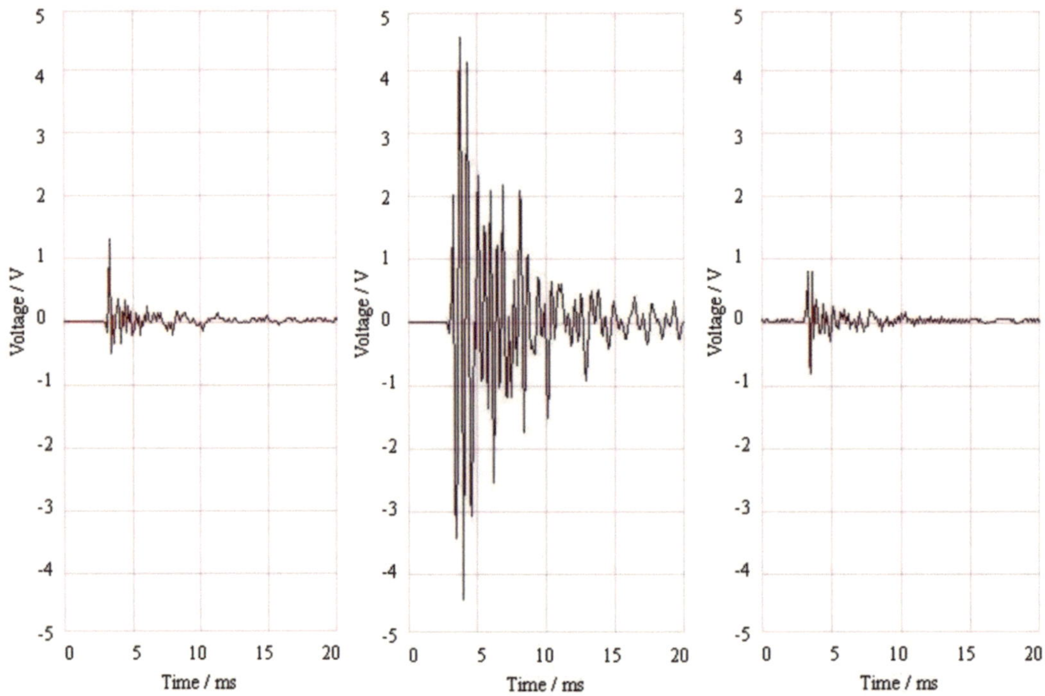

(a) in air (b) under $NaNO_3$ solution(depth=2mm) (c) under $NaNO_3$ solution(depth=20mm)

Fig. 8. Mechanical signals of pulsed infrared laser and metal interaction (150mJ/per pulse)

much higher pressure plasma is formed through the interaction between the laser pulse and metal under neutral solution than in air. The mechanical effect during pulsed laser and stainless target interaction in neutral solution is more than three times higher than in air.

It is also shown that the pressure of the target with infrared laser is decreased with the distance of laser propagation in neutral solution. However, the effect with green laser is stable in the laser transmission range of 20mm. The most probable reason for this is infrared laser is absorbed in neutral solution, but green laser is a case of low absorption. Therefore, the green laser is more suitable for JECM-LD.

5. Modeling of JECM-LD

5.1 Theoretical analysis

Theoretical analysis of JECM-LD is performed by changing machining conditions. There are two conditions in JECM-LD: One is an electrolyte-jet-guided focused pulsed laser with a pulse width of 0.2ms and a frequency of 5Hz. The other is only an electrolyte during the inter-pulse. Fig.9 presents the main idea for the theoretical analysis.

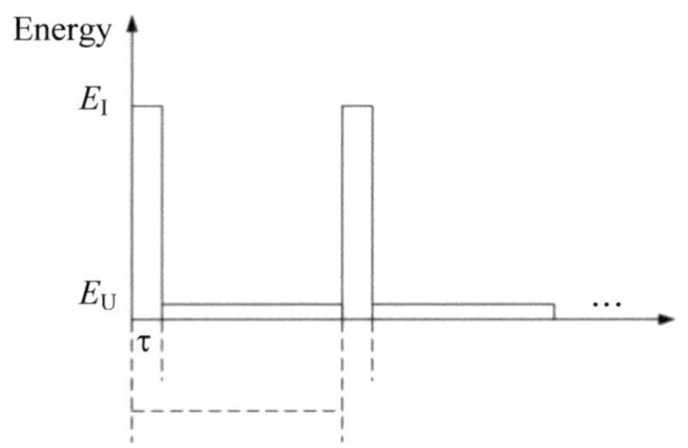

Fig. 9. Schematic diagram of the energy to be used for JECM-LD

As the laser pulse width constitutes only one percent of the inter-pulse and the electrical energy is much lower than laser energy in pulse width, the main process energy E is the heat energy E_I of pulsed laser during the pulse width τ. In contrast, the process energy E of the inter-pulse $(1/f-\tau)$ is the electrical energy E_U of electrical charge flow. The processes comprising both laser drilling and jet electrochemical machining unceasingly proceeds during JECM-LD, so the theoretical model of JECM-LD should embrace the two kinds of reactions.

5.2 Mathematical model of electrolyte-jet-guided laser drilling

Since the electrolyte-jet-guided laser drilling is quite different from laser drilling in air, the following assumptions should be made pertinent to the effects of the electrolyte jet and the temperature dependent material properties:

1. The melting temperature of the material under study is the maximum temperature the work-piece can reach.
2. The material possesses constant thermal properties.

3. The evaporated material does not interfere with the laser beam, and the scattering of laser radiation within the hole can be neglected.

Suppose that the top surface of the workpiece made of stainless plate is located at the plane z = 0 and the plate is exposed at time t = 0 to the electrolyte jet guided laser. Fig.10 schematically illustrates the cylindrical coordinate system.

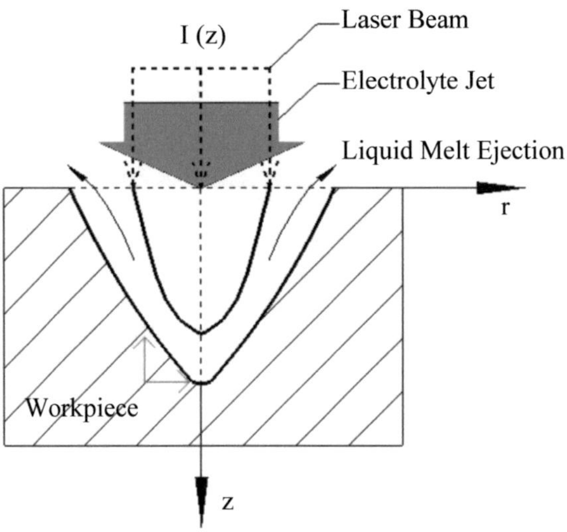

Fig. 10. Schematic diagram of the model for JECM-LD process

To elucidate the evolution of the temperature field, the time-dependent heat conduction equation is solved in the workpiece domain under the appropriate boundary conditions. The governing equation in terms of temperature T is

$$\rho c \frac{\partial T}{\partial t} = \nabla[k \nabla T] + Q \tag{17}$$

Where ρ is the material density, c the heat capacity, t the machining time, k the heat conductivity and Q the laser heat source term. In the cylindrical coordinate system, Eq. (1) can be written into

$$\rho c \frac{\partial T}{\partial t} = k \left[\frac{\partial^2 T}{\partial r^2} + \frac{1}{r}\frac{\partial T}{\partial r} + \frac{\partial^2 T}{\partial z^2} \right] + Q \tag{18}$$

The general laser heat source Q can be expressed by

$$Q = -\frac{\partial I(z)}{\partial z} \tag{19}$$

Where $I(z)$ is the intensity of the laser beam incident upon the top workpiece surface from a given distance, z. In the case of metals, $I(z)$ can be quantified according to the Lambert's Law,

$$I(z) = \varepsilon I(0)e^{-\beta z} \tag{20}$$

Where ε is the surface emissivity, β the absorption coefficient. From Eq. (19) and Eq. (20), Q can be written into

$$Q = \varepsilon\beta I(0)e^{-\beta z} \tag{21}$$

Inserting Eq. (21) into Eq. (17) yields

$$\rho c\frac{\partial T}{\partial t} = k\left[\frac{\partial^2 T}{\partial r^2} + \frac{1}{r}\frac{\partial T}{\partial r} + \frac{\partial^2 T}{\partial z^2}\right] + \varepsilon\beta I(0)e^{-\beta z} \tag{22}$$

As the heat loss due to convection and radiation would take place at the top workpiece surface resulting from the cooling by the electrolyte jet, the boundary condition can be expressed by

$$-k\left(\frac{\partial T}{\partial z}\right)_{z=0} = h_c(T - T_a) + \varepsilon\sigma(T^4 - T_a{}^4) \tag{23}$$

Where h_c is the convection coefficient, T_a the ambient temperature and σ the Stephan-Boltzman constant.

The scraps of the workpiece would be carried away in molten state by the electrolyte jet whenever and wherever the material has reached the melting temperature. Thus, for this model, the isotherm line, which represents the melting temperature of the material, can be considered as the profile of the hole during electrolyte-jet-guided laser drilling.

5.3 Mathematical model of jet electrochemical machining

During the inter-pulse, with the ending of pulsed laser beam, the electrolyte jet creates a contactless electrode and has electrochemical reactions onto the workpiece-anode. In this way, the recast layers and spatters have been effectively removed by the electrochemical reactions.

Based on the cylindrical coordinate system of the model of electrolyte-jet-guided laser drilling, a two-dimensional model for JECM can be developed as follows:

In order to model the shaping process with a free jet, the following assumptions should be made (see Fig.11).

1. The top boundary is the shape formed by the preceding electrolyte-jet-guided laser drilling.
2. The cross-section of the jet remains constant along the whole length of the jet.
3. The diameter of the electrolyte jet is assumed to be the same as that of the nozzle.
4. The jet is axisymmetric, thus the system can be described as 2D in cylindrical coordinates (r, z).
5. The electrolyte has constant conductivity.

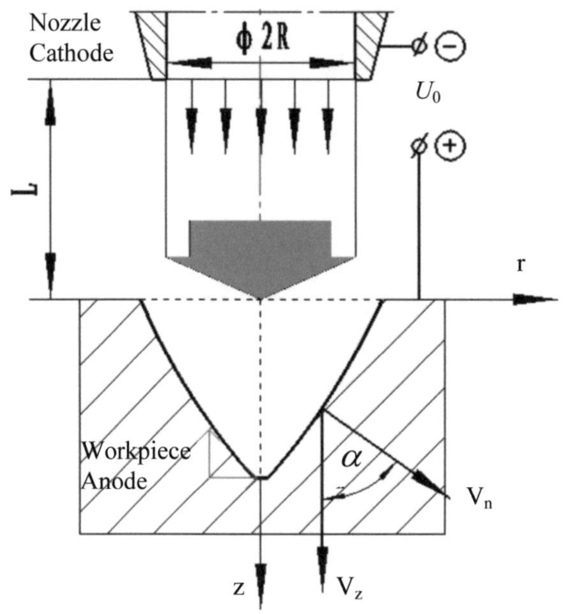

Fig. 11. Schematic diagram of the model for JECM process

Fig.11 shows the change of the shape of the workpiece surface machined by the anodic dissolution as a function of time. During machining, the velocity at which the anodic boundary moves along the *z*-axis is given by

$$V_z = \frac{V_n}{\cos\alpha} \quad (24)$$

Where V_n is the velocity of anodic dissolution, α the angle between the axis of the anode and *z*-axis. The $\cos\alpha$ in Eq.(8), which can be evaluated with the function $z = z_a(r,t)$, describes the shape of the anode as follows:

$$\frac{1}{\cos\alpha} = \sqrt{1 + (\frac{\partial z_a}{\partial r})^2} \quad (25)$$

Substituting Eq. (9) into Eq. (8) yields

$$\frac{\partial z_a}{\partial t} = V_n\sqrt{1 + (\frac{\partial z_a}{\partial r})^2} \quad (26)$$

The velocity of dissolution V_n can be obtained from the Faraday's Law:

$$V_n = \eta\omega i_a \quad (27)$$

Where, η is the current efficiency of anodic dissolution, ω the equivalent volume of material removed by electrochemical processing and i_a the current density of the anode.

The current density i_a on the anode depends on the distribution of electric potential, u, in the electrolyte, and can be determined from the following Ohm's Law:

$$i_a = \kappa E_a \tag{28}$$

Where κ is the electrolyte electrical conductivity and E_a the electric field intensity at anode given by

$$E_a = \left|\nabla u\right|_a \tag{29}$$

Where u is the electrical potential across the gap.

In the case of JECM, the potential between the electrodes, u, can be described by

$$\mathrm{div}(\kappa \,\mathrm{grad}\, u) = 0 \tag{30}$$

By evaluating Eq. (30) in cylindrical coordinates, can be obtained the following equation to define the electric potential distribution:

$$\frac{\partial^2 u}{\partial r^2} + \frac{1}{r}\frac{\partial u}{\partial r} + \frac{\partial^2 u}{\partial z^2} = 0 \tag{31}$$

Eq. (31) can describe the shape of the surface to be machined with the following boundary conditions:

For the cathode-nozzle: $u = 0$, and $r = r_0$;

For the free surface of jet: $\frac{\partial u}{\partial n} = 0$;

For the anode-workpiece: $z = z_a(r,t)$, $u = U_0$, and

$$\frac{\partial z_a}{\partial t} = \kappa \eta \omega \left|\frac{\partial u}{\partial n}\right|_a \sqrt{1 + (\frac{\partial z_a}{\partial r})^2} \tag{32}$$

With the inter-pulse ending, the reaction of JECM comes to a halt and the result of the model becomes the renewed boundary for the next pulse laser drilling. Thus the two parts of JECM-LD model alternatively work during machining.

6. Simulation and experimental results

6.1 Experimental apparatus

As shown in Fig.12, the JECM-LD apparatus consists of three systems, namely, a laser system, an electrolyte supplying system and a power unit. A stream of electrolyte pumped out of the reservoir flows into the jet cell and meantime joins with a coaxial focused laser beam to form a jet 0.5mm dia. The jet together with the beam strikes against one spot on the workpiece surface. By means of the pump, the pressure of the electrolyte jet can be adjusted within the range of 0-1.5Mpa.

A jet cell with an annular cavity is used to obtain a stable electrolyte jet and slim down the transmitting distance of laser in electrolyte. Fig.13 shows the construction of a jet cell.

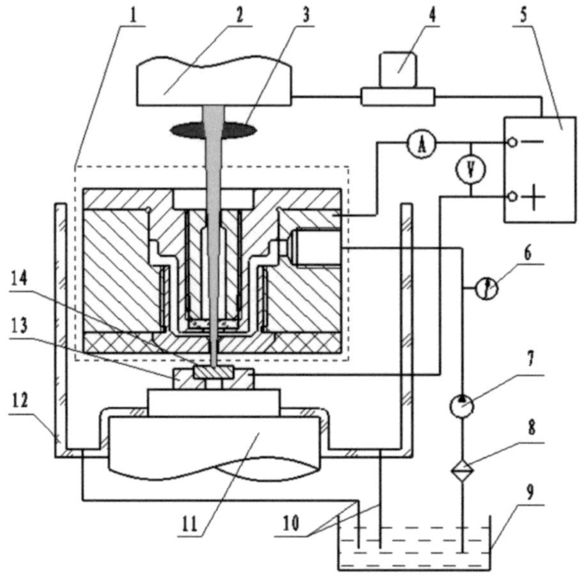

1 Jet cell; 2 Laser system; 3 Focusing lens; 4 Control unit; 5 DC power unit; 6 Pressure gauge; 7 Pump; 8 Filter; 9 Electrolyte reservoir; 10 Back pipelines; 11 CNC Table; 12 Work cell; 13 Holder; 14 Workpiece

Fig. 12. Schematic diagram of JECM-LD system

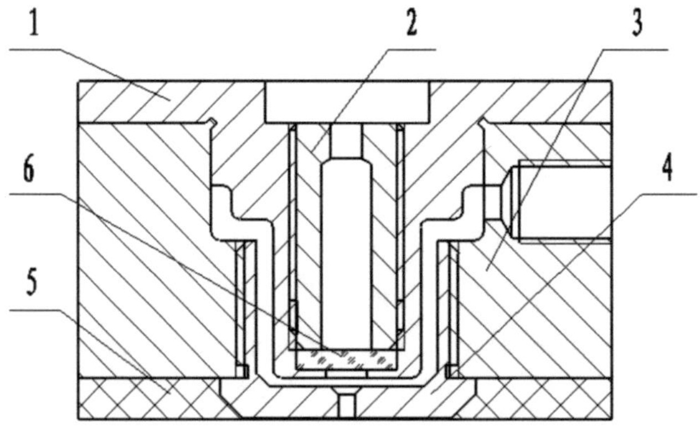

1 Impacted plank; 2 Lock bolt; 3 Base; 4 Nozzle; 5 Insulated layer; 6 Quartz window

Fig. 13. Structure of a jet cell

6.2 Experimental parameters

As a pulse Nd:YAG laser at second harmonic wavelength, it has parameters as follows: wavelength of 532nm; pulse length of 0.2ms; frequency of 5Hz and energy per pulse in the range of 0-300mJ. The workpiece is made of 321 stainless steel 0.5mm thick. Table1 lists the material properties.

A DC power unit with a working voltage in the range of 0-50V and current 5A is adopted in the experiments. The electrolyte from the reservoir on the manifold enters a plunger pump, where it is pumped out at a pressure of 1.5MPa.

Property	Value
Density, ρ [kg·m^{-3}]	7900
Heat conductivity, k [J·m^{-1}·s^{-1}°C^{-1}]	28.5
Heat capacity, C [J·kg^{-1}°C^{-1}]	502
Melting temperature, T_m [°C]	1400
Convection heat transfer coefficient, h_c [W·m^{-2}°C^{-1}]	100
Surface emissivity, ε	0.68
Absorption coefficient, β [m^{-1}]	5×10^3

Table 1. Properties of 321 stainless

In the experiments is used a sodium nitrate electrolyte with a percentage concentration of 18% in weight and a conductivity of 12.2 $(\Omega \cdot m)^{-1}$. The volume electrochemical equivalent of 321 stainless steel is 2.1x10^{-9} $m^3/(A \cdot min)$. The current efficiency is about 0.6. The distance between the nozzle and the workpiece surface is 0.8mm.

6.3 Simulation results

The model of JECM-LD includes two sub-models at different machining stages. One is the sub-model of electrolyte-jet-guided laser drilling and the other of jet electrochemical machining. With the machining stage changing, the simulation is computed with different sub models and the results are summarized step by step.

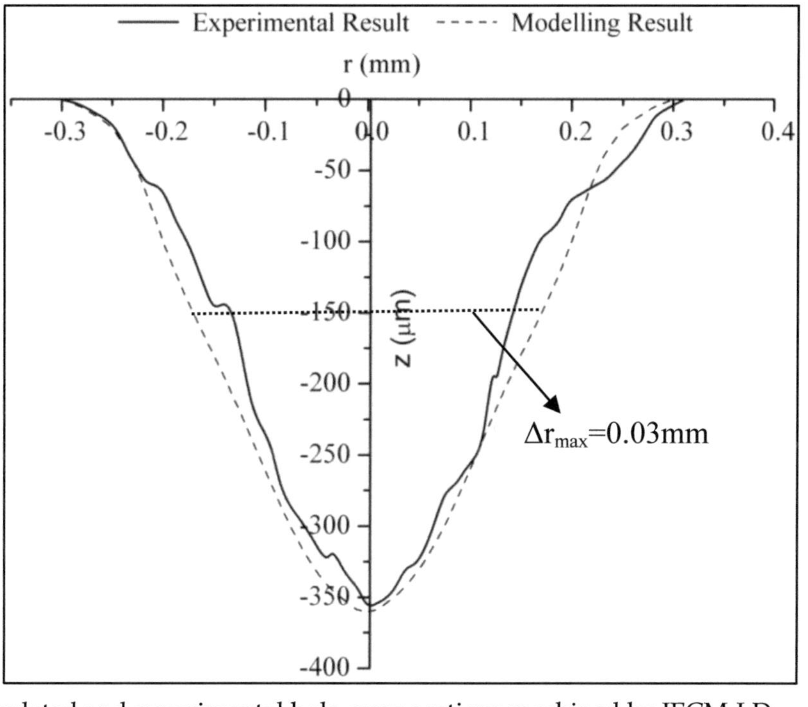

Fig. 14. Simulated and experimental hole cross-sections machined by JECM-LD
(Energy per pulse: 200mJ; ECM voltage: 40V; Electrolyte pressure: 1MPa; Machining time: 10s)

Fig.14 illustrates the shape of the vertical cross-section along the hole-axis from the simulation and from the experiment, which is plotted by means of the three-dimensional profilemeter (MicroXAM, ADE, USA). The comparison of Fig.14 displays a good agreement between the simulated and experimental results in terms of both the hole depth and the shape of the cross-section. However, a noticeable difference exists between the bottom shapes of the simulated cross-section and the experimental one. The former contrasts markedly with the latter by its greater flatness. This might be attributed to the more serious attenuation of laser energy by the scattering and absorption of electrolyte jet.

Therefore, the feasibility of the two-dimensional model is evidenced by the satisfied compliance of the simulated results with the experimental ones.

6.4 Experimental results

In order to make comparison with the laser drilling in air, a millisecond second green laser was used for JECM-LD to work out penetrated holes. The energy per pulse was 200mJ, and the electrochemical machining voltage 40V. The optical microscope was used to examine the experimental results.

As can be seen from Fig.15, for the holes laser-drilled in air, the surfaces both at the entrance and the exit are all encircled by a large irregular area comprising spattering deposits and re-solidified molten layers. In contrast, from Fig.16, it is clear that few spatters could be unveiled on both sites around the hole drilled with JECM-LD. This may be chiefly ascribed to the high-speed electrolyte jet, which effectively cools the material to be processed and discharge scraps. Fig.16 also exhibits better surface quality and a smoother hole periphery, however, an obvious annular electrochemical overcut can be observed at the entrance surface.

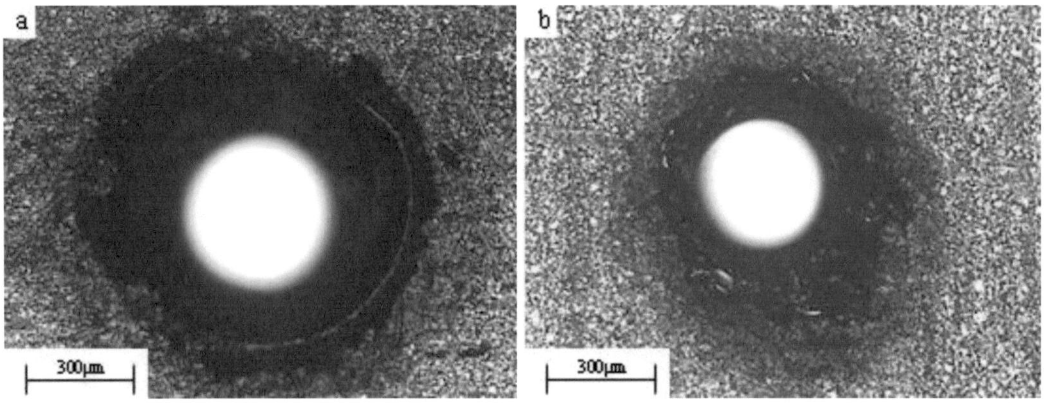

Fig. 15. Micrograph of a penetrated hole drilled with a green laser in air (200mJ, 10s).
(a) Entrance surface; (b) Exit surface

Fig.17 (a) and (b) separately display the details of the exit surface of Fig.15 and Fig.16. Unlike the laser drilling in air(see Fig.17(a)), neither spatters nor recast layers could be discovered on the periphery of the hole drilled with JECM-LD (see Fig.17 (b)). This might be attributed to the electrolyte jet, which rids the recast layers of the surface to be processed.

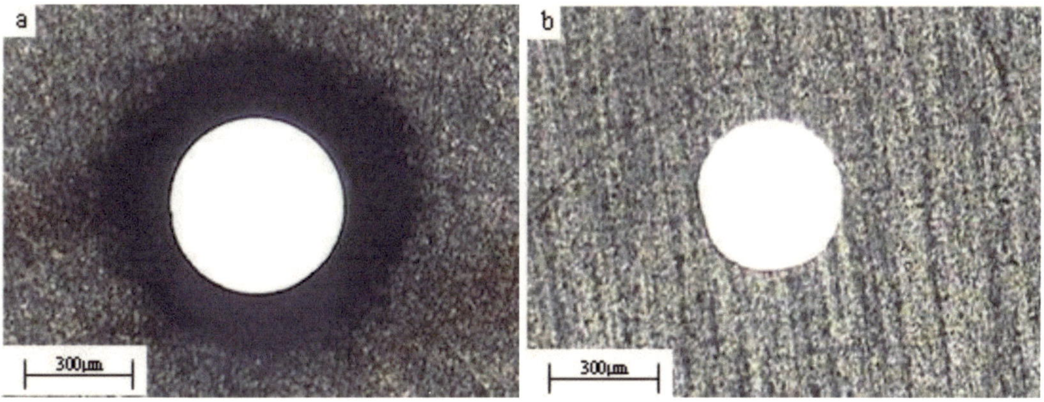

Fig. 16. Micrograph of a penetrated hole drilled with JECM-LD (200mJ, 40V, 20s).
(a) Entrance surface; (b) Exit surface

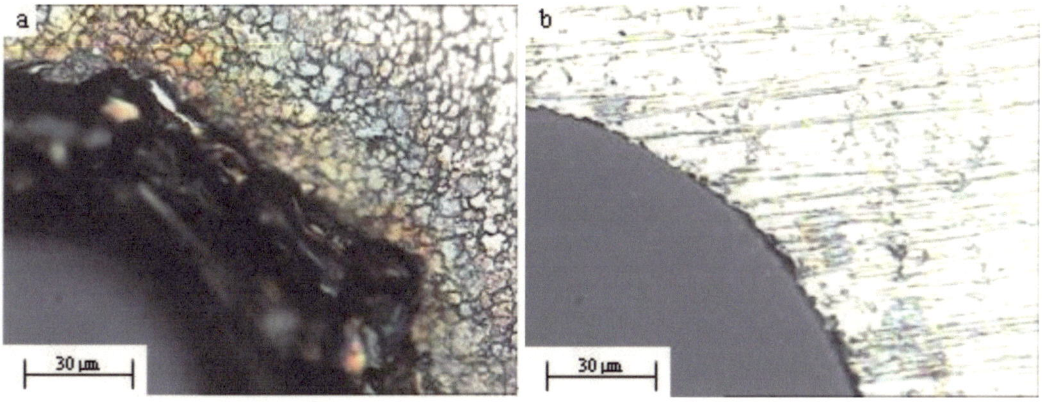

Fig. 17. Micrograph of the peripheral surface of a penetrated hole drilled with different processes.
(a) Laser-drilled in air (200mJ, 10s); (b) JECM-LD (200mJ, 40V, 20s)

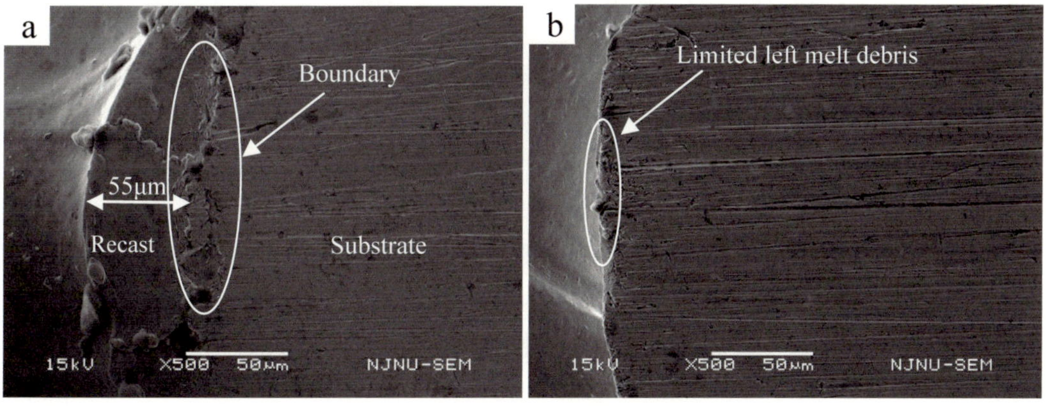

Fig. 18. SEM image of the sidewalls of hole drilled with different process.
(a) Laser drilling in air (200mJ, 10s); (b) JECM-LD (200mJ, 40V, 20s)

The typical cross-sectional view of holes machined with JECM-LD and laser drilling in air are shown in Fig.18. It is shown that there is an obvious boundary between the recast and substrate in Fig.18(a). The thickest recast layer is about 55μm. It is confirmed that recast is an inherent defect associated with holes produced with laser drilling. As can be seen in Fig.18(b), neither boundary nor recast layer can be seen on the sidewalls of hole machined with JECM-LD. There are only few of limited left melt debris adhering to the sidewalls. With comparing the Fig.18(a) and Fig.18(b), the new hybrid process really reduce the recast more than 90% on line.

As the most popularized criterion for defining the cutting efficiency in laser drilling, Material Removal Rate (for short MRR) can be defined as

$$\mathrm{MRR}=\frac{\frac{1}{n}\sum_{i=1}^{i=n}\Delta m_i}{t} \tag{33}$$

Where Δm_i is the lost mass of workpiece in the i-th experiment, n the number of experiments and t the machining time.

Fig.19 evinces the effects of laser pulse energy on MRR and Fig.20 the electrochemical machining voltage on it. These data come from Eq.33 after five repeated experiments with the machining time of 10s each.

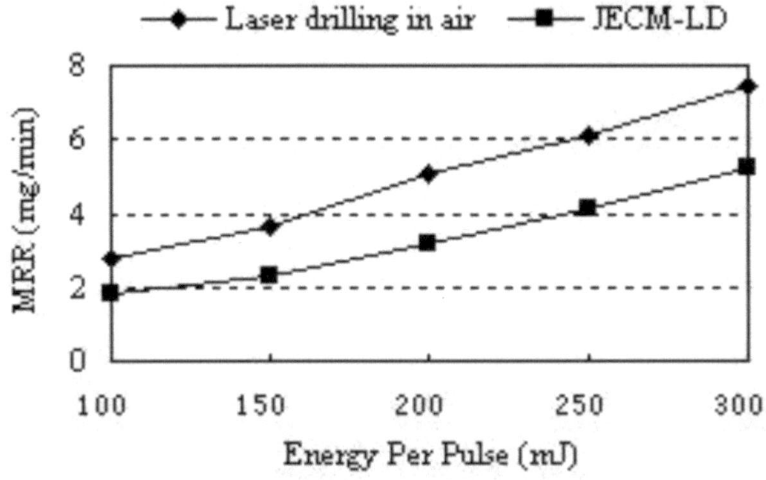

Fig. 19. Effects of energy per pulse on MRR

From Fig.19, it can be seen that MRR increases with the energy per pulse rising. Since a portion of the laser beam has been absorbed and scattered by the electrolyte, the MRR of JECM-LD appears lower than the laser drilling in air, the efficiency of JECM-LD with millisecond green laser is about 70% of laser drilling in air.

Fig.20 shows that MRR remains almost un-changed with the electrochemical machining volt-age increasing. This is because of the insignificant anodic dissolution provided that the voltage is low and the working distance long. It stands to reason that laser beam takes the chief responsibility for removing scraps from the processed surfaces during JECM-LD, and the jet electrochemical machining plays an auxiliary part in reducing the recast layers and spatters.

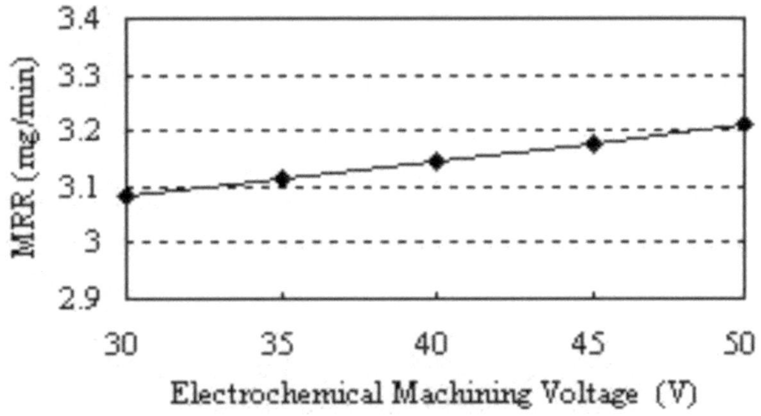

Fig. 20. Effects of ECM voltage on MRR with JECM-LD

7. Conclusions

This work has developed and investigated a novel hybrid process of laser drilling assisted with jet electrochemical machining to improve the quality of laser-drilled holes. The main conclusions can be summarized as follows:

1. The attenuation of green laser in the electrolyte mainly is scattering and the infrared laser is absorbed in electrolyte.
2. Mechanical effect during pulsed laser and stainless target interaction in neutral solution is more than three times higher than in air.
3. Able to effectively reduce the recast layers and spatters, the JECM-LD has an annular electrochemical overcut left behind surrounding the entrance surface after the completion of the process.
4. The laser beam takes the chief responsibility to remove the scraps material during JECM-LD while the jet electrochemical machining serves the auxiliary function of eliminating the recast layers and spatters.
5. The efficiency of JECM-LD with millisecond green laser declines to about 70% of the laser drilling in air.

It is evidenced that the JECM-LD that combines the laser drilling and jet electrochemical machining is on the position to obtain high machining quality with reduced recast layers and spatters. Having great potentiality in eradicating the defects inherent in laser-drilled holes, the hybrid process for hole drilling is expected to become a versatile tool finding broader applications in aerospace and aircraft industries.

8. References

[1] K.T. Voisey, T.W. Clyne. Laser drilling of cooling holes through plasma sprayed thermal barrier coatings. Surface and Coatings Technology, 2004, 176:296-306.

[2] W.S. Rodden, S.S. Kudesia, D.P. Hand, J.D. Jons. Use of "assist" gas in the laser drilling of titanium. Journal of Laser Applications, 2001, 13:204-208.

[3] J.C. Verhoeven, J.K. Jansen, R.M. Mattheij. Modelling laser induced melting. Mathematical and Computer Modelling, 2003, 37:419-437.

[4] Hua Zhang, Jiawen Xu, Jiming Wang. Investigation of a novel hybrid process of laser drilling assisted with jet electrochemical machining. Optics and Lasers in Engineering, 2009, 47(11):1242-1249

[5] H. Zhang, J.W. Xu, J.M. Wang. Green Laser Drilling Assisted with Jet Electrochemical Machining of Nickel-based Superalloy. Key Engineering Materials, 2010, 426-427:75-80.

[6] Zhang Hua, Xu Jiawen. Modeling and Experimental Investigation of Laser Drilling with Jet Electrochemical Machining. Chinese Journal of Aeronautics, 2010, 23(4):454-460.

[7] Hua Zhang, Jiawen Xu, Jianshe Zhao, Guoran Hua. Mechanical effects during pulsed laser and metals interaction in neutral solution. Key Engineering Materials, 2011, 464:623-626.

[8] Hua Zhang, Jiawen Xu, Jianshe Zhao, Guoran Hua. Mechanism of Recast Removal During Laser Drilling Underwater. Advanced Science Letters, 2011, 4(6-7):2071-2075.